Mammoths and Neanderthals in the Thames Valley

Mammoths at Stanton Harcourt as recreated by K. Scott

# Mammoths and Neanderthals in the Thames Valley

## Excavations at Stanton Harcourt, Oxfordshire

Katharine Scott
Christine M Buckingham

Archaeopress Archaeology

Archaeopress Publishing Ltd
Summertown Pavilion
18-24 Middle Way
Summertown
Oxford OX2 7LG
www.archaeopress.com

ISBN 978-1-78969-964-7
ISBN 978-1-78969-965-4 (e-Pdf)

# Contents

# List of Figures

# List of Tables

# Acknowledgements

Countless people assisted in the collection, analysis and processing of the data upon which this book is based. Between 1989 and 1999 this included colleagues and university students with a professional interest in the project, but also dozens of volunteers whose help was indispensable. We are especially indebted to the volunteers from the Earthwatch Institute for their tireless enthusiasm, often working in the most difficult conditions. Their assistance made it possible to retrieve a vast collection of bones, insects, molluscs and other organic material and we have been able to draw on invaluable detailed records kept in hundreds of their notebooks. Large teams of volunteers, some with little or no experience in such work, need advice and supervision and in this respect, we were very fortunate to have the frequent assistance of Jeffrey Wallis (a long-standing active member of the Abingdon Archaeological Society) and students Jim Campbell and Chris Gleed-Owen (Coventry University), Stephen and Anette Lokier (Oxford Bookes University), and Julie Cormack (University of Liverpool). We were also fortunate to have regular help locally from R.J. 'Mac' Macrae, Terry Hardaker, Sally Moyes and John Cooper. John's additional skill as a digger driver enabled us to create essential exploratory trenches and move mountains of spoil.

As will be seen in the Chapters on the fauna and flora, many specialists over the decades examined specimens and analyzed samples from the site. The authors are acknowledged in the relevant sections and listed here with reference to the institutions at which they were based at the time: Prof. Derek A. Roe (Donald Baden-Powell Quaternary Research Centre, University of Oxford); Prof. Russell G. Coope (Birmingham University); Dr David H. Keen, Dr Mike Field, Dr Jim Campbell and Dr C. Gleed-Owen (Coventry University); Mr Terry Hardaker (Oxford Cartography); Prof. Julia A. Lee Thorpe and Dr Ian Gourlay(University of Oxford); Dr Jon G. Hather and Dr Brian G. Irving (Institute of Archaeology, University College, London); Dr Mark Robinson (Oxford University Museum); Dr Joanne Cooper (Natural History Museum, London); Dr. D. Marc Dickinson (University of York) and Dr Rowena Gale (Royal Botanical Gardens, Kew).

In addition to the contributors listed above, several colleagues have provided invaluable advice over the years, been generous with their own data, and commented helpfully on various draft sections of this book. We especially thank Dr Nick Ashton (British Museum), Prof. Martin Brasier (Oxford University), Dr Clark Friend (Oxford Brookes University), Prof. Gary Haynes (University of Arizona), Prof. Adrian Lister (Natural History Museum, London), Prof. Richard Klein (Stanford University), Prof. Danielle Schreve (Royal Holloway, University of London), Prof. Tony Stuart (Durham University) and Dr. Roger Suthren (Oxford Brookes University).

The conservation of so many bones and tusks post excavation took many years and we are especially grateful to Sally Moyse, Clark Friend and Lucia Pinto who dedicated very many hours to help prepare and restore the specimens. Comparative fossil material was essential and no-one could have been more generous with his time and his willingness to make other collections available than Andy Currant at the Natural History Museum. We also thank Suzanne Thompson for her assistance in the field and for collating hundreds of field plans, and Greg Scott for digitizing the extensive collection of excavation slides and for packing 400 crates of fossils and other material for transfer to the Oxford University Museum.

Keeping such a large and valuable collection of Pleistocene material safe from thieves and the elements was never easy. Friends with farms in Oxfordshire kindly offered their barns for storage. We thank Stuart and Gillian Hamilton, Guy Pharaon, and Malcolm Hastings for providing short-term shelter in the early days. We are especially grateful to Nina and Nick Ritchie who put up with our comings and goings on their farm for more than 20 years. A permanent home with display potential was always the goal, to which end several people in the village of Stanton Harcourt, led by Charles Mathew, endeavoured to locate suitable premises for a museum with storage facilities. We are pleased to report that everything is now in the care of the Oxford University Museum of Natural History due mainly to the enthusiasm and perseverance of the Collections Manager, Dr Hilary Ketchum. We are also grateful to the Director, Prof. Paul Smith, and the Head of Earth Collections, Eliza Howlett, for facilitating this event and to the Curry Fund of the Geologists' Association for covering the costs of packing and re-boxing the material. Conservation and curation of the fossils has continued thanks to the expertise of Neil Owen and Neil Adams and the generosity of the Street Foundation.

The entire excavation project hinged on the co-operation of the quarry owners who were unfailingly supportive of our efforts to extract material in very tough conditions. In this respect, Hanson Aggregates (formerly ARC) and the landfill operators Greenways Landfill together with their on-site staff were instrumental in our ultimate success. In particular, we were fortunate to have the interest and co-operation of Hanson's technical and external affairs director, John Mortimer, as the result of which Hanson generously contributed to excavation funds. Invaluable financial and other support was also received by the British Academy, Earthwatch (UK), the Hanson Environment Fund, the Leakey Foundation, the Society of Antiquaries (London) and the Quaternary Research Association (QRA). The project also benefitted greatly from the institutional and personal support given to KS through her Fellowship at St Cross College, Oxford.

The final compilation of this volume was frustrated by the Covid pandemic as one of us (KS) had to remain in South Africa for the year of 2020. Apart from the inherent difficulty of remote collaboration, essential documents in Oxford were inaccessible. KS is particularly grateful to Edward Cropper and Tahli Betteridge for their persistence in locating and scanning many documents and photographs from UK files. Thanks are also due to Polly Courtice and Valda Führ for their editorial assistance, and to Robin Orlić for page-setting the text and illustrations.

Abundant illustrations are essential to describe a site so rich in finds as Stanton Harcourt. The maps and plans in Chapters 1 and 2 describing the geological context were drawn by CMB. As regards the rest of the illustrations, it is hard to do justice to the contribution made by Greg Scott to this volume. Not only did he finalise all KS's line drawings of animal remains, but he produced the finished photographs of over 100 fossils and 74 artefacts, created the vegetation reconstructions in Chapter 6, and designed the cover for the book.

Last but not least, we express deep gratitude to our husbands, Richard Cropper and Keith Buckingham, for their forbearance, support and encouragement over so many years.

# Preface

This book describes an unusual situation near the village of Stanton Harcourt, Oxfordshire. In 1989, fossils came to light in a quarry that had been the focus of gravel extraction a decade previously. The discovery, in a disused pit, presented an opportunity to carry out extensive fieldwork for 10 years rather than the more usual 'rescue' excavation and resulted in one of the most remarkable Pleistocene assemblages in Britain.

At the base of the pit, below the extracted gravel, was a metre or so of 'uneconomic' gravel, so-called because it contained a variety of large stones and organic material. This gravel was left in place, the pit was abandoned awaiting its use for waste disposal, and vegetation took root across the quarry floor. In 1989, in the course of drainage maintenance, the tusk of a mammoth was discovered. By coincidence, the authors visited the quarry shortly thereafter and agreed to return to salvage the tusk. Within a short time, it was clear that there were abundant *in-situ* animal and plant remains. In 1992, there was great excitement when the first of many stone artefacts was found. For the rest of the decade, it was possible to excavate, record and analyse the context of a very large assemblage of Pleistocene bones, artefacts, shells and wood in the sediments of a meandering river.

Such a situation is extremely rare in Britain. The majority of bones and stone tools donated to museums by fossil collectors over more than a century have come from gravel pits. In most cases, there is little or no accompanying contextual information as there is rarely an opportunity to carry out detailed fieldwork at working quarries. Hence fossils and artefacts are generally found in the wake of quarrying at the base of excavated pits or on piles of extracted gravel. Gravel is the accumulation of millennia of ancient river beds and so, even in cases where it has been possible to record the context of fossils, it is a common perception that such material is of limited value in reconstructing the past because, by its very nature, a river is a dynamic environment with the potential to transport and redistribute such material.

Stanton Harcourt proved to be an exception. The excavated deposits revealed a buried channel, a former course of the River Thames. Material that found its way into this channel had evidently been rapidly buried, preventing oxidization and erosion, with the result that organic material was extraordinarily well preserved. In such a situation, it had a better chance of surviving into the fossil record than equivalent material on the ground surface. Apart from more than 1500 animal bones, teeth and tusks, there were molluscs, insects, and vegetation including seeds, nuts, branches and trunks of trees. A date of c.200,000 years makes this assemblage unique for this period in Britain.

Over the course of the decade, it was possible to document the course of the river over a wide area and, through the detailed analysis of the sedimentary environment, to identify a variety of depositional histories within this fluvial setting. The particular significance of this site is that it is possible to describe in unprecedented detail a temperate environment approximately 200,000 years ago (marine isotope stage 7). Analyses of the vegetational remains reveal this region of the Upper Thames to have been an area of open woodland and grassland bordering a river where large mammals grazed. Of particular interest is the most common species at the site – the mammoth. Initially thought to be the woolly mammoth *Mammuthus primigenius*, it was soon recognised as an earlier species – a small form of the steppe mammoth

*M. trogontherii*. This mammoth was living alongside the much larger straight-tusked elephant and other animals – bison, horse, bear, wolf and lion.

The discovery of stone tools in the same context as the large vertebrates was highly significant because, at the time of excavation, it had long been believed that people were not present in Britain during interglacials. The lack of archaeological evidence was taken as an indication of the reluctance or inability of Pleistocene hominins to occupy forested habitats. It has become increasingly clear that the more open environment of MIS 7 was not a limiting factor to human movement in Britain.

As will be seen in the Acknowledgments, a substantial number of people and institutions contributed to the success of the excavation both on- and off-site. Access to several acres over a period of ten years generated a vast amount of data: crates of fossils, buckets of sediment samples, and thousands of photographs, plans and section drawings. Apart from the enormous task of analysing and collating these after the excavation, the vertebrate fossils required conservation. Although generally well-preserved when unearthed, they were not fully fossilized which meant that most larger items required to be encased in fibreglass or Plaster of Paris before they could be lifted. Reversing this process was time-consuming and often difficult especially in the case of the mammoth skulls and tusks two metres in length and curved. Apart from the technical difficulties encountered in such a conservation task, suitable workspace was always a problem. Derek Roe generously allocated us an office at the Donald Baden-Powell Quaternary Research Centre in Oxford but the University lacked the extensive space required for the excavated material. The Oxford University Museum of Natural History had long expressed interest in acquiring the Stanton Harcourt material (as well as that from several other Upper Thames sites at which we had worked) but had not sufficient space to house it. Friends on farms provided temporary accommodation in their barns for the first few years and then Hanson (the quarry owners) gave us access to a large disused glasshouse. This offered everything we needed and had always lacked – ample space, light, and water. However, the publicity that resulted from media interest in the discovery of mammoths near Oxford and the excavation having featured on television in the *Time Team* series made the collection extremely vulnerable. Sporadic thefts of fossils from the excavation were a continual problem but, when thieves broke into the glasshouse and stole some of our best, fully prepared specimens, it was a major blow. The theft triggered an immediate and onerous move to another farm where conservation went on in secret until 2018. At this point, the University Museum took possession of buildings at a former airbase and acquired funds to employ a curator. It was a great relief to spend the summer of that year cataloguing and crating up the entire collection to be donated to the OUM, knowing that it would be curated, eventually made available to future researchers, and displayed.

With the future of such a large and valuable collection assured, it remained to us to complete our analysis of the large vertebrates, put the finds within their geological context, and collate all other contributions in preparation for this volume. It is a matter of great regret to us that four of the contributors are no longer with us. Derek Roe, Russell Coope, David Keen and Terry Hardaker were all very generous with their support and expertise during the excavation years, we benefitted greatly from their enthusiasm and presence on site, and from innumerable discussions with them. The reports they prepared make a significant contribution to this volume and we trust that the way we have incorporated their research gives these authors the recognition they deserve.

# Chapter 1

# Introduction

As the title denotes, this book concerns the excavation of 200,000-year-old fossiliferous deposits at a site known as Stanton Harcourt in Oxfordshire (SP413051). More accurately, the excavation site was Dix Pit, a former gravel quarry near the village of Stanton Harcourt (Figure 1.1). Oxfordshire is rich in mineral resources. Those which are used for primary aggregate production comprise extensive alluvial sand and gravel resources along the River Thames and its tributaries. River terraces occur at several levels above the modern floodplains within the Thames, Evenlode, Windrush and Thame valleys and their minor tributaries. The sands and gravels within these terraces comprise mainly unconsolidated materials laid down by rivers and streams and are an important resource in the county. Once the gravel is removed, the pits are flooded and used for recreational purposes.

In the immediate vicinity of Stanton Harcourt, the sand and gravel deposits are attributed to the meltwater of the penultimate cold stage. Dix Pit, quarried in the 1970s by Hanson Aggregates (formerly ARC), was unusual for two reasons. Firstly, below the 5-6m of quarried sands and gravels was a less well sorted deposit approximately 1m thick. This was considered uneconomic and not quarried. Secondly, as Dix Pit had been ear-marked as a waste disposal site, the abandoned quarry was not flooded, and vegetation grew on the quarry floor. Over the next decade, sporadic visits by Quaternary scientists led to the conclusion that this lower gravel represented a former river channel cut into the Oxford Clay and it became known as the Stanton Harcourt Channel (Briggs *et al.* 1985). Within these channel deposits were warm adapted molluscs, insects and plant remains that were not typical of the bulk of overlying quarried gravel which represented cold conditions.

In 1989, during drainage operations, the tusk of a mammoth was unearthed at the base of the pit. The authors' initial interest was in trying to retrieve this tusk but it was soon apparent that it was in sediment containing fresh water shells of a species normally found in much warmer climates than in Britain today and that this area of the pit might be another exposure of the Stanton Harcourt Channel reported by Briggs *et al.* (*op. cit.*). Mammoths are generally associated with a cold climate but the associated molluscs indicated temperate conditions. However, the gravel deposits were predominantly of fluvial origin so the question that arose was: had the tusk and shells originated in different climatic episodes and become mixed together by river action or was the mammoth a survivor from a previous cold episode that had become adapted to a warm climate? This apparent anomaly of creatures of cold and warm habitats in the same deposit led the authors to explore the site further.

As the pit had been designated for waste disposal in the foreseeable future, field work at the site was initially in the nature of a rescue operation. As time passed and as the finds and their importance increased, funding was applied for and field work was undertaken between 1990 and 1999 on a more systematic basis (Figure 1.2). For several years, funds facilitated three two-week excavations with volunteers and field assistants. At other times, the authors and various local helpers made regular visits to the site. The excavations became known as the 'Mammoth Project' and later as the 'Oxford Mammoths'.

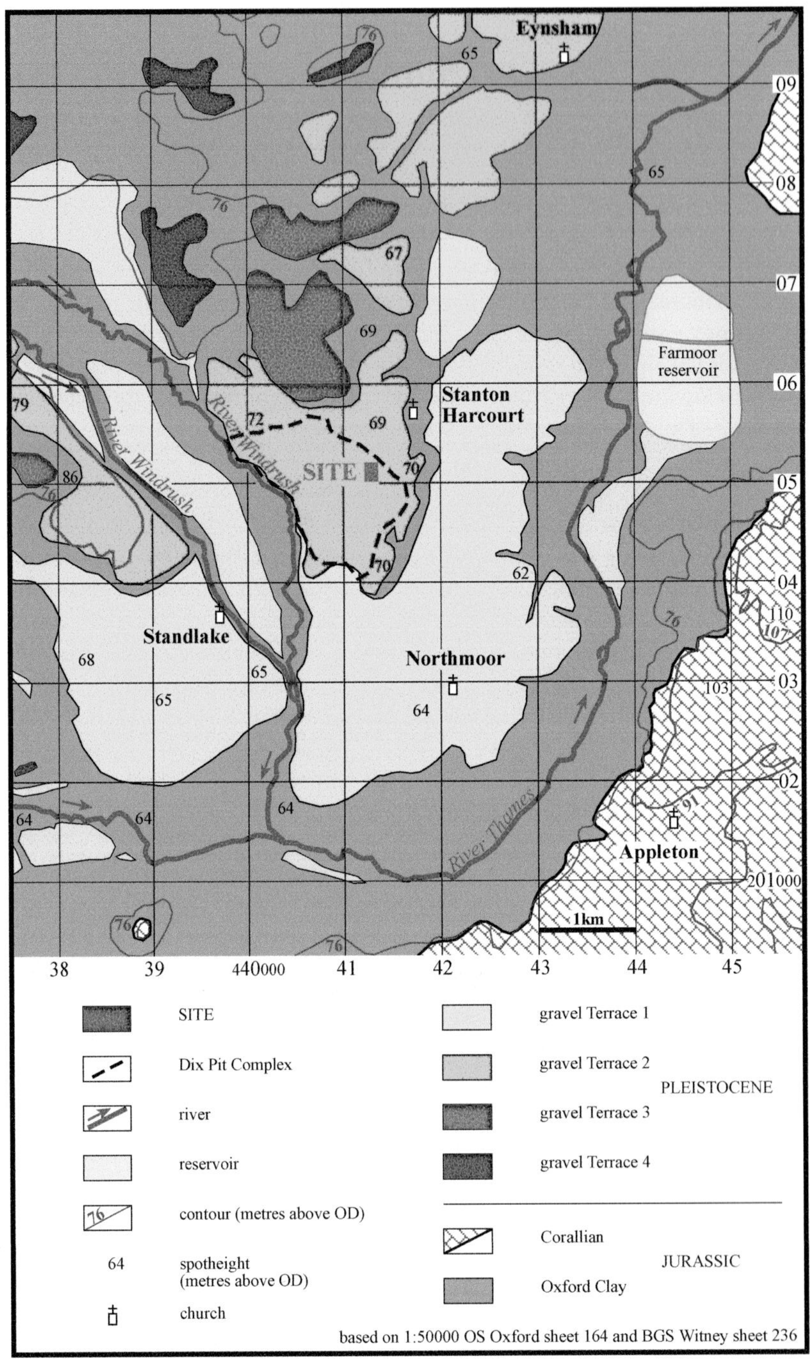


Figure 1.1 Site location map with simplified geology

Figure 1.2 Looking SE at part of the excavation site showing the Pleistocene fluvial sediments that had been left at the base of the pit after gravel extraction and, below these, the Oxford Clay.

Preliminary reports on the early years were given by Buckingham *et al.* (1996) and Scott and Buckingham (1997). Further descriptions of the excavations and the importance of the finds were published by Scott and Buckingham (2001), Scott (2001), Jones *et al.* (2001), Buckingham (2007) and Scott (2007).

It was a very interesting time to undertake this excavation as new research on the application of oxygen isotope stratigraphy to oceanic sediments was indicating a far greater number of warm and cold periods within the last 2 million years than had previously been thought. In the absence of suitable material for absolute dating, previous attempts to distinguish between British interglacial deposits had depended solely on the terrestrial record. A means of distinguishing between interglacial deposits based on botanical remains had been proposed by West (1963, 1968). Mitchell *et al.* (1973) applied West's palaeobotanical interpretations of the temperate deposits, together with geological evidence for the deposition of other sediments under extremely cold or even glacial conditions, to create a chronostratigraphic framework that would enable Quaternary specialists to identify glacial and interglacial deposits. However, it soon became apparent that the 1973 scheme as a chronostratigraphic tool was problematic in the case of deposits that were widely separated and had little or no pollen. Furthermore, the fossil mammals from the Lower Thames seemed not to fit into the scheme and Sutcliffe (1975, 1976) argued for a hitherto undocumented temperate phase between the Hoxnian/Holsteinian Interglacial and the Last Interglacial to account for anomalies within the large vertebrate assemblages from these localities.

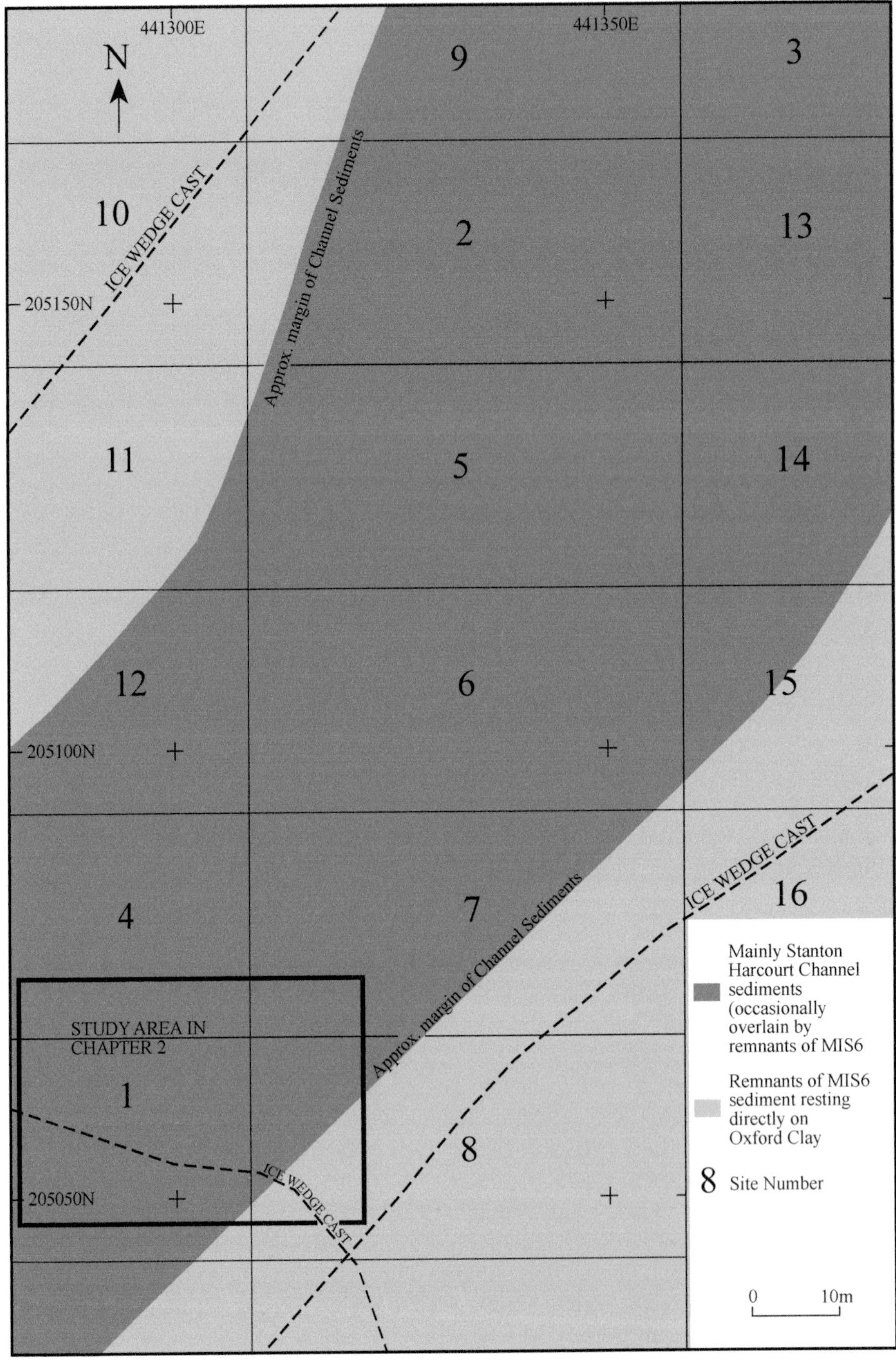


Figure 1.3 Plan of the excavation grid at Dix Pit, Stanton Harcourt

The emergence of the oxygen isotope record to become a stratigraphic standard with which terrestrial sequences may be correlated was thus widely welcomed. Of particular relevance to this site was the proposal that the temperate deposits of the Stanton Harcourt Channel should be equated with the so-called 'new' interglacial (MIS 7) between the established Last Interglacial (Ipswichian) and the Hoxnian (Briggs *et al.* 1985). Amino acid racemisation of molluscs from the site supported this view and indicated a date of c.190,000-200,000 years BP (Bowen *et al.* 1989).

## The excavations

The area covered by fieldwork between 1990 and 1999 was approximately 150m x 100m and was divided into a number of separate excavations (Figure 1.3). From the outset, bones, teeth and tusks were numerous and frequently associated with Pleistocene molluscs and vegetation (including oak) clearly indicative of a warm climate. Although many larger bones and tusks appeared to be in poorly sorted coarse gravels, there were also other bones in silt and sand layers. Sedimentary structures such as ripple laminae within these sand or silt layers indicated relatively undisturbed sedimentary deposits within the river channel. There was no evidence of major floods or mixing of material of different Pleistocene environments and this suggested that the basal sediments at Stanton Harcourt were all deposited during the same interglacial by the same river, gradually, over a period of time (See Chapters 2 and 3).

Of the numerous large vertebrate remains at the site, mammoth was overwhelmingly the best represented (Table 1.1; Figures 1.4 and 1.5). Initially, this was identified as the cold-adapted woolly mammoth *Mammuthus primigenius*, yet the associated environmental evidence indicated that these mammoths were living in a warm climate in the vicinity of deciduous woodland. This was further supported by the discovery in 1993 of the molar of a straight-tusked elephant *Palaeoloxodon antiquus*, a true temperate forest species. There was much discussion as to whether the mammoths represented populations from the previous cold stage that had been isolated in Britain by the MIS 7 sea-level rise and adapted to a temperate habitat. However, during the course of preparing the fossil material for publication, new research into the evolution of mammoths indicated that, on the basis of certain distinctive dental characteristics, some assemblages of mammoths previously described as woolly mammoths were more correctly a late form of *M. trogontherii* - the steppe mammoth (Lister and Sher 2001). Based on their criteria, the Stanton Harcourt mammoths were also identified as *M. trogontherii* (Scott 2007). The significance of this

| Number of specimens | |
|---|---|
| **Carnivores** | |
| *Canis lupus*, wolf | 1 |
| *Ursus arctos*, brown bear | 10 |
| *Felis spelaea*, lion | 3 |
| **Herbivores** | |
| *Palaeoloxodon antiquus*, straight-tusked elephant | 57 |
| *Mammuthus trogontherii*, steppe mammoth | 922 |
| *Proboscidean unidentifiable post-cranial* | 274 |
| *Equus ferus*, horse | 34 |
| *Cervus elaphus*, red deer | 4 |
| *Bison priscus*, bison | 125 |
| *Bovid/equid unidentifiable post-cranial* | 21 |
| *Other post-cranial unidentifiable to species* | 126 |
| **TOTAL** | **1577** |

Table 1.1 Summary of identifiable large vertebrate remains from Stanton Harcourt

Figure 1.4 Excavating a mammoth tusk

Figure 1.5 A mammoth mandible with dentition being uncovered

finding was that the steppe mammoth is a common element in interglacial faunal assemblages and thus the apparent anomaly of mammoths in a temperate environment was resolved. A particularly diagnostic feature of the Stanton Harcourt mammoths is their small size relative to steppe mammoths from earlier interglacials (Figure 1.6). As described in Chapter 4, this is now recognised as an important marker for MIS 7 (Lister and Scott in press).

Unexpectedly, soon after the excavations began, the first of more than 30 stone artefacts was discovered, many of them in good (unrolled) condition. Until the emerging realisation of the 'new' interglacial c.200,000 years ago, it had been generally accepted that the north European interglacials were characterised by heavy forestation. The apparent absence of archaeological evidence from known sites of interglacial age supported the conclusion that forested habitats were unsuitable for hunter-gatherers. Although this remains true for the Last Interglacial, the evidence at Dix Pit indicates that the terrain during MIS 7 (at least as represented by the excavated deposits) was predominantly open with some woodland in the vicinity and was evidently favourable to hominins (See Chapter 8).

Continued excavations at Stanton Harcourt throughout the decade enabled detailed documentation of this MIS 7 environment, a habitat where hominins co-existed with lion, bear and various large herbivores, including the steppe mammoth. This was a unique opportunity to excavate an ancient riverbed in 3-dimensions and to establish the nature of the fossil accumulation in its sedimentary deposits.

## Geological context of the Stanton Harcourt Channel

Understanding the sediment that is deposited by a river requires an appreciation of a large number of variables. The amount and particle size of the sediment carried by a river is mainly a feature of its discharge. This in turn depends mainly on the rainfall, the river gradient and the size of the catchment. The type and structure of the rock over which the river flows are

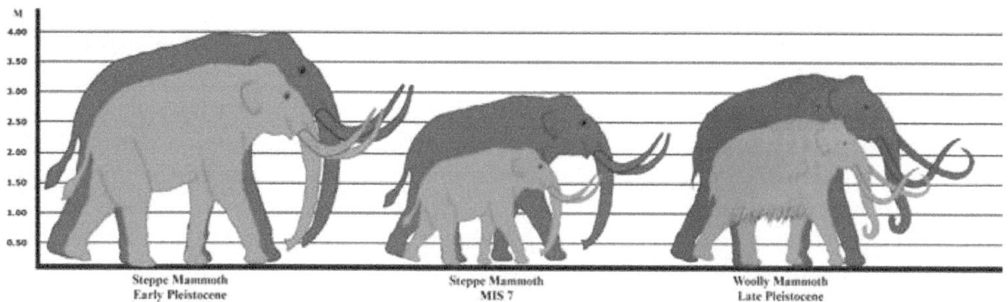


Figure 1.6 Shoulder height in British Pleistocene mammoths estimated from fossil post-cranial remains. Minimum and maximum shoulder heights for the Early Pleistocene steppe mammoths and the Late Pleistocene woolly mammoths are calculated from the skeletal remains of males (data from Lister and Stuart 2010). Females are likely to have been smaller. The remains of the MIS 7 steppe mammoths were not sufficiently complete to determine the sex of individuals. Thus the figure simply represents the shortest to tallest individuals represented (Scott and Lister in press).

also significant factors in determining how easily material can be entrained and transported by the water and how the size and nature of that river evolves. Disarticulated bones and pieces of wood can also be regarded as potential sediment to be transported if the discharge is high enough and deposited when the energy wanes.

### *The regional setting of the Stanton Harcourt Channel*

At the time of the existence of the Stanton Harcourt Channel the catchment area of the Upper Thames and its tributaries was probably of a similar size to that of today. The palaeo-Thames had a meandering route on low lying relatively flat land along a wide valley between Cheltenham and Oxford, north of its current route. This area is mainly defined by the NE/SW strike of the Jurassic Oxford Clay, north of the Corallian limestone escarpment. Throughout the Middle and Late Pleistocene, the River Thames has shifted laterally down-dip to the SE, in response to both channel migration and regional uplift (Maddy 1997).

The main tributaries drained NW to SE down the limestone dip slope of the Cotswolds, joining the palaeo-Thames on its left bank, when looking downstream to the NE (Figure 1.1). The gradient of the tributaries would have been higher and the potential for both incision and deposition would consequently have been greater than that of the main river, especially during Glacial Stages of the Pleistocene. As a result, previously deposited fluvial sediments remain as discontinuous river terraces on the NW side of the main valley, mainly at the confluences with, and in, the lower reaches of the tributaries. The fluvial deposits form a series of terraces which become progressively younger and at a lower elevation towards the current R. Thames. Bridgland *et al.* (2004) have stressed the importance of terrace sequences for providing a semi-continuous record of Pleistocene fluvial deposition, with periodic incision isolating individual deposits with unique faunal and floral characteristics.

The tributary of particular interest to the Site is the River Windrush. In the area near Stanton Harcourt, sediment mainly from the R. Windrush has encouraged the R. Thames to migrate south eastwards and to undercut the Jurassic Escarpment. The Pleistocene sediments discussed here underlie the Second Terrace surface (approx. 70m above OD) and at the time of deposition were located near the confluence of the River Thames and its tributary the R. Windrush (Fig 1.1). These sediments are separated from the First Terrace by an outcrop of Oxford Clay which underlies the church at Stanton Harcourt village.

At Dix Pit, Stanton Harcourt, the Second Terrace is comprised of two distinct gravel deposits representing two climatic events within the Pleistocene. The upper deposit, mainly composed of limestone dominated sand and gravel, was the deposit of commercial interest and reached a depth of 5-6 metres. Underlying this, the bone-bearing sediments formed about 1m of sediment that remained at the bottom of the gravel pit after quarrying ceased, mostly below 64m above OD. These basal deposits described as the Stanton Harcourt Channel are formally known as the Stanton Harcourt Bed of the Summertown-Radley Member of the Upper Thames Formation and are correlated with the interglacial at Marine Isotope Stage 7 (MIS 7) (Bowen, 1999).

The presence of the earlier gravel terraces and fossils within the Oxford Clay provided the raw material for potential incorporation into the sediment of the Stanton Harcourt Channel. The gradient of the channel would have been low where it flowed over the clay. At the excavation site the bone-bearing channel deposits were found to lie between clear geological boundaries.

*Lower Stratigraphic Boundary*

Mapping of the Stanton Harcourt Channel showed that the MIS 7 bone-bearing sediments of interest had a lower stratigraphic boundary with the Upper Jurassic Oxford Clay (Buckingham *et al.* 1996; Buckingham 2004 and Figure 1.). The excavation area was in the Stewartby Member (Middle Oxford Clay), near the transition from the *Kosmoceras spinosum* sub-zone of the *Peltoceras athleta* zone to the *Quenstedtoceras lamberti* zone (Cox *et al.* 1992). The Stewartby Member is a stiff, light blue grey, plastic clay, which is occasionally silty (Hollingworth and Wignall 1992). Ammonite zone fossils in this clay indicate that the base of the Stanton Harcourt Channel was near the boundary with the Lamberti limestone, a distinctive fossiliferous, calcareous mudstone bed within the Oxford Clay. 'Turtle stones' (large pillow-shaped concretions from the Oxford Clay) were also commonly *in situ* near the top of the clay, at the base of the Pleistocene sediments. Scours and grooves below the sediments of interest, together with local fossils and lumps of clay within them, suggest that this boundary was erosive (Buckingham 2004).

Fieldwork suggested that the harder mudstone layers and turtle stones within the Oxford Clay were instrumental in inhibiting the lateral migration of the Stanton Harcourt Channel. An environment was created where there was net sedimentation rather than erosion which resulted in the location of this exceptional fossiliferous site (See Chapter 2).

*Upper Stratigraphic Boundary*

At the excavation site it was observed that the upper stratigraphic boundary of the MIS 7 bone-bearing beds was also predominantly erosive with scours and cobbles at the base of overlying limestone dominated sand and gravel. Evidence indicated that the overlying sediment belonged to periglacial conditions of the succeeding glacial at MIS 6. Patterned ground with ice wedge casts was mapped. One cast bisected two mammoth tusks and a bison jaw within the MIS 7 sediments (Figures 1.7 and 1.8). A reindeer shed antler was also excavated from the MIS 6 basal sediment. In areas of Dix Pit beyond the margins of the MIS 7 channel, the MIS 6 sediments directly overlie Oxford Clay.

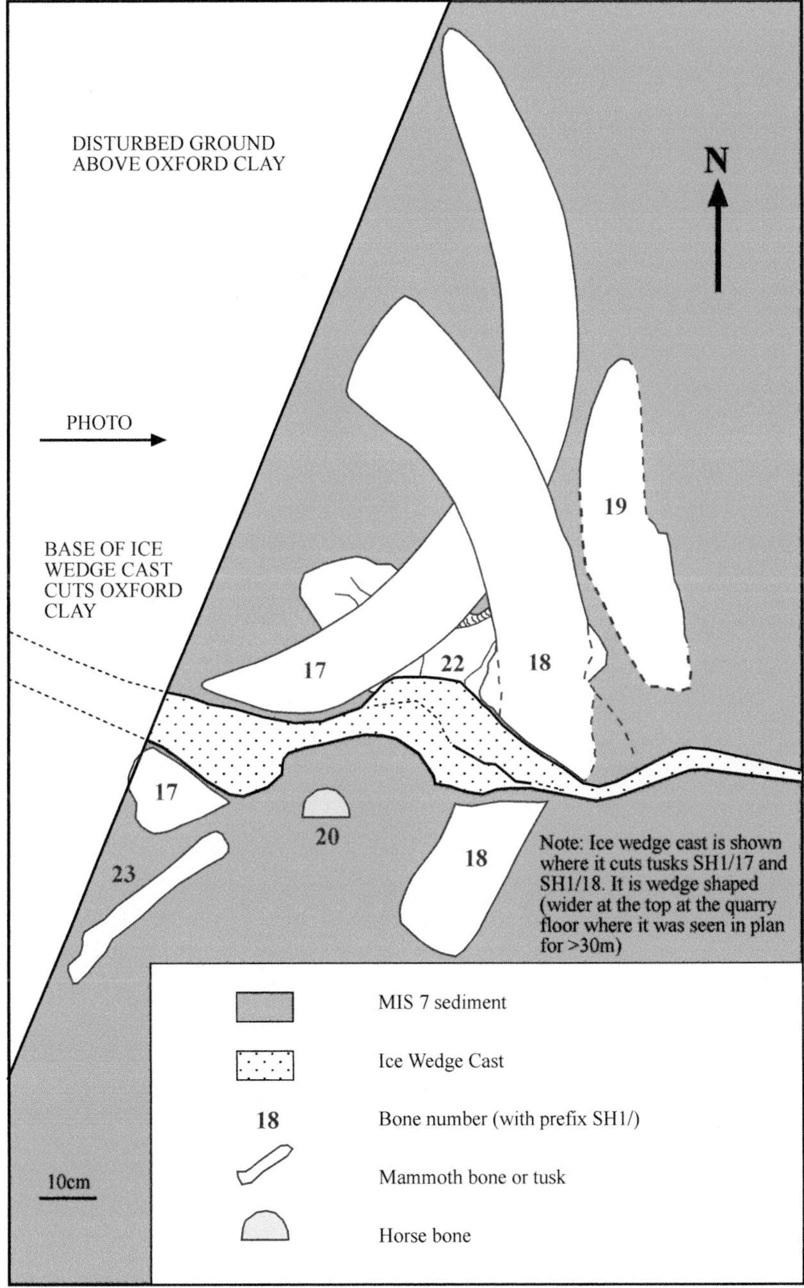


Figure 1.7 (a) Plan of ice wedge cast at Site 1 bisecting mammoth tusks SH1/17 and SH1/18

Figure 1.7 (b) Photograph of the ice wedge and tusks taken from west

Figure 1.8 Ice wedge bisecting bison mandible SH1/340). Inset: the mandible once restored

# Chapter 2

# Evidence for the Contemporaneity of Bones, Wood, Molluscs and Artefacts

There are many instances in which fossils retrieved from fluvial deposits might be regarded as not *in situ* and therefore of limited stratigraphic importance. At Stanton Harcourt, however, a large area was available for field work for an unusually long time. This facilitated extensive excavation of the Channel sediments and detailed recording in three dimensions. As a result, it has been possible to demonstrate various types of preservation (and therefore of stratigraphic relevance) of both animal and plant remains indicating whether they were *in situ*, almost *in situ*, or transported.

Although all the excavated finds were in the Channel sediments, they fall into two categories representing different habitats: the animals (i.e. molluscs, fish) and plants that lived in the water, and the animals (vertebrates, insects, snails, etc.), trees and other vegetation that originated on the landscape bordering the river. As will be discussed in further Chapters, the assemblage as a whole is indicative of interglacial conditions during the latter part of MIS 7. Thus, it could be argued that the thousands of plant and animal remains accumulated randomly in the river over many millennia.

The purpose of this Chapter is to show through the detailed analyses of some areas of the site that a significant proportion of the fauna, flora, and artefacts were contemporaneous with one another. Furthermore, it is evident that some animals and plants had remained undisturbed in their life/death locations in the river, and that some of the animal remains and artefacts originated from the river bank close by.

Two factors are crucial to such interpretations: the condition of the material and a comprehensive understanding of the sediments. Assessing the state of preservation of the fossil material was relatively straightforward. The bones and tusks near the surface of the excavation were usually in a fragile state due to crushing by quarry vehicles and subsequent exposure to the elements for many years thereafter. However, where bones were excavated well below the quarried surface of the site, the majority were not severely desiccated or cracked, indicating that they had not been subjected to prolonged exposure on a land surface and had been buried quickly. Although frequently excavated from gravel layers, many of the bones show little evidence of the kind of damage that would be expected if they had been transported any distance in a fluvial environment (see Chapter 4). Assessing relative contemporaneity in fluvial deposits requires meticulous recording in three dimensions and, at this site, resulted in a dataset of great complexity. As the importance of this cannot be underestimated, the recording process is described here prior to presenting examples of contemporaneity.

When animals die on the flood plain or in the river, there is the potential for remains of the skeleton to be incorporated into the fluvial sediment. During flood events burial could be rapid or the process may be gradual as the carcass decays, with bones becoming buried in different layers of sediment near the death site or they may be dispersed and transported by the current.

In fluvial deposits the type of sediment is controlled by the material available for the river to erode and transport and the energy of the river (discharge). The latter is the result of many factors such as the amount of rainfall, the gradient and the size of the catchment area. When trying to correlate individual patches of sediment over an old river bed, it is important to appreciate that a river can deposit similar looking sediment on a number of occasions or various types of sediment may be deposited in different parts of the channel at the same time. It is not unusual for gravel to be transported in the faster flowing parts of a channel with sand or silt being deposited in quiet water near the bank. If the channel position shifts by erosion then the area of a particular sediment type will also change. Discontinuous beds and layers which grade laterally or vertically are therefore likely in old river deposits.

During the excavations of the large vertebrates at Dix Pit a detailed record was kept of the sediment in which the bones were found. Vertical sections were recorded at approximately 1 metre intervals at Site 1 and parts of Sites 4, 7 and 8. Elsewhere the sections were generally more widely spaced. Although there was clear bedding seen in the sections it was often very difficult to follow individual layers when excavating large bones and tusks especially when discontinuous lenses were encountered. Each section drawing was therefore labelled individually in reverse stratigraphic order following standard archaeological procedure (layers as dug) to allocate a bone or other item of interest, such as wood, to a layer on that section drawing. Layers between sections were correlated at a later date to give a stratigraphic order (sedimentary beds numbered in order of deposition). These beds were the smallest unit that could be correlated between sections and sometimes represented more than one depositional event, especially if they consisted of multiple very thin layers of alternating sediment type. Notebooks and plans were used to record extra information as each square was excavated.

## Stratigraphy and sedimentology

Sedimentary evidence collected from previous visits to the area and the excavations under the direction of the authors, suggests that during MIS 7 there was a meandering river at the site (Briggs *et al.* 1985; Buckingham *et al.* 1996). The overall distribution of the MIS 7 sediment and the dominant current direction across the excavation area appeared to be from SW to NE with some local variations (Chapter 1, Figure 2). This river transported a variety of sediment types, mainly gravel with some cobbles at the base (the bed load) with sand and silt comprising the suspended sediment load. Scours and grooves in the surface of the Jurassic Oxford Clay suggest that it was being actively eroded by the MIS 7 river and therefore the channel probably had clay banks. There was unlikely to be much alluvium on the floodplain at this time as this is largely a feature of modern environments after cultivation began in the Neolithic Period.

The excavations revealed that large Pleistocene vertebrate bones, teeth and tusks are an integral part of this fluvial deposit, together with *in situ* deciduous tree roots, other wood and fresh-water Mollusca. The latter are abundant and the bivalves are often articulated so that the distribution of warm water species such as *Potomida littoralis* (Cuvier) and *Corbicula fluminalis* (Müller) were used to help map the distribution of the MIS 7 sediment across the excavation site. These bivalves, in a similar way to mussels, would become disarticulated soon after death. Their frequent completeness indicates that they were located near to their life/death position (see Chapter 6).

The environmental analyses in the following chapters are based on samples taken from areas of the site where bones, wood and molluscs are associated. In other parts of the excavation site better sorted sand and gravel deposits were encountered and the majority of this sediment is believed to date to the following cold stage (MIS 6) and is not discussed further here.

Detailed analysis of the sediments and their biological content has enabled the identification of different depositional environments within the MIS 7 river. Low energy sediments deposited marginal to the flood plain are represented as well as higher energy channel deposits.

In a meander the '*thalweg*' is faster flowing water which moves coarse bed-load material in a roughly 'S' shape from one bank to the other. Areas marginal to the '*thalweg*' have slow moving or still water allowing silt or sand to accumulate. There is also likely to be faster flowing water between meander bends in straighter sections of the river. In the latter there would mainly be through-flow of fine sediment with a lag of coarse material at the base. Throughout the excavation site, in the higher energy gravel deposits, single large vertebrate teeth, disarticulated bones and mammoth tusks are common. Some of the teeth have clearly been rolled by the current and are interpreted as lag deposits. However, even within the main channel, many of the tusks are complete and appear to have remained a long time in the same position while a succession of different sedimentary layers was deposited in and around them.

Large bones and tusks are commonly associated with abundant cobble-sized *Gryphaea dilatata* (Sowerby) shells. These Jurassic oyster shells have been eroded from the Oxford Clay, from levels in the clay stratigraphically above that found under the MIS7 sediment in the excavation area. At the present day these shells outcrop on the Corallian escarpment near Appleton, to the SW of Stanton Harcourt (Chapter 1, Figure 1). There is a variety of other cobbles and the occasional boulder, probably a lag from earlier climatic events, but the *Gryphaea* dominated the bed load at the time of bone deposition. These shells often form an armour layer on the channel bed. This is a relatively coarse surface layer which forms when fine material is winnowed away by the current causing improved packing of individual grains. The *Gryphaea* are often concave down, protecting a layer of finer sediment and tiny molluscs or are stacked together (imbricated) against bones, tusks, wood or other large items and dip upstream. Some bones have flat elements facing upstream. Once imbricated it is more difficult for the current to pick up those items again. It appears that even in the faster flowing parts of the river, the water current was often of insufficient strength to move the *Gryphaea* and large bones far.

The bones are also associated with 'turtle' stones: pillow shaped concretions found locally within the Jurassic Oxford Clay at the excavation site. When uncovered these mudstone or siltstone boulders dry and crack into sub-angular pieces resembling a 'turtle' shell. The MIS 7 river has uncovered these septarian nodules, many of which are complete and are on clay pillars with *Gryphaea* against them (Figure 2.1).

At Site 7, where there is a straight section of the river, an *in situ* 'turtle' stone was observed below the base of the channel, within the Oxford Clay. In the same area other complete

Figure 2.1 A complete 'turtle' stone partially resting on the bluish-grey Oxford Clay (left) with *Gryphaea* shells against it on the upstream side. The distance between pegs 13 and 14 is 1 metre

specimens were only partially uncovered. The 'turtle' stone in Figure 2.1 is also shown lower centre in Figure 2.2. Broken pieces of 'turtle' stones were frequently excavated within the basal layers of the channel sediment. The SW to NE trend of the 'turtle' stones closely mirrors the current direction at Site 7. The area of Site 7 shown in Figure 2.2 has mostly large disarticulated bones, single teeth, mammoth tusks and large pieces of wood associated with abundant *Gryphaea* valves. It is likely that the 'turtle' stones and thin discontinuous mudstone beds found locally within the Oxford Clay were more resistant to erosion than the softer clay and helped to inhibit the lateral migration of the river.

Imbrication, grooves and other indicators support a dominant SW to NE current direction at Site 7. Wood and bones have their long axis mainly at right angles or elongate to the current. It is suggested that Site 7 was probably a fast flowing, shallow, turbulent part of the channel (a 'riffle' section). Cobble-sized *Gryphaea* shells formed the bed of the river. In this environment, there would have been winnowing taking place and any small bones would have been transported further downstream. The tusks are mostly complete. Some of them have one end embedded in lenses of clay or silt or they are in multiple layers. Their awkward shape may have inhibited movement by the current. Some of the bones and teeth show evidence that they have been rolled. The condition of the bones and the body parts recovered is discussed further in Chapters 4 and 5. Other areas of the excavation site at Stanton Harcourt appear to have less disturbed bone collections and some articulated specimens were excavated. There are several instances where parts of a mammoth carcass appear to have remained close to the site of death of the animal.

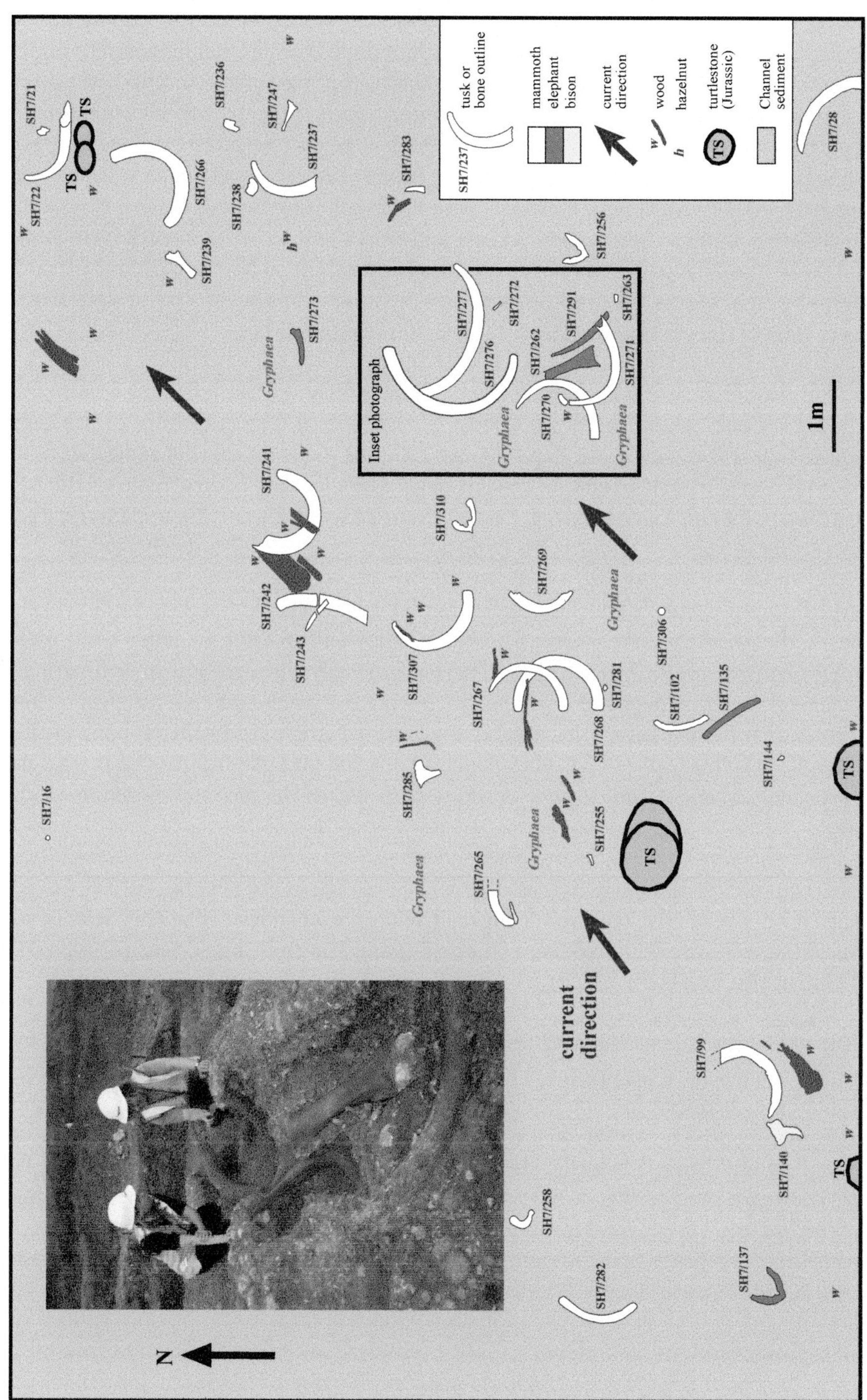


Figure 2.2 Mammoth tusks, teeth and bones with wood and 'turtle' stones at Site 7

As part of a doctoral thesis CMB selected an area at Site 1 which appeared to have a concentration of relatively undamaged bones, stone artefacts, wood, seeds, nuts and molluscs for a sedimentological study to document the sequence of deposition in that area and to place the finds in context. Very thin layers of silt, sand and clast-supported gravel implied a gradual accumulation of sediment here and the sequence seemed relatively undisturbed compared to some other parts of the site. The layers between sections were correlated to produce a series of plans to illustrate the 3-dimensional build-up of sediment near the left bank of the river (Buckingham 2004, 2007). A variety of methods were used including field notes, overlays and tables. Since 2007 this area has been extended East and North East to include sections in Sites 4, 7 and 8 and these plans have therefore been modified and extended to give a clearer picture of the stratigraphy right across the Stanton Harcourt Channel from the left to the right bank in the southern part of the excavation site. A summary is presented here.

The Oxford Clay profile at Site 1 is asymmetric, sloping gently on the left bank and more steeply at the right bank, where there was probably a deep pool. The curvature of the clay contours implies that this is a meander bend (Figure 2.3). The plans reveal that sediment first accumulated on the inside of this bend on the left bank when looking downstream to the NE and developed as a point bar. Anastomosing grooves in the Oxford Clay surface and imbricated gravel support water movement towards the bank as the current rotated anti-clockwise.

The first bed contained some cobble-sized sub-angular silt blocks which may be a broken 'turtle' stone. At this time the main current probably followed the tight meander bend with the outside right bank near 1R (Figure 2.3) and there was mostly through flow of sediment in the '*thalweg*'. Gravel was then deposited as bars which were transverse (at right angles) to the current and rotated anti-clockwise. They extend at an oblique angle from the left bank at the downstream end of the meander bend. The moderate sorting and general lack of matrix in these beds, especially in Bed 4, indicates re-working of gravel in the channel, probably close to the '*thalweg*'.

At the downstream end of the early bar sediment in Site 4 there are several 'turtle' stones which appear to be where the '*thalweg*' changed direction and moved into the next bend downstream. These were probably resistant to erosion and the right bank near position 1R was located here at this time.

After the early gravel there was a waning of energy and silty clay was deposited followed by SW to NE trending small sand bars (scrolls) forming an undulating surface. Cross-lamination in the sand and cross-bedding in very fine gravel layers, have steeper faces to the NW and indicate a dominant SE/NW current of low to medium energy up the bar surface. Water movement was initially inward towards the left bank, probably by secondary currents to the '*thalweg*'. There are sand scrolls of different ages, formed by gentle currents re-working sediment on the bar top. The original meander bend was then filled with stacked layers of silt or sand. At this time, it is likely that the right channel margin gradually shifted from position 1R to 2R as the outer bank was eroded (Figure 2.3).

The early bar sediment and the meander infill of fine sediment are below the layers containing bones at Site 1. The bar at this time had an undulating profile due to the sandy scroll bars on

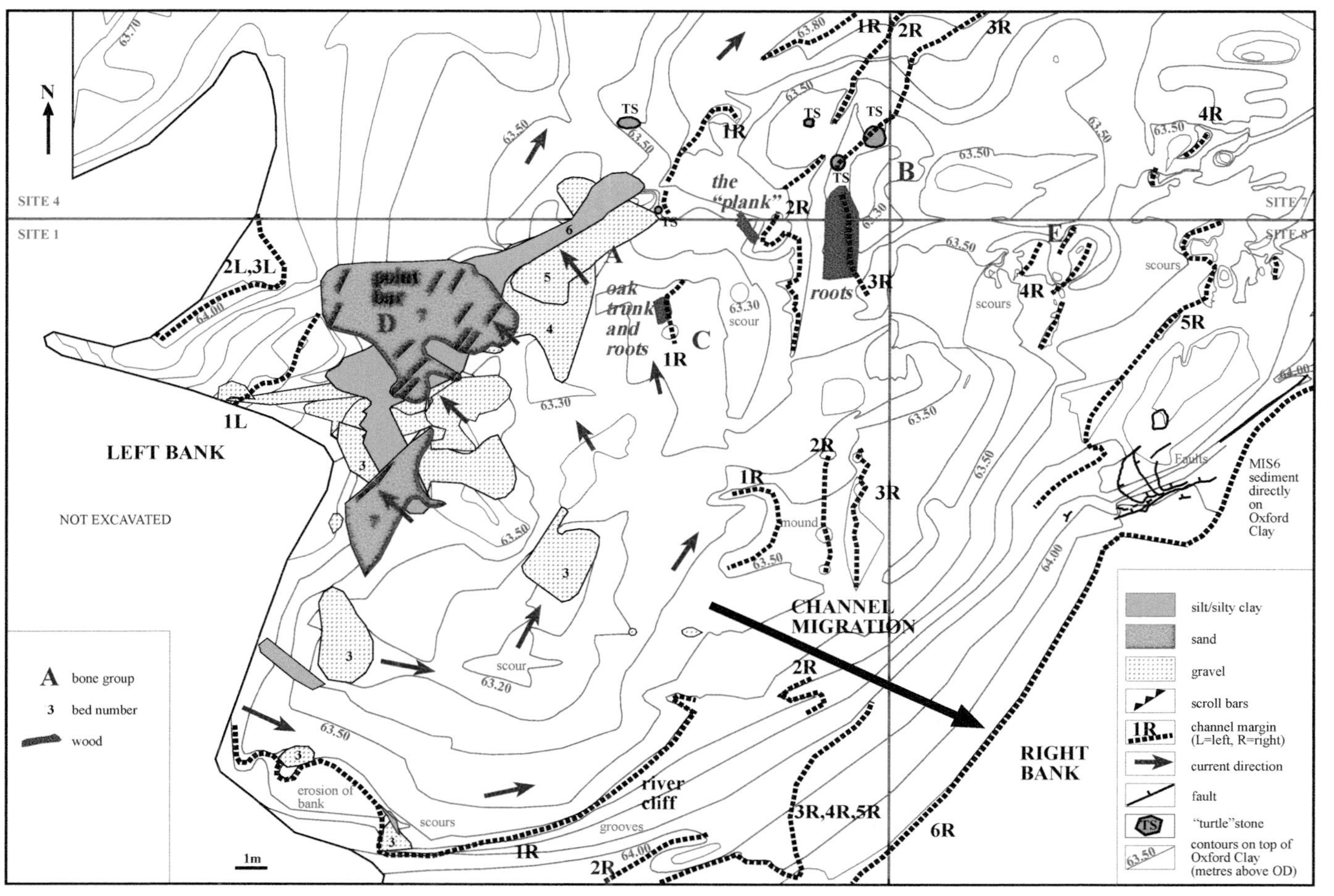


Figure 2.3 Early bar sediment on the left bank of a meander bend

the top and the sediment pile was up to 50cm in depth. It had a ridge and swale topography. Some fine sediment accumulated in lenses between the scrolls.

Excavated organic material indicates that the point bar was well established with some grassy vegetation before the first bones appear in the sedimentary sequence at this locality, in predominantly gravel beds. The first of several stone artefacts also only appears on the bar at the same time as the bones. These are discussed later. Warm water Molluscs and wood are seen throughout the bar sediments and it is likely that it was channel migration which changed the type of sediment that was being deposited at this locality, rather than a change of climate. There is no evidence of major flooding but changes in the sediment type do imply some seasonal variations in river discharge. The predominantly quiet water area in the meander bend where silt or sand had collected was later an area where gravel was being moved.

Beds became more complex as the bar grew. Many beds thin towards the left bank over the earlier sediment and become finer grained upwards. This supports the conclusion that the sediment on the bar was mainly deposited by lateral accretion. This is an important conclusion as it was first thought that some of the sequence was missing due to the quarrying activities. Although the MIS 7 sediment at the excavation site was only about a metre in depth, the pit floor was almost exactly at the interface with the overlying commercial MIS 6 sediment.

Large bones and tusks which were excavated from the basal part of the gravel frequently rest on silt or sand of the early bar sediments (Figure 2.4 and Figure 2.8). As fine-grained sediment accumulated as a point bar on the left bank, when looking NE, the stratigraphic evidence at Site 1 suggests that the *thalweg* (the fastest flowing part of the channel) changed its position and migrated across the Oxford Clay as the channel eroded its right bank. There would have been movement of gravel and some cobbles, the bed load material, on this side of the channel. This sediment was moved as transverse bars which gradually rotated towards the left bank to be deposited at the downstream end of the point bar. Evidence suggests that occasionally, probably during seasonal flooding, coarser material from the bed load was washed on to the bar. Some of this sediment collected with

Figure 2.4 Bones and tusks stacked against sandy silt at the base of the bar (scale = 30 cm)

environmental material between the ridges of the earlier scroll bars. Coarse sediment was also deposited downstream on the Oxford Clay. The presence of large bones and tusks on the bar encouraged further channel migration as the *thalweg* shifted laterally to preferentially erode the softer Oxford Clay.

Scours and grooves in the Oxford Clay indicate that there was probably an eroded river cliff on the right side of the channel. Small irregular ridges or mounds between scours in the Oxford Clay surface and the presence of *in situ* tree roots in the overlying channel sediments indicate possible positions of the right bank as the channel migrated south-eastwards (1R, 2R, 3R, 4R, 5R and 6R) on Figure 2.3 (R = right). There only remains a faint indication of the earlier right banks as they would have periodically collapsed and then the clay would have been washed away or incorporated in the matrix of any deposited sediment. The sediment in these areas often includes lumps of clay or has a clay matrix. The proposed channel margins are discontinuous on Figure 2.3 to illustrate the initial areas of partial destruction of the right bank. Erosion would have been sporadic and remnants of earlier banks would have been a feature at the channel margin. Several 'Turtle' stones are present near the downstream end of the point bar where the river changed direction into the next bend (the inflexion point) as the bank remained intact here for longer than other parts of the right bank.

At Site 8 there is the indication of a river cliff on the SE and E side of the channel where the clay had been undercut by erosion (Figure 2.3). There are tight curvilinear faults in the clay near the last position of the right bank (6R). Weathered Oxford Clay and channel sediment here has been moved after deposition, by slumping or slip movement downward along these fault planes. Similar modern slips in undercut clay are shown in Figure 2.5. In contrast, the left

Figure 2.5 Modern curvilinear faults in clay. Rotational sediment movement is downward into the void on the left.

bank was an area of net sedimentation so it was less eroded and probably just degraded with time (1L,2L,3L) on Figure 2.3, L=left).

## Bone assemblages at their death sites

Although many of the bones in the Stanton Harcourt Channel Deposits are disarticulated, there are some clusters of bones or mammoth skulls which appear to have had minimal disturbance. These are discussed below as Groups A-E. Before describing them, it is informative to consider an example of the taphonomic processes affecting an African elephant after death that might be applicable to the interpretation of the Stanton Harcourt bones.

A young adult female died on a Zambian lakeshore during the hot dry season of early October 2010 (White and Diedrich 2012). Lions were first to scavenge on the carcass, feeding on the intestines and inner organs; spotted hyenas followed, scavenging on the fresh carcass with an emphasis on the feet and leg bones. Following this initial scavenging phase, where the fresh meat and softer material was eaten and the majority of the bone damage occurred, the desiccated remains were abandoned on the lakeshore as a more or less intact carcass with the thick hide covering the mostly articulated skeletal elements. During the seasonal floods from December 2010 through May 2011, the carcass was submerged. Nearly one year after death, skeletal material from this elephant lay scattered over an area of 20 x 25 m but the main concentration of bones, including most of the larger bones and two articulated sections of the vertebral column, remained within a 10 x 10 m area (Figure 2.6). Additional bone damage (cracks and flaking) was not attributable to further scavenging by large predators but rather to changes in temperature and humidity. However, the dispersal of the bones had been caused by the seasonal flooding and most of the smaller bones were missing.

Although the environment is not exactly comparable, similar processes were likely at Stanton Harcourt. After the selective removal by predators, any remaining bones within the channel banks could have been moved by water and sorted with other sediment of a similar size, shape and weight. Seasonal flood events may also have moved bones lying on the surface if water spilled on to the flood plain. Even though they may have moved from their death site, the transport may be very limited before they came to rest again, became buried and thus part of the sedimentary record. If there is relatively little evidence of damage caused by rolling and bouncing in the current, or if they are still articulated, then these bones are likely to be representative of the environment in which other associated environmental material is found.

The bone groups A - E discussed here are all in the southern part of the excavation area and are described in stratigraphic order. The bone descriptions are by KS and the context is provided by CMB. Plans of the sedimentary context show the minimum extent of any particular bed based on the available section drawings. Only remnants of earlier beds are likely due to later erosion as the channel position changed with time.

A photograph is included of the modern River Evenlode near Combe Mill, in an area with a similar scenario to that presented in Figure 2.3, to indicate the suggested locations of these bone groups within the MIS 7 channel. The right bank of the River Evenlode, a tributary of the

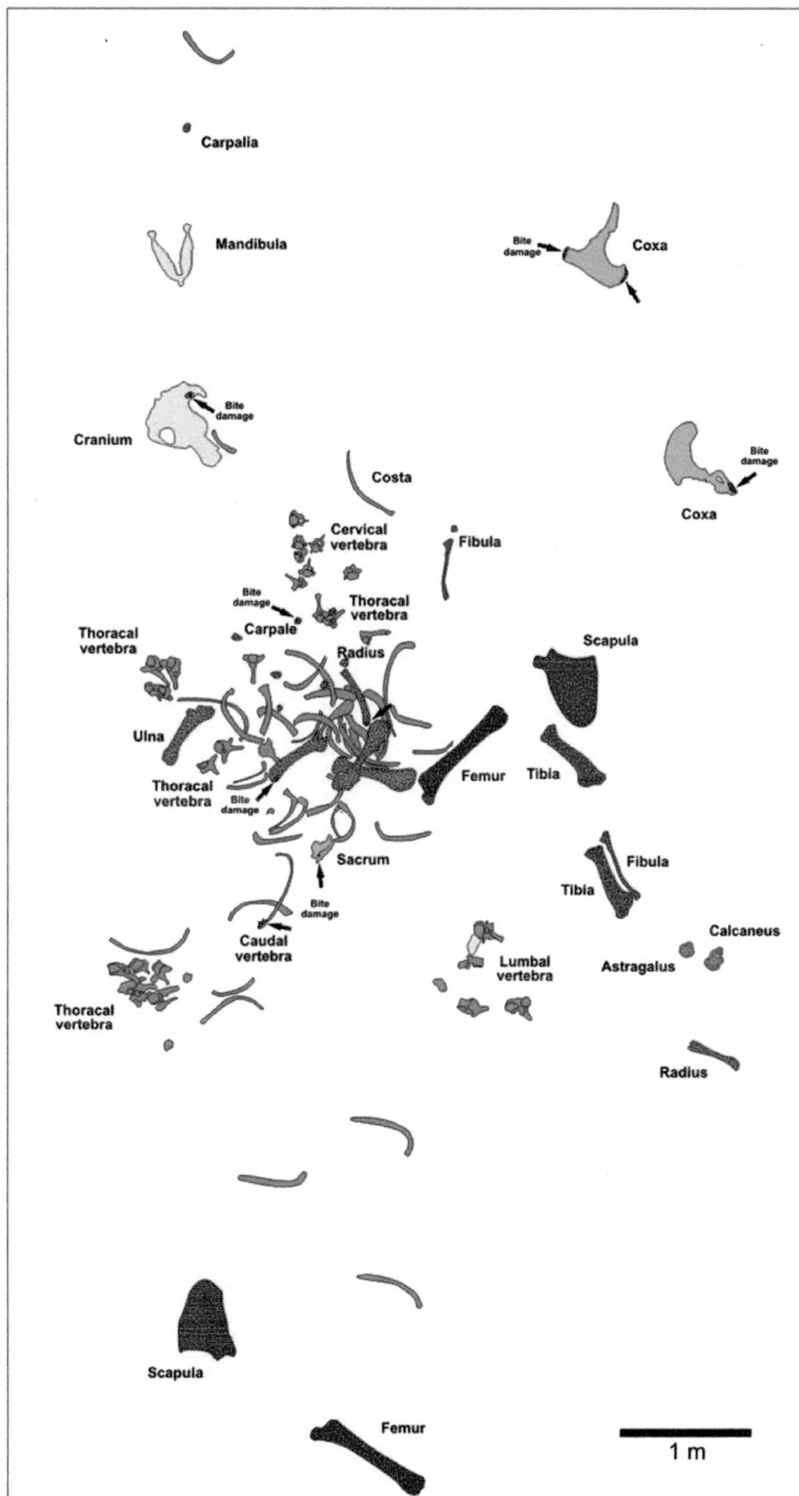


Figure 2.6 Map of the bones of an African elephant *Loxodonta africana* one year after death Skull = yellow, fore limbs = green, hind limbs = brown, axial skeleton = grey, pelvis = blue (from White and Diedrich 2012).

Figure 2.7 Proposed locations of bone groups in the channel
(Photograph taken at Combe Mill, Oxfordshire, on the River Evenlode)

River Thames at the present day, shows erosion in several places on the outside bend of the river and a point bar is developing on the inside bend (Figure 2.7).

### *Group A*

The mammoth skull SH1/242 forms the focal point of a possible associated group (Figure 2.8). This centres on the skull of a very old individual (c.55 years of age: see Chapter 4). Two tusks (SH1/170 and SH1/264) of similar size lie to the south and the north-west of the skull. Their relatively undamaged condition, particularly in the case of SH1/170, suggests that they scarcely moved once they dislodged from the premaxillae (tusk sockets) of the skull. There is also a scapula (SH1/261) and ulna (SH1/260) of a mature mammoth. Other possible associated bones are a thoracic vertebra (SH1/255) and two ribs.

*The context of the mammoth skull SH1/242 in Group A*

There is a very small bird bone SH1/273 immediately under the edge of this skull on the east side. The base of SH1/242 is mainly in a trough of mixed grey silt/silty clay and poorly sorted gravel with cobble sized *Gryphaea* above Oxford Clay (Bed 37 of the sedimentary sequence in Figure 2.9). This layer thins W over the earlier sediment pile and there are many *Potomida littoralis (Pl),* freshwater bivalves indicating an MIS 7 age.

The skull is partially on the Oxford Clay and appears to be against ridges of earlier gravel, which trend SW to NE. These are probably transverse bars deposited at the base of the channel. This sediment was part of the early development of the point bar to the west of the

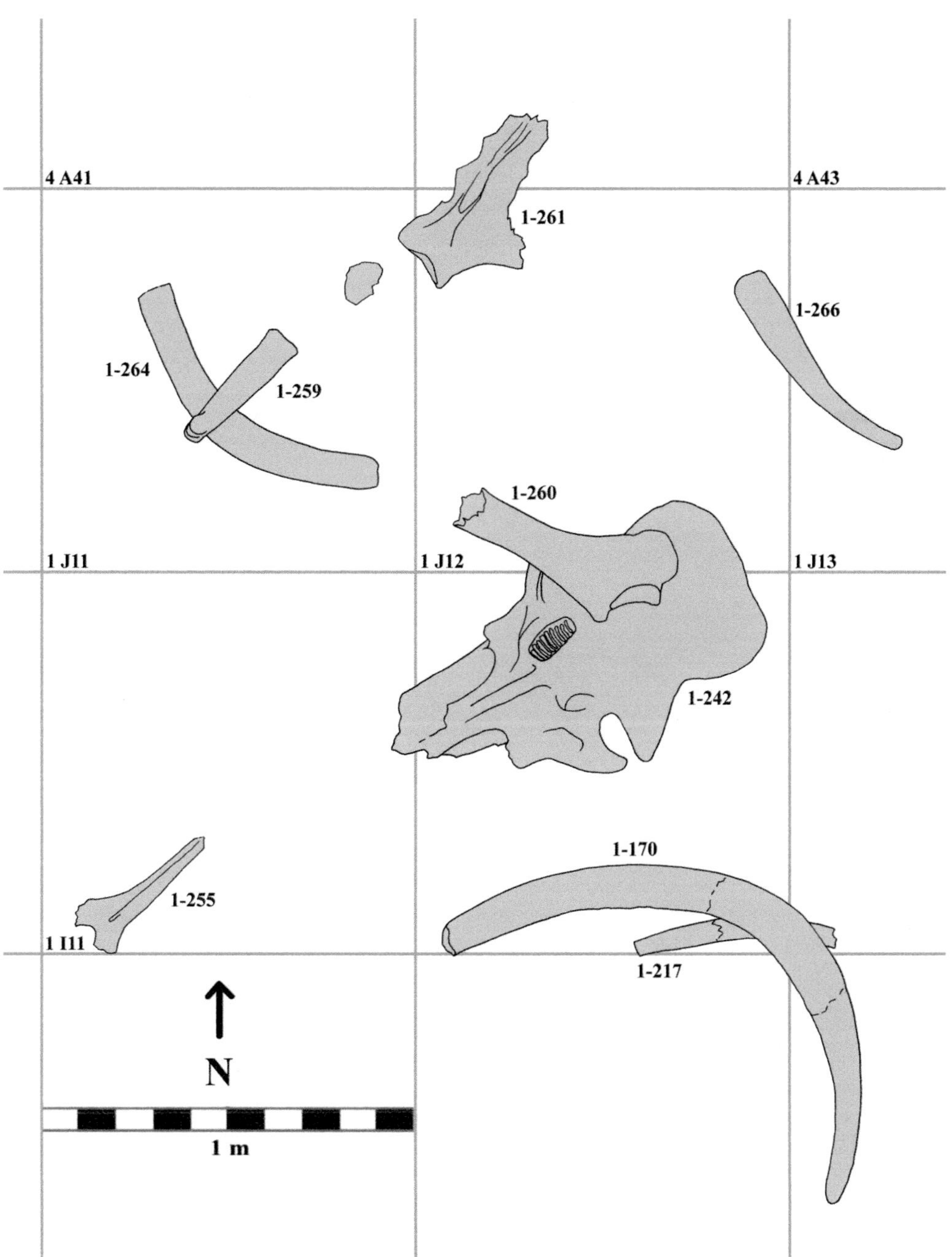


Figure 2.8 The bones of Group A: centred around mammoth skull SH1/242

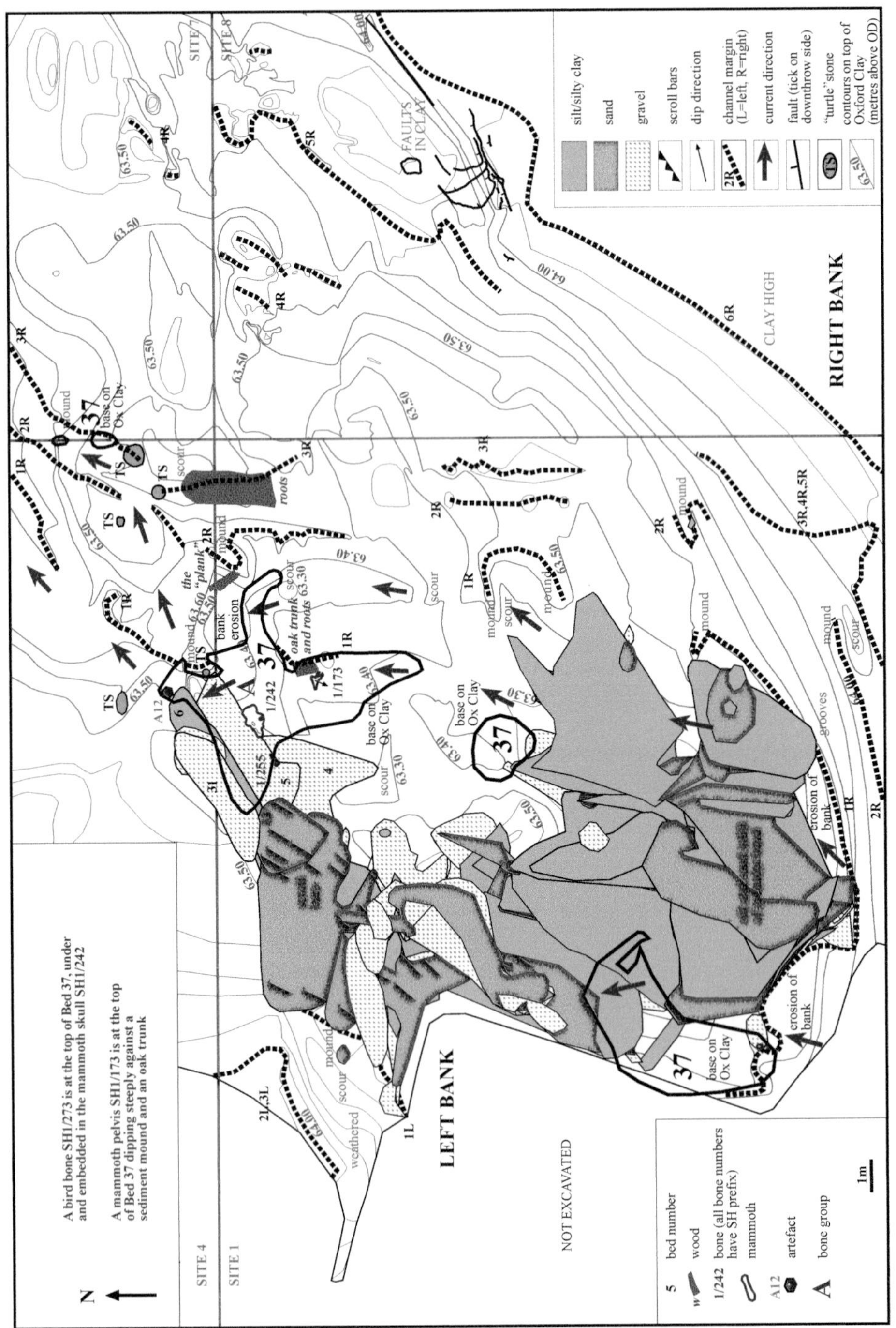

Figure 2.9 Bed 37. Bone Group A. Mammoth skull SH1/242 in the channel in Bed 37

skull, projecting from the left bank at the downstream end of the anti-clockwise meander bend at Site 1 (Figure 2.3).

It appears that the skull SH1/242 was also close to the right bank when the latter was at position 1R, just before the next clock-wise bend into Sites 4 and 7. Sedimentary evidence suggests that poorly sorted gravel of Bed 37 was partially banked up against a 'turtle' stone and a clay mound when the right bank was at position R1 (Figure 2.9). This bank had been partially eroded, especially between the 'turtle' stone and an *in situ* oak trunk with roots, to form the start of a new bank eastwards near 2R. There was probably a river cliff where the current had undercut it. The tree roots and the remains of the tree trunk imply that this bank had mature vegetation (as in Figure 2.7). The area to the east of the trees would have been a remnant of the floodplain. The bank at position 1R is also degraded in the southern part of Site 1. Some gravel of Bed 37 was moved over fine sediment in the original meander bend and it is probable that the bank had also moved southwards at this time.

The skull may be *in situ* or may have been moved slightly by the strongest current (*the thalweg*) as the latter veered from the right bank across to the left bank (Figure 2.7). This skull is upside-down but it is not clear if the current would have been sufficiently strong here to overturn it as it is considerably larger than the *Gryphaea* and other cobbles associated with it. An artefact (A12) is in Bed 37 just north and downstream of the skull (see later discussion on artefacts).

There are also small patches of Bed 37 in Site 7 which would involve sediment movement to the NE through another small breach in the bank 2R between three other 'turtle' stones. This is at the downstream end of the original meander bend near the start of the next bend, between the right bank positions 2R and 3R at this locality.

There were other breaches of the right bank into Sites 7 and 8 soon after the skull SH1/242 became partially buried. Bed 38 in Site 7 (Figure 2.10) represents two areas of deposition, either side of another upside-down skull (SH4/124 in Group B, which is discussed later). The main area of deposition is downstream of the group of 'turtle' stones and is an elongate deposit trending SW to NE on the right bank near the start of the next meander curve downstream. Bed 38 is very thin here and probably represents only moderate erosion of the bank at 3R. There is a smaller area of Bed 38 where a new right bank was developing closer to position 4R. Scours and small potholes in the Oxford Clay surface indicate turbulent flow on this side of the channel. SW/NE grooves in the clay and imbricated *Gryphaea* on the SW side of a clay mound indicate a moderately strong current from the SW. The current probably flowed from a small breach in the bank at position 3R, south of *in situ* tree roots and wood (Figures 2.3 and 2.10).

There are *Gryphaea* banked up on the west side of the skull SH1/242 with moderate to poorly sorted cross-bedded, sandy, iron stained gravel above. This is Bed 39 in the sequence (Figure 2.10). Cross-sets are to the N of the skull, dipping mainly NE, indicating a current in this direction. The large skull had clearly influenced sedimentation. Bed 39 is more extensive than Beds 37 and 38 but also appears to have been influenced by the presence of the 'turtle' stones at the downstream end of the bar near the Site1/Site 4 boundary. Bed 39 is near the right bank 1R here near the inflexion point between meander bends.

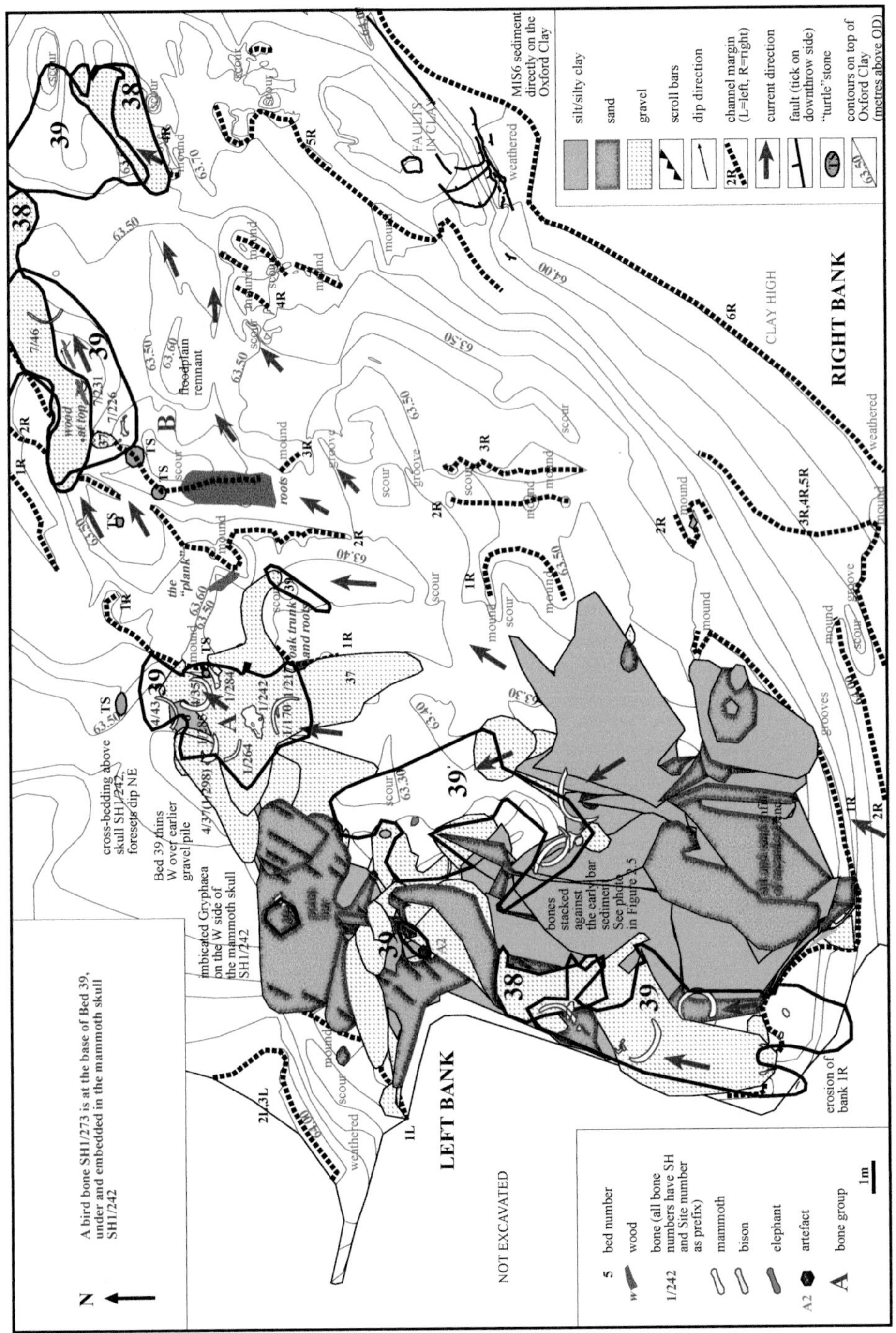

Figure 2.10 Beds 38 and 39. Erosion and deposition near 'turtle' stones

Bed 39 is also seen in Site 7 to the NE, where it is near to the bank as it curves into the next bend, following a similar SW/NE trend to Bed 38. This is immediately downstream of a cluster of 'turtle' stones and is mainly on the Oxford Clay. Long axes of large pieces of wood indicate a SW/NE current here.

There are many bones in Bed 39. Near the boundary of Sites 1 and 4, large mammoth bones and tusks are near the right bank associated with the 'turtle' stone mentioned in Bed 37. Some of these may be parts of the same carcass as the skull SH1/242, which had originally been close to the right bank.

To the NE of the skull is a mammoth ulna SH1/284 which trends SW/NE. Its NE end is against a 'turtle' stone on the W side, separating it from a mammoth tusk SH4/35 which is concave downstream. There are also bones downstream to the NW of the skull SH1/242 including a mammoth tusk SH1/285 above a mammoth mandible SH4/37. These are stacked on top of each other above a large piece of wood and a large cobble at the top of a mound of Bed 31 near the southern end of the point bar. Bed 39 thins against this mound to the west. The base of tusk SH1/264 is against a low mound of grey or iron stained poorly sorted gravel to the NW of the skull, which is slightly downstream. Wood and a horse metatarsal SH1/259 are immediately above this tusk but in a later bed (Bed 58). In this area it is likely that most of the bones in Bed 39 were originally on the bar surface or close to the right bank before being buried immediately by sediment or transported a short distance by the current.

The mammoth ulna SH1/260 is above SH1/242 in a very thin layer of silty gravel with involutions which appears to be later sediment (Bed 59) which was banked up with wood over the skull. If it is part of the same carcass it has hardly been moved. SH1/261, a mammoth scapula (shoulder blade), is in the same silt/silty gravel as SH1/260 but is nearer the Site1/Site 4 boundary downstream. Bone fragments immediately above the shoulder are labelled SH1/376 and are probably part of it.

The levels on the mammoth thoracic vertebra SH1/255 suggest that it is very close to the top of the Oxford Clay with the base resting in very iron stained gravel of Bed 4. This is below most of the main bone beds. This is interesting as it could mean that this was part of the mammoth skull SH4/124 which also partly rests on the Oxford Clay at a similar level. It is possible that both were resting on the clay before the point bar started to accumulate. Bed 4 is close to the *thalweg* here and this bed has little matrix showing evidence of re-working, so this bone may belong to Bed 37 and came to rest near the edge of Bed 4. In either case it is likely to be contemporaneous with the skull SH4/124.

In the southern part of Site 1, some bones and tusks in Bed 39 have been stacked by the current near the base of the bar sediment (Figures 2.4 and 2.10). In this area the right bank of the early meander bend 1R had probably been mostly eroded away by this time with a new bank location near position 2R. There is an artefact (A1) in Bed 39 which is partially embedded in the top of the early bar sediment (see later discussion on artefacts).

After the cross bedding of Bed 39 there are mainly very thin beds and lenses of sediment with much wood and molluscs above SH1/242. Some involutions possibly indicate some turbulent flow over the skull, which would have been a large object in the channel.

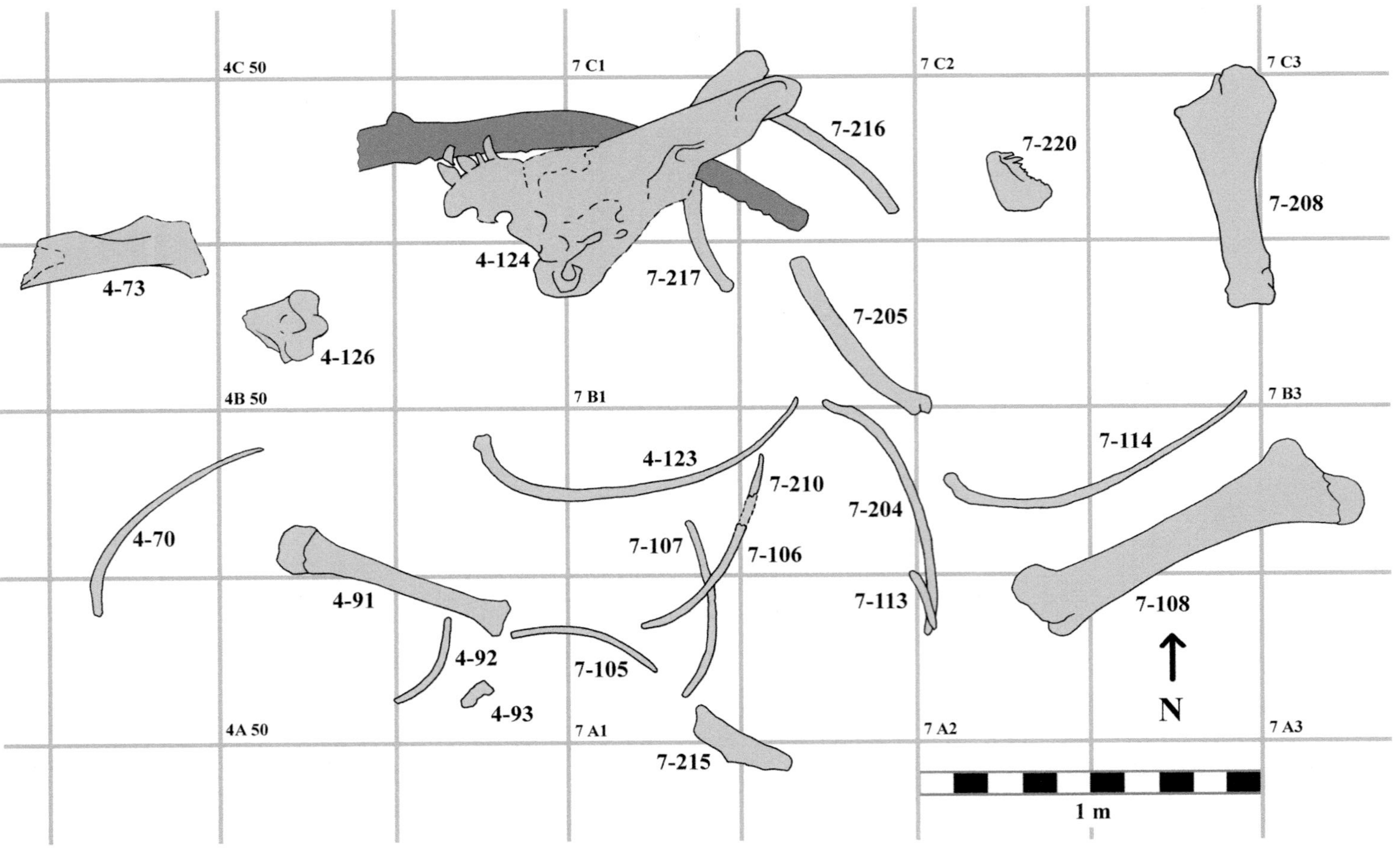


Figure 2.12 The bones of Group B: centred around mammoth skull SH4/124

Figure 2.11 Mammoth skull (SH4/124) resting on oak logs and large ribs

***Group B***

This group of bones possibly represents one mammoth (Figures 2.11 and 2.12). SH4/124 is an almost complete skull although somewhat crushed and distorted. An axis vertebra (SH4/126) lies a short distance to the south west of the base of the skull. All around are many complete ribs and a sternum, suggesting that the rib cage was entire within its skin at this location. The dentition in the skull is the right last molar M3 which indicates a mature animal. Nearby was a left upper M3 (SH7/220) in excellent condition with delicate, undamaged roots suggesting that it might have come from the skull. This tooth gives an estimated age at death of 47 years (see Chapter 4). The three limbs with fused epiphyses (indicative of maturity) may also belong to this individual: a complete right ulna and radius (SH7/208 and SH4/91) and right femur (SH7/108).

*The context of the mammoth skull SH4/124 in Group B*

The mammoth skull SH4/124 trends roughly SW to NE between the eastern edge of Site 4 into Site 7. Warm water Molluscs including *Corbicula fluminalis (Cf) and Potomida littoralis (Pl)* indicating an MIS 7 age are in sedimentary layers below and above this skull.

The skull is upside down with the base of the eastern part at the top of a small mound of grey sandy, silty gravel (Bed 44 of the stratigraphic sequence). Bed 44 thins eastwards over a low Oxford Clay mound and then thickens again. The exact level of the base of SH4/124 in the central area was not recorded due to the difficulty of lifting it but, when finally removed, grey silt was adhered to the underside implying that the skull partially rested in the top of Bed 43.

Figure 2.13 Bed 43. Bone Group B. Mammoth skull SH4/124 at the wooded right bank

There are large pieces of wood below the skull which are at the interface of Bed 43 with Bed 44 (Figures 2.13 and 2.14). This wood and the skull are shown in Bed 44 on Figure 2.14. Other wood and ribs were excavated at this location from just above the Oxford Clay within Bed 43 (Figure 2.13).

Bed 43 is a composite bed with a very variable composition. The base is on the Oxford Clay (Figure 2.13). The main feature is an area of *in situ* roots of deciduous trees, to the east of the earlier oak trunk, and small pieces of drifted wood which trend N to S near the bank position 3R. The roots are in grey silt, gravelly silty clay or clayey gravel just above the Oxford Clay. Above and between them there are small scours filled with well sorted sand or silt. On the western side there are involutions and scours in the Oxford Clay, with many derived Jurassic fossils including crocodile and ichthyosaur vertebrae, implying active erosion of the clay. In this area Bed 43 frequently forms low mounds of silty, poorly sorted gravel. There is much re-worked clay and gravel in an Oxford Clay scour to the southwest of the skull SH4/124 and N of the roots.

Immediately west of the skull, there are many small twigs in Bed 43 and it fines and thins upwards to the north over earlier sediment with broken 'turtle' stones, which indicates a probable near bank location. Grey silty gravel with some cobbles grades southwards into slightly sandier gravel with many molluscs. This area is to the east of a 'turtle' stone which rests on 5-10cm of poorly sorted gravel and has been eroded from the Oxford Clay. The molluscs include articulated *Corbicula fluminalis* (*Cf*) and *Potomida littoralis* (*Pl*). These molluscs prefer a sheltered area in close proximity to fairly deep, moderately fast flowing, warm, fresh water.

Bed 43 represents the partial destruction of the river bank at position 3R near the junction of Sites 1,4,7 and 8 (Figures 2.3 and 2.13). The river would have been undercutting the clay bank on this side of the channel and the sediment is likely to be a mixture of slumped clay mixed with marginal channel sediment. This is just before the next bend. The presence of the 'turtle' stones probably inhibited lateral migration of the channel eastwards and there was probably a remnant of floodplain here.

The mammoth skull SH4/124 has been moved from its death site as it is upside-down on two sturdy oak logs and an elephant rib in Bed 43. The second piece of wood is not seen in the photograph (Figure 2.11), but it is at right angles and under the one on the left (see Figure 2.12).

It is proposed that the original location of the skull was on or very near the undercut, wooded, right bank of the river, at the downstream part of the anti-clockwise rotating meander, just before the next clock-wise rotating meander bend. It is possible that erosion around the skull area isolated part of the bank as a small island (just east of position 3R). More than one very thin sedimentary layer is banked up against the mammoth skull SH4/124 especially on the S and W side.

A number of possible scenarios can be implied for its final upturned position. Firstly, that it was turned over by the current. There is a SW/NE scour in the Oxford Clay on the west side of the skull, to the east of a 'turtle' stone. The main part of the scour is infilled with poorly sorted silty or sandy gravel with some cobbles and re-worked clay implying some erosion here. A

Figure 2.14 Bed 44. Bone Group C. Meander expansion and migration

very strong current would have been required to move this large object. Molluscs indicate a marginal channel location just off the main current. The skull is also on the landward side of the preserved tree roots in Bed 43 and in this vicinity is unlikely to have been transported far by the current, if at all. Small scours around the roots, infilled with silt or sand and drifted wood indicate some water movement and erosion, but by gentle currents. It is possible that the skull SH4/124 was overturned when, during a seasonal flood event, a strong current first breached the bank between the 'turtle' stone and the *in-situ* roots. The orientation of the wood, ribs and skull do appear to be roughly at right angles or elongate with the suggested water current as it overtopped the bank here. As mentioned earlier, the skull rests partially on wood in silt/silty clay of Bed 43 and partially on a very thin lens of moderately sorted gravel (Bed 44 in Figure 2.14) so this scenario seems unlikely unless a bloated carcass floated on to sediment and wood near the bank. Bed 43 does not extend far eastwards and Bed 44 is very thin at this locality. Water overtopping the bank probably slightly rearranged this bone group but did not transport them far.

There are other bones in or on Bed 43 which also do not appear to have been transported far. The western end of a mammoth tusk SH1/356 (called SH8/49 in Site 8) is in Bed 43 on the eastern side of the tree roots (Figure 2.13). This is still in part of its tusk socket which is fragile and would easily have broken if transported far. This tusk is nearly complete with some minor damage at the distal end. It may have been protected from transport by the roots. There is some erosion around the tusk with sand lenses in small scours, which probably occurred while the tusk was *in situ* at the channel margin.

An alternative explanation for the skull being upside down is that other mammoths in the vicinity upended the skull. Elephants are known to pick up or move the bones of dead elephants; such behaviour on the part of mammoths might have resulted in the skull rolling down slope from the bank on to wood at the channel margin.

The unusual layout in Figures 2.11 and 2.12 leads to the intriguing possibility that hominids overturned the mammoth skull SH4/124. Their presence, approximately contemporaneously, is indicated by two of the excavated stone tools: A11 in Bed 44 to the south of the skull near the original bank at 2R (Figure 2.14) and another was located in Bed 48, just above the *in-situ* wood and the mammoth tusk SH1/356.

Immediately below the skull SH4/124 is a W to E trending oak branch on which it rests (Figures 2.11 and 2.12). The wood rises steeply westwards. This oak has an angled joint which is the sort of wood that might be selected for a lever. Oak is particularly favoured for this purpose due to its strength and flexibility. Under this wood and at right angles to it is another N-S trending flatter piece of wood which was longer when first uncovered and extended further south. This has a trimmed side branch at the northern end which would make it easier to handle if also using it as a lever. The southern end of this wood rests on the articulated (western) end of a very sturdy elephant rib (SH7/216) which could also be useful as a tool.

The layout in Figures 2.11 and 2.12 is comparable to a similar arrangement at the archaeological site Gesher Benot Ya'aqov in Israel (Goren-Inbar *et al.* 1994) where it was suggested that people had overturned an elephant skull to get at the brain case (see discussion in Chapter 8). The association with mammoth ribs may be significant as ribs were found

Figure 2.15 The 'plank'. The distance between peg A48 and the peg in the centre of the photo is 1 metre.

embedded in the parietal region of a mammoth skull at La Cotte de Saint-Brelade, Jersey (Scott 1980, 1986). Brains were a valuable commodity for early hominids, useful for food and in the tanning process of hides. Other large ribs below the wood, near the interface of Bed 43 and Bed 44, and close to the skull SH4/124 may have been used as tools to scoop out the brain and chosen for this purpose. These include mammoth ribs SH7/105, SH7/106, SH7/107, SH7/114, SH7/205, SH7/210, SH7/217 and SH7/227.

Other bones in Bed 43 are a mammoth ulna SH7/208 (the NW end rests in silt near the top), another mammoth ulna SH8/46, an epiphysis SH7/219, and a mammoth radius SH4/91. Some of these bones partially rest on the Oxford Clay. These were all bones that had probably originally collected with wood in slack water near to the right bank. An interesting find a few metres west of skull SH4/124 was a large section of smoothed oak, described herewith as 'the plank'.

*The 'plank'*

This large flat piece of wood trends NW to SE and is immediately west of a ridge of Oxford Clay (Figure 2.15). It is in Bed 44 but appears to have been located close to a remnant of the bank in position 2R (Figure 2.3).

This 'plank' had a very smooth surface when first uncovered. It is uncertain whether this is the result of erosion by sediment laden water flowing over the top of it or whether this plank could have been shaped and polished by humans. A flat polished plank was also recorded at the archaeological site of Gesher Benot Ya'aqov, Israel (Goren-Inbar *et al.* 2002: 91). A plank would be useful for the preparation and drying of hides during tanning processes and local oak bark could also have been gathered for this purpose. This is described further in Chapter 8.

As described above, skull SH4/124 was partially resting on Bed 44 (Figure 2.14). This shows a clear shift in the position and dimensions of the initial meander bend, mainly following the bank defined by the *in-situ* roots at position 3R. A series of gravel bars composed of bed load material rest directly on an eroded Oxford Clay surface. The distribution of these bars indicates a similar anti-clockwise rotation to the earlier meander seen in the Oxford Clay surface but by this time the bend had expanded and shifted downstream to the NE and sideways to the E and SE. Erosion to the SE is supported by lenses of poorly sorted gravel with cobbles and *Gryphaea* infilling scours in the Oxford Clay bank. Involutions and some loading of the gravel into fine sediment possibly indicate some turbulent flow at the channel margin.

Sediment in the main part of Bed 44 consists of channel gravel with many cobble sized *Gryphaea* and some cross bedding. There are many places where it is directly on the Oxford Clay. In other areas it is above thin layers of earlier sediment. The matrix is variable in amount and type depending partly on the sediment immediately below it, but the gravel is mostly clast-supported. It is interpreted that Bed 44 largely represents water movement in the *thalweg*, the area of the fastest current in the channel. The sediment forms bars which are mostly at right angles to the current direction. Scours and grooves in the underlying Oxford Clay suggest the water movement was rotating from the right towards the left bank near the right channel margin R3 before veering NE into Site 7 where the channel was straighter (see the River Evenlode in Figure 2.7 and the plan of the tusks and 'turtle' stones plan (Figure 2.2).

Many of the ribs and limb bones in Bed 44 have their long axes aligned with or at right angles to the water direction indicating that they had been moved by the current. Some of the larger bones and tusks remained roughly in the same place while later sedimentary layers were deposited. The meander had evidently expanded in size and moved downstream from its initial position at this time.

The mammoth skull SH1/242 in Bed 37 is located between the gravel bars of Bed 44 which were deposited after this skull, when the channel had shifted eastwards.

During the deposition of Bed 44 there was a small breach of the right bank 3R south of the skull SH4/124, between the 'turtle' stone and the tree roots into Site 7 and another one south of the roots into Site 8. There are potholes, scours and grooves at the base of Bed 44 in Site 8 suggesting turbulent flow and there are imbricated *Gryphaea* which are stacked against a clay mound. The grooves and imbrication indicate a moderately strong current mainly from SW to NE to the south of the *in-situ* roots. Bed 44 is banked up against Bed 43 near the mammoth ulna SH8/46 which appears to have protected Bed 43 from erosion here. This bone is at right angles to the current direction.

As the right river bank at 3R further deteriorated it is possible that the remaining sediment under the skull SH4/124 became a small island with water following both the original direction around the meander bend and other water following a straighter course into Sites 7 and 8. Discontinuous small ridges of clay indicate a new clay bank position at position 4R then later 5R.

In a similar way to the mammoth skull SH1/242, the upturned mammoth skull SH4/124 was a large object which influenced later sedimentation in the channel.

There are many cobble-sized *Gryphaea* stacked against it particularly between the W side and the western end of the large oak log on which it rests. Most of these appear to be in Bed 48. The *Gryphaea* dip SW, implying a current from this direction when they came to rest. There are also small pieces of wood, many small molluscs, articulated *Potomida littoralis (Pl)* and *Corbicula fluminalis (Cf)* confirming an MIS 7 age when the skull was partially buried. A mammoth axis SH4/126 is also near the top of Bed 48. There are many thin sedimentary layers after bed 48 before this large object was finally buried.

The occurrence of two such upturned skulls within a few metres of one another, both with sediment banked up against them, would indicate that neither of the mammoths they represent had moved far from the site of their death.

***Group C***

SH1/245 is comprised of four neck vertebrae of a mammoth. From right to left in Figure 2.16 they are as follows: the axis and three cervical vertebrae. The first vertebra in the spine, the atlas, is absent. Their proximity to one another, their condition (colour and degree of weathering), and their measurements suggest that they represent a small section of the neck of one individual. The fact that the intravertebral discs are fully fused to the centrum of each vertebra indicates a mature animal. There was some degree of movement of this section of the spine after the death of the mammoth but while flesh still encased the bones; the axis and the 2nd and 3rd cervical vertebrae were anterior side down but the 1st cervical vertebrae had twisted so that the anterior was facing upwards. Some of the vertebral processes are damaged, possibly as stones, molluscs and other bones moved with the flow of the river. No damage could certainly be identified as carnivore gnawing.

Figure 2.16 A section of the neck of a mammoth (SH1-245). From right to left are the axis, 1st, 2nd and 3rd cervical vertebrae

*The context of the semi-articulated mammoth vertebrae SH1/245*

These bones are located just above the Oxford Clay at the downstream end of Bed 45. Gravel in Bed 45 is mainly located between the bars of Bed 44. The semi-articulated vertebrae SH1/245 partially rest on the Oxford Clay but have been moved slightly by the current against a mound of Bed 44 (Figure 2.14). A large piece of wood, at right angles to the current is above them. The mammoth mandible SH1/239 also appears to have come to rest against the wood.

The position in the channel is just to the east of the *in-situ* tree trunk at the earlier bank position 1R. It is possible that remnants of this degraded bank originally sheltered these bones, preventing them being transported far. A mammoth pelvis SH1/173 at the top of Bed 37 (Figure 2.9) is also mainly in Bed 45 and dips steeply against the SW side of the remains of the *in-situ* tree trunk.

Beds 44 and 45 formed an undulating surface throughout the meander bend defined by the right bank at position 3R. Sediment continued to build in this area with mainly very thin undulating gravel layers which thinned both north and west over the earlier sediment pile. Some sand and silt lenses represent occasional periods of lower energy. Cross bedding is more common especially at the downstream part of the bar and there is much re-working. It appears that the mammoth skulls discussed earlier were obstacles in the channel where sediment banked up. On the left bank the bar continued to grow in complexity especially at the downstream end, where wood and bones became stacked on top of each other.

***Group D***

At Site 1 is the complete lower mandible of a mammoth, a partial skull and two tusks that possibly represent the head of one individual. The mandible (SH1/180) has both molars (M3). The estimated age at death of this individual is 47 years (Chapter 4). Nearby are a crushed partial skull with a damaged but *in situ* tusk (SH1-162) and a complete tusk (SH1/179) (Figure 2.17). The skull was almost at the junction of the floor of the quarry and had suffered extensive machine damage and the effects of weathering. A partial tusk (approximately 40 cm in length) remained in the premaxilla (tusk socket) but both the skull and tusk were too deteriorated to retrieve. However, the proximal diameter of the tusk was 7 cm, the same as the proximal diameter of the nearby tusk SH1/179, suggesting that they may have been a pair. The complete tusk is slender and only slightly curved, indicating a female.

*The context of the mandible SH1/180, skull SH1/162 and tusk SH1/179*

The mammoth skull and tusk SH1/162 are located on the west side of the main sedimentary deposit of the Stanton Harcourt Channel at Site 1. These became part of the sedimentary record at a late stage in the point bar development. They are on the east side of an eroded SW/NE ridge of Oxford Clay which is interpreted as a former edge of the channel on the west bank, when looking NE (2L on Figure 2.3). These bones are near the top of a low mound of poorly sorted iron stained gravel with some cobbles especially *Gryphaea* valves in Bed 55 of the sedimentary sequence (not illustrated). This was deposited later than the channel shift at Bed 44 and is predominantly bed load material from the faster flowing part of the channel

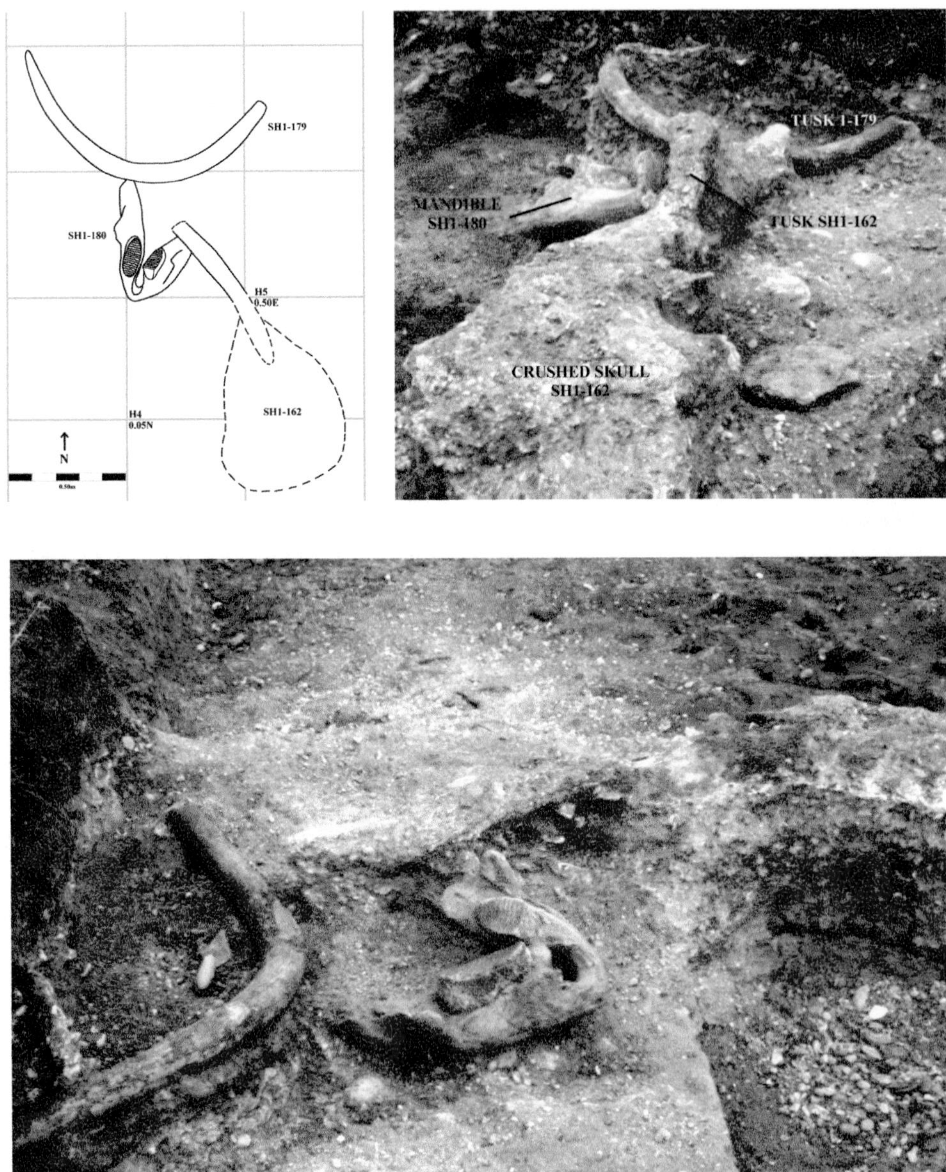


Figure 2.17 The bones of Group D. Top left: Plan of mandible SH1/180, tusk SH1/179 and the skull SH1/162 with part of a tusk still in its socket. Top right: photograph of the group showing extensive damage to the skull. Below: after the removal of the skull

mixed with variable amounts of fine-grained bar sediment. This sediment forms low ridges and would have been moved up the point bar surface by the current.

Bed 55 trends S to N to the east of the large tusks in Bed 39, which are stacked against fine-grained bar sediment (see tusks in Figure 2.4 and Bed 39 in Figure 2.10) and were clearly obstacles which the current went round. Sediment was moved over the point bar surface from the SSE or SE, possibly during a seasonal flood event. This material was then deposited between the earlier sand scrolls and also to the NW of a small gravel bar.

It is uncertain when or how the skull and tusk SH1/162 came to be in this area but it is possible that the carcass was originally on the surface of the silt/sand point bar but was then moved up the bar surface, with bed load material, by the current at high stage. The long axis of the tusk and skull is aligned approximately with the current direction. However, the fact that the tusk was still in the skull suggests little movement by the current.

The skull and tusk SH1/162 are extremely weathered and it is possible that they had been on the bar surface for some time before being buried. There are a number of other bones and tusks at a similar height in the vicinity of this skull (the mammoth mandible SH1/180, the tusk SH1/179, a mammoth rib SH1/161 and a mammoth femur SH1/163) and also appear to be in Bed 55. They may be from the same carcass. SH1/179 is aligned approximately at right angles to the current and is concave downstream. A bison vertebra and limb fragment (SH1/160, SH1/164) are also found in this bed.

Other nearby bones and tusks have been buried in other layers on the bar with a different current direction. The sedimentary sequence indicates some erosion of the upstream side of the bar at a late stage of bar development. This appears to be water taking a 'shortcut' across the bar from SW to NE which is typical of a complex bar deposit.

The mammoth tusk SH1/159 is near SH1/162 but was deposited with *Gryphaea* on the west side of the clay ridge at a later stage, but still in MIS 7. Layers with articulated and disarticulated *Potomida littoralis* become thinner against the tusk SH1/159 on the west side and may have been washed on to the upstream side of the bar during seasonal flooding. Some later imbricated gravel beds imply more sustained SW to NE currents eroding this side of the bar as the channel position shifted. The northern end of another mammoth tusk SH1/88 also roughly follows the clay ridge.

It should be noted that the final resting place is not necessarily where the animal died. All the bones near the skull SH1/162 may have initially been from the same death site on the bar close to the west bank but became incorporated in to the final sedimentary sequence at different times. They have not been moved in the '*thalweg*' of the river.

### *Group E*

This is part of a mammoth skeleton found at the river margin (Figure 2.18 and 2.19). The focal point is a virtually undamaged right pelvis (innominate SH7/119), articulated bones of the lower spine (lumbar and sacral vertebrae SH8/4), and a right femur (SH8/5). East of this central group is a left femur of the same size (SH8/25) and a complete right upper molar M3

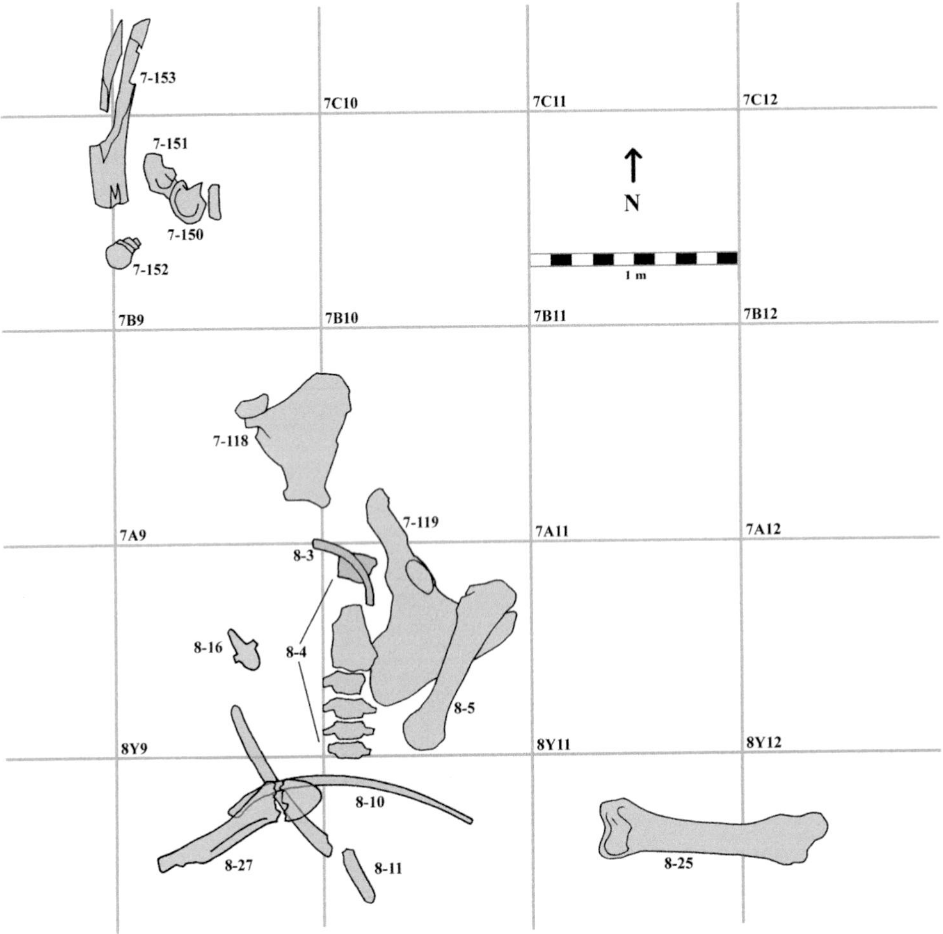

Figure 2.18 The bones of Group E: the post-cranial remains of a mammoth

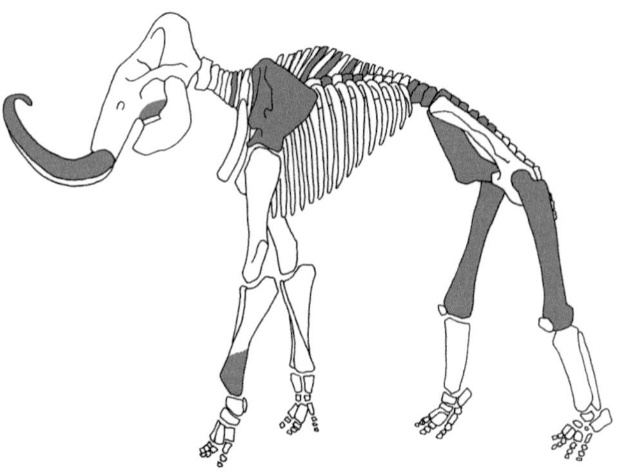

Figure 2.19 Tusk, molar and post-cranial remains (shaded) of the mammoth at Site 7/8 (Group E)

Figure 2.20 Three lumbar vertebrae (part of SH8/4). These and the adjoining sacrum were all dorsal side down in the silt. The spines (although cracked in the process of excavation) showed almost no damage, suggesting little post-depositional disturbance.

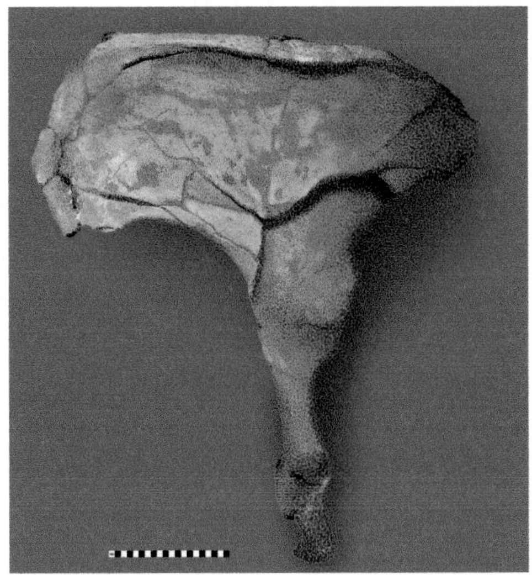

Figure 2.21 Right innominate (pelvis) SH7/119. The long margin of the iliac crest is virtually undamaged, indicating little post-depositional movement of this bone. This delicate long edge is incomplete in all other examples of this bone (see Chapter 4).

(SH8/22). To the northwest is a left scapula (SH7/118), two consecutive thoracic vertebrae (SH7/151,152), a cervical vertebra (SH7/150) and a tusk (SH7/153). Not far from these is another thoracic vertebra (SH8/18), a fragmentary distal ulna (SH8/27/SH8/28) and a couple of ribs. To the north and south of the main group are several almost complete as well as fragmentary ribs. All the epiphyses of long bones and the distal margin (the iliac crest) of the pelvis are fused which indicate a fully mature animal. The tooth (M3) suggests an age of approximately 43 years (see Chapter 4) which corresponds with the mature status of the postcranial remains.

Several lines of evidence support the impression that these are the remains of one individual mammoth whose death occurred locally. The condition of the central group of bones indicates that relatively little post-depositional movement took place. During their excavation, it was noted that the articulated lumbar and sacral vertebrae were embedded dorsal side down in very fine silt and sand (Figure 2.20). This protected and preserved the delicate transverse spinal processes which were so completely undamaged as to suggest that, once embedded in the silt, they had remained undisturbed. This is also true of the scapula, the delicate dorsal spine of which was embedded in the silt and remained intact. The superb condition of the right innominate is indicative of minimal post-depositional disturbance (Figure 2.21). It and the right femur were found in close anatomical relationship to one another and to the sacrum. As regards the condition of the femur, it appears that once the ligaments holding the head of the femur in the socket of the innominate had disintegrated, the length and weight of the distal end caused it to drop down into the silt. This resulted in the perfect preservation of the distal epiphysis, but the exposed proximal end became damaged, possibly at a considerably later date. This damage is in line with the junction of the MIS 7 channel deposits and the overlying MIS 6 deposits indicating that the proximal femur was damaged by the force of the incoming MIS 6 gravel.

*The context of the bones of Group E*

Group E is the last group of bones that do not appear to have been moved far. These bones are partially resting against a small mound of Oxford Clay near the boundary of Sites 7 and 8. The lower parts are mainly embedded in sandy silt of Bed 58 (Figure 2.22).

Bed 58 is composed of irregular lenses of grey silty clay or silt which are sandy and laminated in places and represent the deposition of suspended sediment in low energy conditions. Lenses are found on the Oxford Clay to the NE of the earlier sediment pile near a remnant of the bank at position 4R or along the channel margin to the SE between 5R and 6R (Figures 2.3 and 2.22), above earlier sediment.

It is likely that Bed 58 was originally more extensive. Mammoth ribs SH8/8, SH8/9, SH8/10, SH8/11 and the S end of SH8/3 are in this bed. The articulated mammoth vertebrae SH8/4, a mammoth pelvis SH7/119 and a mammoth femur SH8/5 appear to partially rest on or in this layer. Artefact A21 was excavated from the top of this layer near the SE corner of SH7/119. The mammoth vertebra SH8/16 also appears to partially rest on this bed.

The bases of some of the bones in Group E are partially in the top of the Oxford Clay on the S side of a small clay rise. There is also a very eroded Oxford Clay surface to the SE of this bone collection and some faulted and slumped clay. These are probably remnants of the right bank when it was at position 4R. These bones are in a marginal channel location. They rest on each other, moved together by later currents (mainly Beds 60 and 62 in Figure 2.23) but they do not appear to have been moved far.

To the SE, Bed 58 is very thin and there are small-scale faults at the top and base. A slab has become detached from the main deposit and has moved NW over earlier sediment. The latter is mostly composed of very thin silt, sand or gravel layers with some molluscs and is interpreted as marginal channel sediment. The faults at this locality are tight and slightly concave in plan and section. Fault surfaces are striated and there appears to have been downward, slightly rotational movement in several directions but mostly away from the channel margin at position 6R. The Oxford Clay has steep slopes in this area. Meteoric water or groundwater, at the interface of the very thin silty clay layers with other more permeable layers, probably lubricated the clay surfaces and promoted bank collapse where the sediment had been undercut at the right bank (Figures 2.3, 2.5 and 2.21).

The pile of bones in Bed 58 appears to have influenced later sedimentation. Bed 60 is mainly to the NE of the gravel of Beds 44 and 49 and forms a linear deposit trending roughly W to E. It is concave N, forming a horse-shoe shape around Group E (Figure 2.23). It forms mounds of sediment on the Oxford Clay which are mostly lenses or thin layers of grey or iron-stained f/m gravel (pea grit) with a sandy or sandy silt matrix, with some graded bedding and very thin silt/silty clay lenses. Fine grained sediment is characteristic of the western edge of Bed 60 where low mounds are stacked against each other to the W or SW and thin eastwards. The sediment type implies some re-working of earlier deposits. There are also some NW/SE

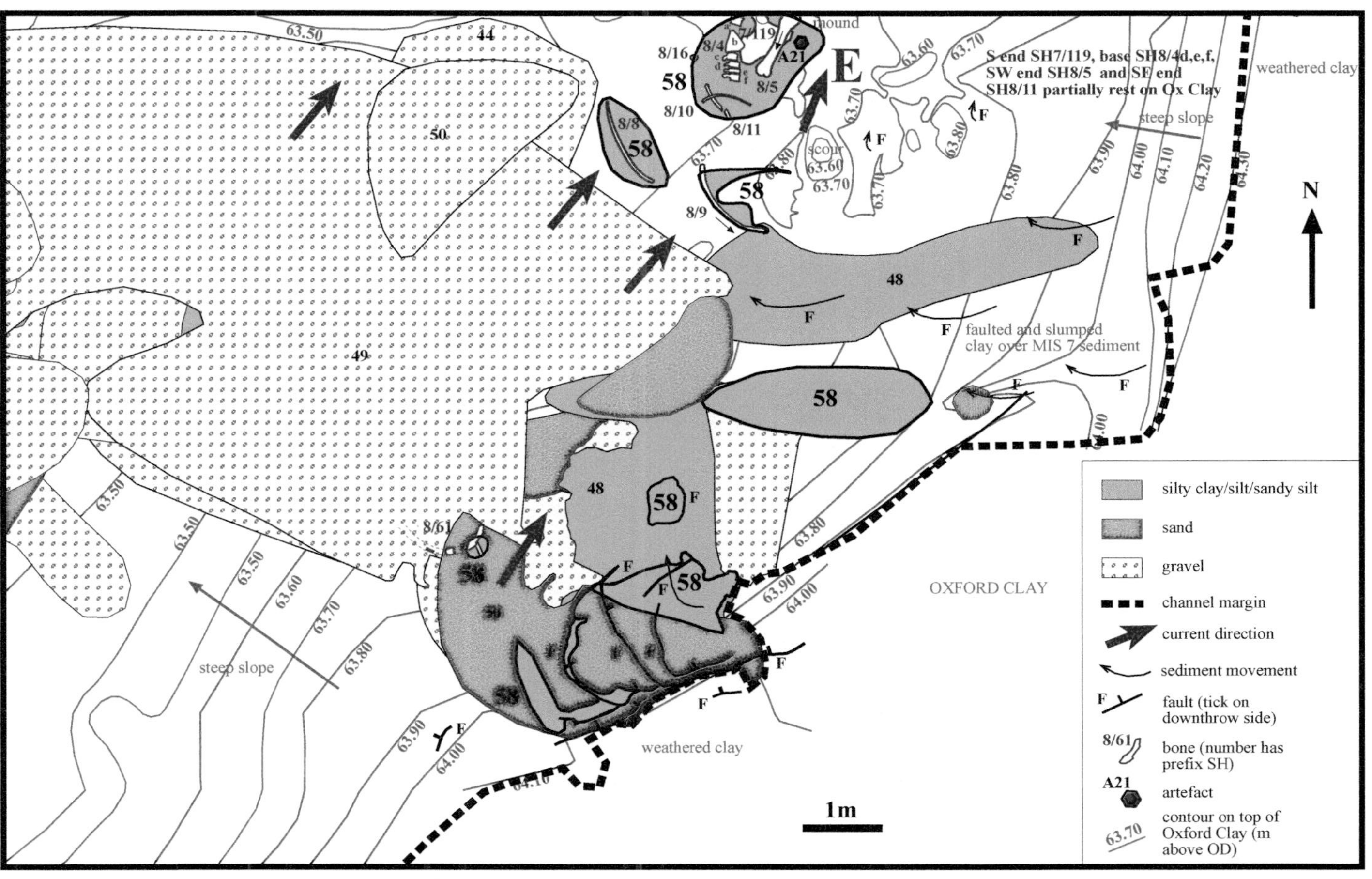


Figure 2.22 Bed 58. Bone Group E. An undisturbed mammoth carcass near the undercut right bank at Site 8

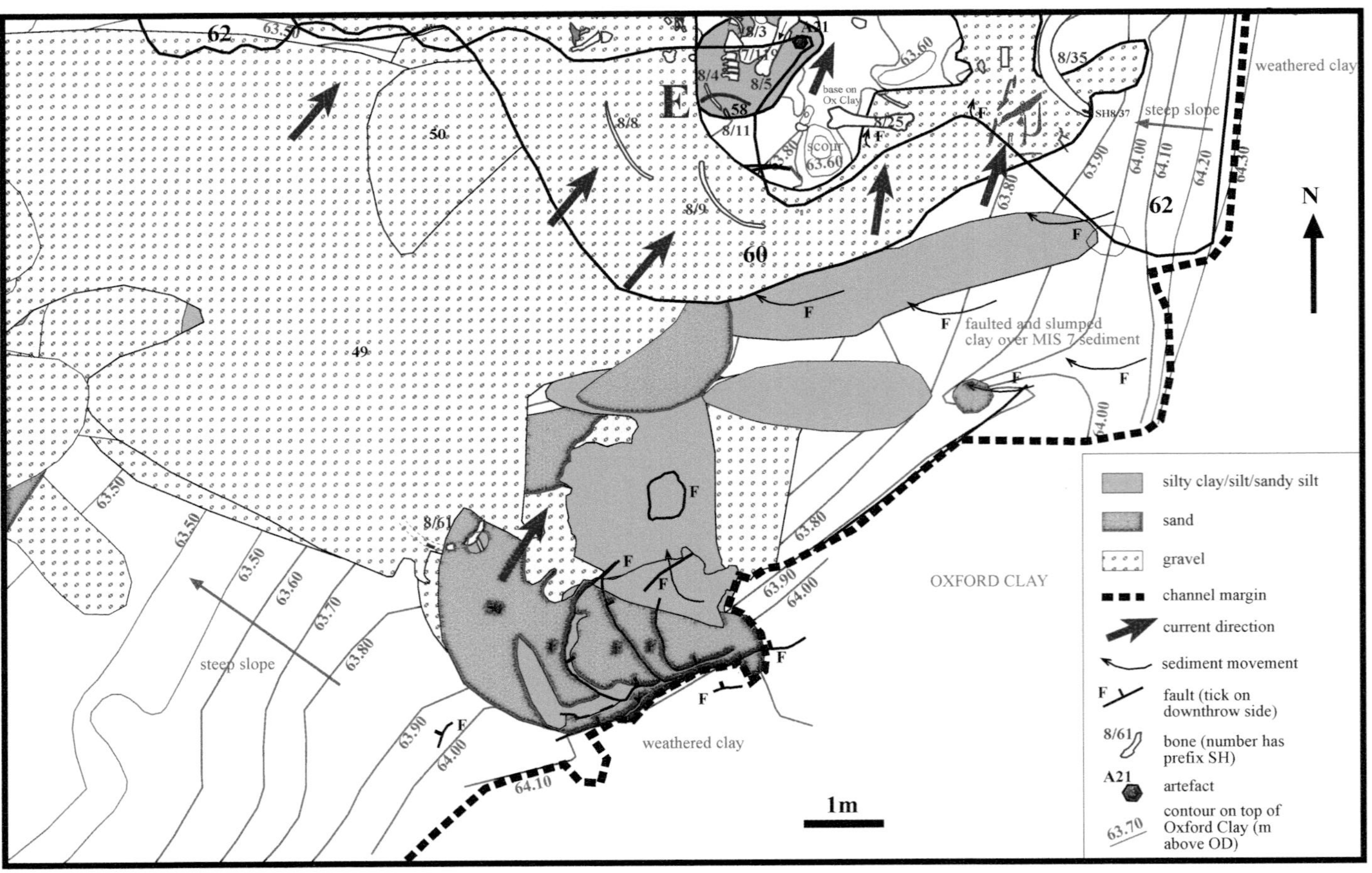


Figure 2.23 Bed 60 and 62. Sediment movement to the NE at Site 8. Carcass burial

trending, shallow, coarse sand filled grooves in the top of laminated sandy silt of Bed 60. These may be burrows as there are many *Potomida littoralis* (*Pl*) present at the base of the overlying gravel (Bed 62).

There are a few cobbles at the base of Bed 60, at the interface with the Oxford Clay, and some poorly sorted sediment was deposited nearer the faster flowing part of the river. Scours and small potholes in the Oxford Clay surface and involutions indicate some turbulent flow, possibly as water flowed over the earlier gravel pile. Long axes of wood are parallel or at right angles to the Oxford Clay contours. This bed is interpreted as a transverse bar with sediment mainly from the re-working of the earlier gravel mound collecting downstream to the NE.

Bed 60 covers the mammoth ribs SH8/8 and SH8/9 which appear to be orientated with their long axis at right angles to the current direction and are concave downstream to the NE. A mammoth femur SH8/25 and mammoth rib SH8/29 also have their long axis approximately transverse to the current direction. The NW end of the femur is partially embedded in the Oxford Clay surface. The tusk SH8/35 is concave downstream and the S end is in this layer. Hazelnuts and other nuts collected in organic silty clay at the top of this bed as energy waned.

The mammoth vertebra SH8/16 is at the interface of Bed 58 and 60. SH8/16 is part of SH8/18 which is also in this Bed. It is likely that Bed 60 involved some re-working of Bed 58.

There is a possible rotational fault above the mammoth femur SH8/25 at the top of Bed 60/ base of Bed 62. Sediment movement above the fault appears to be downward to the N. The femur does not appear to have been affected.

The upper part of bone Group E is partially covered by poorly sorted gravel from Bed 62. Some re-working of the sandy silt/silty clay of Bed 58 is implied and the bones have been stacked against each other by the current from a south or south westerly direction. The long axis of SH8/5 is in this direction.

Bed 62 also infills scours and potholes in the Oxford Clay and continues into Site 7. Involutions suggest that there was some turbulent flow possibly as a result of water flowing over the earlier sediment pile from the S or SW, although scours and ridges in the Oxford Clay are possibly remnants of the degraded bank at 4R. The northern part of the tusk SH8/35 is in this layer. Near the boundary between Site 8 and Site 7 this bed is composed of undulating layers of grey poorly sorted gravel, with some *Gryphaea* overlain by silty clay which drape the earlier mounds of Beds 60 and 61. A moderately strong current followed by a waning of energy is implied.

The channel current direction in the region of the central group (referred to henceforth for convenience as the 'carcass') flowed from SW to NE before veering slightly westwards some 8 meters downstream. The significance of the current observations lies in how they affect the interpretation of both the carcass and the more widely dispersed bones. The disarticulation of the bones would have occurred through the gradual decay of the carcass, moving some bones slightly downstream. Two further elements that might belong that have been carried downstream: part of a humerus (SH7/139), a lower M3 (SH7/96) of the same age as the upper M3 near the carcass, and a complete tusk (SH7/80).

Figure 2.24 The decaying carcass of an African elephant on a river bank. The mammoth carcass (bones of Group E) is interpreted as having originated similarly at the edge of the river and become incorporated into the Channel sediments (photograph reprinted with permission from The Illustrated Encyclopaedia of Elephants by S.K. Eltringham. London: Salamander Books).

As mentioned earlier, linear NW/SE coarse sand filled grooves are at the base of Bed 62 in sandy silt/clay of the underlying Bed 60. Many articulated and single *Potomida littoralis (Pl)* and small articulated *Corbicula fluminalis (Cf)* and some flattened wood were excavated from the base, indicative of MIS 7 when the bone Group E was first buried.

The NE end of the mammoth femur SH8/5 is partially covered by a low mound in Bed 62 but the top of the bone is truncated by loose very sandy gravel by remnants of MIS6 sediment at the base of the gravel pit. In this area Bed 62 is very iron stained and/or cemented in places, especially at the top of the MIS7 deposits, immediately below the MIS6 sediment. The femur must have been partially uncovered by erosion during the succeeding periglacial climate.

There are various reasons to explain the situation of this mammoth at the channel margin during MIS7. This is an old individual and may have been nearing death when it came to the river. When elephants are ailing or too old to feed, their last recourse is frequently to water where they might be killed by lions or hyaenas or die of natural causes. As in the case of the elephant shown in Figure 2.24, the old mammoth probably died at the river. Although scavengers would then have had access to the flesh, the generally good condition of the bones of this individual suggests that the carcass might soon have been buried by fluvial sediment, probably during seasonal high water. All but one of the bones of the lower limbs and feet of this individual are missing (Figure 2.19). Although the presence of lion, wolf and bear at the site (Chapter 5) does not rule out the selective removal or destruction of the smaller bones by predators, none of these carnivores habitually transports bones away from kill/death sites and very few of the bones from the excavation show evidence of gnawing. Thus, the general scarcity of smaller elements at Stanton Harcourt and the predominance of tusks, and of larger bones of all species, is considered to be the likely result of winnowing by the river current.

Lighter material would have been more easily transported downstream while the heavier items would have become stacked together or not moved at all.

The last sediment in the sequence at Site 8 was deposited about the same time as material also accumulated on the upstream side of the bar at Site 1.

Figure 2.25 Examples of the preservation of wood at the site. Clockwise from top left: logs and tusks at Site 4; roots at Site 2; branches at Site 7; log and branches at Site 1

## The context of wood, fresh-water molluscs and other environmental material at the excavation site

One of the extraordinary features within the MIS 7 sediments at Stanton Harcourt is the abundance of well-preserved wood, including huge logs of oak. Not only do we have an instant picture of the local woodland but the presence of *in situ* roots can demarcate areas of the riverbank where these trees grew.

Figure 2.26 Log jams across the site concentrated and preserved molluscs, seeds and small bones.

Some wood was desiccated and rotten near the old quarry surface but most of the wood was in remarkably good condition especially when within siltier layers, which indicate lower energy areas of the channel. In addition to the wood at Site 1 mentioned earlier, *in situ* roots and drifted wood were excavated from Site 2. A tree trunk with a tusk against it also occurred at Site 6. Tree branches, some more than a metre in length, were excavated from the MIS 7 sediment across the site (Figure 2.25).

Mapping of sedimentary features in the MIS 7 sediments during the excavations, gave an indication of the strength and direction of the current. As at Site 7 (Figure 2.2) it is clear that many large pieces of wood had been transported by the faster flowing parts of the channel (as indicated by associated imbricated or cross-bedded coarse gravel). In these areas the wood is mainly orientated with the long axis in the direction of the current or at right angles to it. Sometimes pieces of wood have been stacked on top of each other, often with large bones or tusks.

Concentrations of wood and other material do not necessarily indicate long-distance transport in the fastest part of the channel. For example, near the Sites 1,4,7 boundary at the downstream end of the bar just before the next bend, there was a pile of wood, bones and tusks. However, the generally good state of preservation of this collection of wood and bones suggests that, although transported, it had not moved far. It is likely that material which had been lying on the bar or near the right bank, was later pushed together by the current, probably at high stage. Many of the large branches are only slightly downstream from where the bank was being undercut (Figure 2.3).

Figure 2.27 Above: clusters of articulated *Potomida littoralis* with commissures pointing upwards. These are interpreted as an in-situ population that had been rapidly buried. Below: single *P. littoralis* (not in life situation) but showing typically excellent preservation of these molluscs at the site.

Wood also frequently occurs with other material in relatively undisturbed contexts. For example, Gourlay (Chapter 6) identified a 120-year-old oak log which is part of the pile up of wood, bones, tusks immediately north of a bison skull and overlying the sequence of articulated mammoth neck vertebrae in Group C described above.

It appears that wood, tusks and large bones often combined to create a 'log' jam. These items, together with the occasional rocks and boulders, formed sediment baffles for organic matter and fine-grained sediment particularly in areas marginal to the active *thalweg*, providing niches for mollusc populations. Molluscs, seeds and small bones are frequently found in the lee of such log jams. (Figures 2.26).

The significance of such concentrations lies in the microenvironments that were created. For example, at Site 1, in the lee of the 'plank' (Figures 2.3 and 2.15) there were very thin layers of silt or sand with some shell debris overlain by a bed of articulated *C. fluminalis*. Nearby clusters of articulated *P. littoralis* with their commissures all pointing upwards occur in a similar situation (Figure 2.27). It is likely that these were also an *in-situ* population, which had been rapidly buried. In such cases, these were probably communities which had selected a locality in quiet water protected by the wood, but close enough to the *thalweg* for the nutrients necessary for their filter feeding. Some articulated *P. littoralis* at the site were also excavated from shallow coarse sand-filled burrows in the top of silt or clay lenses and are likely to have been in their original life position. A bulk sediment sample containing *C. fluminalis* from this locality, used in the isotope study (Buckingham 2004; see also Chapter 6), is predominantly of articulated and undamaged specimens of all sizes suggesting that this is an *in-situ* assemblage or has been subjected to minimal re-working (Figure 2.28). As mentioned earlier bivalves would become disarticulated soon after death. The insides of these *Corbicula* are all undamaged and shiny with the only noticeable weathering being iron stains where the mollusc body was attached. This particular bulk sample also contains a variety of seeds, hazelnuts and wood, indicative of temperate conditions.

### The presence of hominins

The artefacts from the Stanton Harcourt Channel are described in Chapter 7 where they are discussed with reference to other artefacts from the immediate region. Typologically they are of late Middle Pleistocene age. Although some of the artefacts are damaged as to suggest they might have been transported by the river, the condition of the majority suggests that they have not been moved any distance. Their context strongly indicates that they are contemporaneous with the bones and environmental material.

In the southern part of the excavation area (Sites 1/4/7/8) more intensive excavations were possible than for much of the rest of the site. In this area, closely spaced sections and hand-excavation enabled a more detailed interpretation of the 3-dimensional accumulation of the sediments and the relative position of the artefacts (Buckingham 2004; 2007).

In this area there were 10 excavated stone tools. These were from the stacked sequence of sediment which accumulated mainly as a point bar on the left bank of the river or close to the right bank which was being undercut (Figures 2.3 and 2.6). The sedimentary evidence suggests that there was seasonal fluctuation in the flow of the river. The very early bar sediments

Figure 2.28 Above: Molluscs (*Corbicula fluminalis* and a few snail shells) from Stanton Harcourt illustrating their good preservation. Below: *C. fluminalis* from the excavation (left) compared to a modern Corbicula from Indonesia.

probably accumulated below the water line and apart from molluscs, wood and some vegetation, are devoid of bones and artefacts. Initially the bar would have been a temporary feature with unstable sediment which would have been an unsuitable environment for the preservation of organic material and artefacts. It is believed that any material that might have collected there would probably have been washed away by the next storm. As a bar grows, the potential for environmental material to collect on its surface increases when it becomes semi-emergent and vegetation starts to colonise the surface. Warm water molluscs, particularly *Corbicula fluminalis* and *Potomida littoralis* occurred throughout the bar sediments at this location and it is believed that this entire feature accumulated during part of MIS 7.

The sedimentary evidence indicates that this bar had an undulating profile on several occasions which was due to sand/gravel scroll bars forming on the upper surface (Buckingham 2004, 2007). Bones, artefacts, wood and vegetation are often located between the scroll bars. The detailed stratigraphy of this bar indicates that this material accumulated on the bar surface at different times, often collecting in depressions or at the downstream end of the bar nearer the '*thalweg*'. At times the surface was vegetated and this helped to trap sediment and organic debris. Patches of immature soil were recorded.

The formation of a bar provides conditions for the rapid burial of organic material and artefacts without long distance transport in the channel. Some bones and artefacts are in poorly sorted gravel beds but have still not been moved a great distance in the channel as they are not substantially rolled. It is quite likely that organic material and artefacts were originally on the bar top but were sometimes moved together by the current at high stage. Seasonal floods moved material up and across the bar.

At the downstream end of the bar there is also the potential for artefacts which had been discarded on the right bank to be moved with bones and other material from one bank to the other by the '*thalweg*'. Handaxe A12 was located at the downstream end of the bar (Bed 37 in Figure 2.9 and see discussion on Bone Group A).

A11 was located with large pieces of wood near the right channel margin 2R in Bed 44. A27 was found just above the *in-situ* roots at the right channel margin 3R in Bed 48. A21 rested on the silt/silty clay of Bed 58 on which Bone Group E was located, near the right bank 4R (Figure 2.22). It was near the SE corner of the pelvis SH7/119. Although A16 was excavated from only 15 cm above a mound of Oxford Clay, the trough of silt with some gravel was late in the sequence at Site 1, in Bed 93. It was located near a remnant of the channel margin at position 3R (Figure 2.3).

On the bar top, some of the artefacts are resting on or are just above fine-grained sediment. Artefact A2 was two-tone in colour when first excavated, where one end was embedded in the top of a silt layer of the early bar sediment. The other half was in later, more iron stained layers, including gravel that had collected between the scroll bars (Figure 2.10: Bed 39). After excavation this colour difference was less noticeable, probably due to oxidation on exposure to the air.

Other artefacts were excavated from even higher on the bar top. These have been slightly water polished but have not transported any distance e.g. A4 (Figure 2.29) and A5 Figure 2.30).

Figure 2.29 Handaxe A4. The negative scars are quite sharp, suggesting that the artefact was not exposed for long on the land surface before becoming buried and moved very little once in the streambed.

Figure 2.30 Patinated artefacts excavated from the bar sequence at Site 1. A3 was from just above the early sand scrolls. A5 is an unrolled artefact from a very thin fine sand near the top.

The handaxe A4 was excavated from a shallow scour on the bar top, resting partially on a very thin sand and silt layer in Bed 68 (Buckingham, 2007). The bar was well established by this time. A1 was recovered from a trough of iron-stained sandy gravel, between the scroll bars of the early bar sediment in Bed 72. Artefact A3 was excavated from just above silt and sand of the early meander fill but late in the bar sequence (Bed 76). A5 was excavated from the bar top even later in the sequence, in Bed 114.

Another indication of the lack of transport of an artefact after burial may be seen in the patination of A3 and A5, for example (Figure 2.30). A3 was damaged during excavation and revealed deep white patination with very little of the original dark flint visible. Some flint pebbles in the vicinity also had this deep patination showing a similar post-depositional history in which the flint becomes chemically altered. Slight damage on A5 suggests a similar patination. This is the least rolled artefact and was recovered at a higher level in the bar sequence than the handaxe A4. It was excavated from near the top of the bar in a shallow trough of fine sand just below a remnant of the limestone dominated Stanton Harcourt Member of the Summertown-Radley Formation (MIS 6). It is proposed that percolating meteoric water through both these permeable layers has caused the distinctive patination. This type of patination, when silica becomes unstable in calcareous conditions, is recorded elsewhere (Luedtke 1992). In this state it is extremely brittle, and the artefacts would have broken had they been subjected to much fluvial transport. This is especially true of A5 where the ridges are only slightly polished and the edges virtually undamaged.

The most likely scenario for the artefacts in this area is that they were mostly discarded on the bar during low stage, probably in the summer, when the top of the bar was above water level. At this time the bar was an extension of the flood plain on the left bank and was likely to have been visited by animals and people. Other artefacts were probably left at the wooded right bank near the possible *in situ* mammoth skulls discussed earlier. The evidence suggests that the artefacts are contemporaneous with the molluscs and wood and accumulated during MIS 7. There is no reason to believe that the artefacts could not have been directly associated with some of the excavated bones.

# Chapter 3

# Dating The Stanton Harcourt Channel Deposits

## Absolute dating

Several methods of absolute dating were used on sediment, bones, and molluscs associated with the bones with varying degrees of success.

### *Amino acid dating of shell*

Amino acid geochronology of Valvata piscinalis shells from the material collected by Briggs et al. (1985), suggests an age of about 200,000 years BP for the deposit (Bowen *et al.* 1989; Bowen 1999). Other samples from Sites 1 and 2 have given a range of dates from 190-210,000 years BP (Bowen, pers. comm.). More recently, amino acid results from four opercula of *Bithynia tentaculata* from Stanton Harcourt place them firmly among values for samples from other sites of MIS 7 age (Penkman *et al.* 2011). The chronology provided by the opercula data is interesting in that it indicates a number of sites with artefacts (including Stanton Harcourt) that fall within the age range for MIS 7, but a total absence of sites with artefacts attributed to the Last Interglacial MIS 5e (Table 3.1).

### *Amino acid dating of dental enamel*

Amino acid analyses were carried out by M. Dickinson (University of York) on multiple Elephantidae teeth from the Stanton Harcourt. His report is as follows:

> 'Recent amino acid studies have isolated the intra-crystalline fraction of calcium carbonate based biominerals, such as shells, which provides a closed-system repository enabling amino acid degradation to be used as an indicator of age (Penkman *et al.* 2013; Oakley *et al.* 2017). However, recent developments in the preparative method of calcium phosphate based biominerals have enabled the expansion of the intra-crystalline protein decomposition (IcPD) technique to mammalian remains (Dickinson *et al.* 2019). Most amino acids can exist in two or more optical isomers (an L and D form), and the conversion between the two forms is known as racemisation. In a closed system, the extent of racemisation can be used to infer the relative ages of samples with similar temperature histories, as the progress of the reaction is only dependent on temperature and time. It has been shown that a fraction of amino acids that exhibits closed system behaviour can be isolated from enamel, making it suitable for use as a tool for relative age estimation (Dickinson 2018; Dickinson *et al.* 2019). Three *Mammuthus trogontherii* samples of enamel were collected from Stanton Harcourt for amino acid analysis: SH1/335 (L. upper M1), SH1/365 (L. lower m3) and SH15/53 (R. lower dp4/m1).
>
> A pilot dataset for a UK enamel aminostratigraphy is currently being developed, but the data from Stanton Harcourt can currently be compared to a number of sites with independent evidence of age. The levels of racemisation in enamel from Stanton Harcourt are significantly lower than observed from Barnfield pit at Swanscombe

| 1 | | | 2 | | | | 3 | 4 | | | 5 | | | | | | 6 |
|---|---|---|---|---|---|---|---|---|---|---|---|---|---|---|---|---|---|
| **Site** | | | **D/L** | | | | | | | | **Archaeology** | | | | | | |
| | | | **Ala** | | **Val** | | | | | | | | | | | | |
| Site number (SI Fig. 1) Site | *Bithynia* species | Number analysed | mean | standard deviation | mean | standard deviation | Independent Geochronology | Terrace stratigraphy | *Hippopotamus* / *Corbicula* | *Arvicola* / *Mimomys* | Mesolithic | Upper Palaeolithic | Levallois | Acheulian | Clactonian | Undiagnostic | Consensus MIS |
| 1 Acle (modern) | *le* | 3 | 0.04 | 0.00 | 0.02 | 0.00 | • | | | | | | | | | | |
| 1 Acle (modern) | *te* | 2 | 0.04 | 0.00 | 0.02 | 0.00 | | | | | | | | | | | |
| 2 Enfield Lock | *te* | 4 | 0.05 | 0.01 | 0.03 | 0.01 | | | | | | | | | | | 1 |
| 3 Quidenham Mere, 250-260cm | *te* | 4 | 0.05 | 0.01 | 0.03 | 0.00 | | | | | | | | | | | 1 |
| 3 Quidenham Mere, 640-650cm | *te* | 3 | 0.05 | 0.01 | 0.03 | 0.00 | | | | | | | | | | | 1 |
| 4 Newby Wiske | *te* | 3 | 0.06 | 0.01 | 0.04 | 0.00 | • | | | | | | | | | | 1 |
| 5 Aston-upon-Trent | *te* | 4 | 0.06 | 0.00 | 0.04 | 0.00 | • | | | | | | | | | | 1 |
| 6 Star Carr, 245-250cm | *te* | 2 | 0.07 | 0.01 | 0.04 | 0.00 | • | | | | | | | | | | 1 |
| 7 Sproughton | *te* | 4 | 0.09 | 0.00 | 0.05 | 0.00 | • | | | | | | | | | | 2 |
| 6 Star Carr, 524-528cm | *te* | 3 | 0.10 | 0.02 | 0.04 | 0.00 | • | | | | | | | | | | 2 |
| 8 Cassington | *te* | 4 | 0.12 | 0.01 | 0.07 | 0.00 | • | Th | | | | | | | | | 5c-4 |
| 9 Isleworth | *te* | 4 | 0.14 | 0.01 | 0.07 | 0.01 | | Th | | | | | | | | | |
| 10 East Mersea (R. site) | *te* | 2 | 0.16 | 0.02 | 0.08 | 0.01 | | | Hippopotamus | | | | | | | | 5e |
| 11 Coston | *te* | 3 | 0.17 | 0.00 | 0.08 | 0.00 | | | Hippopotamus | | | | | | | | 5e |
| 12 Maxey | *te* | 4 | 0.17 | 0.03 | 0.08 | 0.02 | | NW | | | | | | | | | |
| 13 Woolpack Farm | *te* | 1 | 0.17 | 0.00 | 0.09 | 0.00 | | | | | | | | | | | |
| 14 Jaywick Sands | *te* | 4 | 0.17 | 0.01 | 0.09 | 0.01 | | | | | | | | | | | |
| 15 Saham Toney, 92 | *te* | 4 | 0.17 | 0.00 | 0.10 | 0.01 | | | | | | | | | | | |
| 16 Itteringham | *te* | 1 | 0.18 | 0.00 | 0.09 | 0.01 | | | | | | | | | | | |
| 17 Tattershall Castle | *te* | 4 | 0.18 | 0.01 | 0.09 | 0.01 | • | TW | | | | | | | | | 5e |
| 18 **Bobbitshole** | *te* | 3 | 0.18 | 0.01 | 0.10 | 0.00 | | | | | | | | | | | 5e |
| 19 Shropham | *te* | 4 | 0.18 | 0.01 | 0.09 | 0.01 | | | Hippopotamus | | | | | | | | 5e |
| 20 Trafalgar Square | *te* | 6 | 0.19 | 0.01 | 0.10 | 0.01 | | Th | Hippopotamus | | | | | | | | 5e |
| 15 Saham Toney, 94 | *te* | 2 | 0.19 | 0.00 | 0.11 | 0.01 | | | | | | | | | | | |
| 21 Bardon Quarry, Area 3 | *te* | 3 | 0.19 | 0.02 | 0.09 | 0.01 | | TW | | | | | | | | | |
| 15 Saham Toney, 93 | *te* | 4 | 0.19 | 0.01 | 0.10 | 0.01 | | | | | | | | | | | |
| 16 Itteringham, Bed d | *te* | 4 | 0.19 | 0.01 | 0.10 | 0.00 | | | | | | | | | | | |
| 22 Eckington | *te* | 4 | 0.20 | 0.01 | 0.10 | 0.01 | | SA | Hippopotamus | | | | | | | | 5e |
| 23 Cropthorne New Inn | *te* | 4 | 0.20 | 0.01 | 0.09 | 0.00 | | SA | Hippopotamus | | | | | | | | 5e |
| 21 Bardon Quarry, Area 2 | *te* | 4 | 0.21 | 0.03 | 0.11 | 0.02 | | | | | | | | | | | |
| 24 Funtham's Lane East | *te* | 25 | 0.22 | 0.01 | 0.12 | 0.01 | | NW | Corbicula | | | | | | | | |
| 25 Stanton Harcourt | *te* | 4 | 0.23 | 0.01 | 0.11 | 0.00 | | Th | | | | | | | | | 7 |
| 26 Crayford | *te* | 4 | 0.24 | 0.00 | 0.12 | 0.01 | | Th | Corbicula | | | | | | | | 7 |
| 27 Barnwell | *te* | 2 | 0.24 | 0.01 | 0.12 | 0.01 | | | Corbicula | | | | | | | | |
| 28 Ailstone-on-Stour | *te* | 4 | 0.24 | 0.01 | 0.12 | 0.01 | | SA | Corbicula | | | | | | | | |
| 29 Histon Road | *te* | 2 | 0.24 | 0.01 | 0.12 | 0.01 | | | | | | | | | | | |
| 30 Strensham | *te* | 4 | 0.25 | 0.02 | 0.12 | 0.02 | | SA | | | | | | | | | |
| 31 Stutton | *te* | 4 | 0.25 | 0.01 | 0.14 | 0.01 | | | Corbicula | | | | | | | | |
| 32 Coronation Farm | *te* | 1 | 0.25 | 0.00 | 0.13 | 0.00 | | TW | | | | | | | | | 7 |
| 33 Somersham | *te* | 4 | 0.25 | 0.01 | 0.12 | 0.00 | | | Corbicula | | | | | | | | |
| 34 Norton Bottoms | *te* | 15 | 0.25 | 0.01 | 0.13 | 0.01 | | TW | Corbicula | | | | | | | | 7 |
| 35 Block Fen | *te* | 4 | 0.25 | 0.02 | 0.13 | 0.01 | | | | | | | | | | | |
| 33 Somersham | *tr* | 4 | 0.26 | 0.01 | 0.13 | 0.01 | | | Corbicula | | | | | | | | |
| 36 Ebbsfleet, Lmst Brickearth | *te* | 3 | 0.30 | 0.01 | 0.15 | 0.01 | | | | | | | | | | | |
| 41 Stoke Tunnel | *te* | 2 | 0.30 | 0.01 | 0.15 | 0.01 | | | | | | | | | | | |
| 40 West Thurrock (Lion Pit), 3 | *te* | 25 | 0.30 | 0.01 | 0.16 | 0.02 | | Th | Corbicula | | | | | | | | 7 |
| 40 West Thurrock (Lion Pit), 6 | *te* | 5 | 0.31 | 0.01 | 0.16 | 0.01 | | Th | Corbicula | | | | | | | | 7 |
| 42 Selsey | *te* | 3 | 0.31 | 0.01 | 0.16 | 0.00 | | | | | | | | | | | |

| | | | | | | | | | |
|---|---|---|---|---|---|---|---|---|---|
| 43 Aveley | te | 4 | 0.32 | 0.02 | 0.16 | 0.01 | | Th | 7 |
| 44 West Wittering | te | 3 | 0.32 | 0.00 | 0.16 | 0.00 | | | |
| 45 Bushley Green | te | 1 | 0.34 | 0.00 | 0.16 | 0.00 | | SA | |
| 46 Cudmore Grove | te | 4 | 0.34 | 0.02 | 0.19 | 0.01 | | Th | 9 |
| 47 Hackney Downs | te | 2 | 0.34 | 0.01 | 0.17 | 0.01 | | Th | 9 |
| 46 Cudmore Grove, sample PE | te | 2 | 0.36 | 0.00 | 0.20 | 0.00 | | Th | 9 |
| 48 Belhus Park | te | 1 | 0.36 | 0.02 | 0.20 | 0.02 | | Th | 9 |
| 49 Purfleet, 1 | te | 23 | 0.38 | 0.01 | 0.18 | 0.01 | • | Th | 9 |
| 50 Barling | te | 4 | 0.38 | 0.01 | 0.20 | 0.01 | | Th | 9 |
| 49 Purfleet, 6 | te | 24 | 0.38 | 0.01 | 0.21 | 0.01 | • | Th | 9 |
| 51 Shoeburyness | te | 5 | 0.39 | 0.01 | 0.20 | 0.01 | | | |
| 52 Grays | te | 2 | 0.40 | 0.01 | 0.20 | 0.00 | | Th | 9 |
| 53 Marks Tey | te | 2 | 0.40 | 0.01 | 0.22 | 0.01 | | | 11 |
| 48 Belhus Park, BP18 | te | 4 | 0.40 | 0.01 | 0.21 | 0.01 | | Th | 9 |
| 54 Trimingham | te | 13 | 0.40 | 0.01 | 0.22 | 0.01 | | | |
| 55 Hoxne, Stratum B2, 64 | te | 4 | 0.40 | 0.02 | 0.22 | 0.01 | • | | 11 |
| 56 Barnham, BEF92 | tr | 3 | 0.40 | 0.02 | 0.21 | 0.01 | | | 11 |
| 57 Elveden, ELV 96 | te | 2 | 0.41 | 0.01 | 0.20 | 0.01 | | | |
| 57 Elveden, ELV 95 | te | 5 | 0.41 | 0.01 | 0.21 | 0.01 | | | |
| 58 Swanscombe (Barnfield Pit) | te | 6 | 0.42 | 0.01 | 0.24 | 0.03 | | Th | 11 |
| 56 Barnham, BEF93 | te | 6 | 0.42 | 0.02 | 0.20 | 0.02 | | | 11 |
| 59 Woodston | te | 4 | 0.42 | 0.01 | 0.23 | 0.01 | | NW | 11 |
| 55 **Hoxne, Stratum E** | te | 4 | 0.42 | 0.01 | 0.23 | 0.01 | | | 11 |
| 60 West Stow (Beeches Pit) | te | 4 | 0.42 | 0.01 | 0.25 | 0.01 | • | | 11 |
| 61 Dierden's Pit (Ingress Vale) | tr | 2 | 0.43 | 0.01 | 0.22 | 0.01 | | Th | 11 |
| 55 Hoxne, Stratum B2, 50 | te | 4 | 0.43 | 0.01 | 0.24 | 0.01 | | | 11 |
| 62 Southfleet Road | te | 7 | 0.43 | 0.01 | 0.22 | 0.01 | | | 11 |
| 61 Dierden's Pit (Ingress Vale) | te | 4 | 0.45 | 0.01 | 0.21 | 0.01 | | Th | 11 |
| 63 Clacton-on-Sea | te | 4 | 0.46 | 0.01 | 0.25 | 0.02 | | Th | 11 |
| 64 Waverley Wood | tr | 6 | 0.48 | 0.03 | 0.27 | 0.03 | | | 13? |
| 65 Sidestrand, Upper Unio-Bed | te | 7 | 0.54 | 0.02 | 0.29 | 0.01 | | | 13? |
| 65 Sidestrand, Lower Unio-Bed | te | 5 | 0.55 | 0.01 | 0.29 | 0.01 | | | 13? |
| 66 Sugworth | tr | 6 | 0.56 | 0.01 | 0.28 | 0.01 | | Th | 15? |
| 67 Little Oakley | tr | 2 | 0.56 | 0.00 | 0.28 | 0.00 | | | 15? |
| 68 Pakefield, PaCii | tr | 4 | 0.57 | 0.03 | 0.30 | 0.01 | | | 17/15? |
| 68 Pakefield, PaCi | tr | 3 | 0.58 | 0.01 | 0.35 | 0.02 | | | 17/15? |
| 69 **West Runton** | tr | 11 | 0.61 | 0.02 | 0.33 | 0.01 | | | 17/15? |
| 70 **Bavel** | b/t | 4 | 0.81 | 0.02 | 0.54 | 0.04 | | | 31? |
| 71 Weybourne (Weybourne Crag) | tr | 2 | 0.89 | 0.01 | 0.69 | 0.03 | | | |
| 72 **Tegelen** | tr | 4 | 0.90 | 0.01 | 0.75 | 0.03 | | | |
| 71 Weybourne (Weybourne Crag) | te | 1 | 0.92 | 0.00 | 0.79 | 0.01 | | | |
| 73 Thorpe Aldringham (N. Crag) | sp. | 1 | 0.94 | 0.00 | 0.84 | 0.04 | | | |
| 74 Frechen (Pliocene) | te | 4 | 0.93 | 0.01 | 0.91 | 0.02 | | | |

Table 3.1 Intra-crystalline amino acid date from the opercula of Bithynia from sites in southern Britain (Columns 1 and 2). In Column 4 data are shown alongside terrace stratigraphy (NW – Nene/Welland, SA – Severn/Avon, TH – Thames, TW – Trent/Witham) and occurrences of important biostratigraphic indicator species. Sites with in situ archaeology are indicated in Column 5 and existing consensus views on their correlation with the MIS record are indicated in Column 6. Table modified from Penkman *et al.* 2011

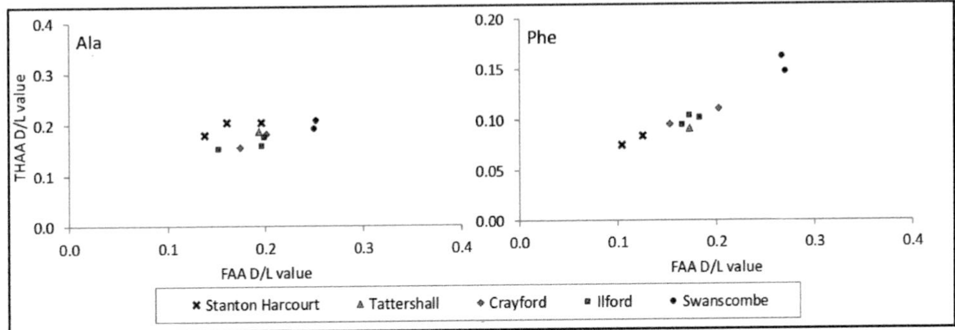


Figure 3.1 Free amino acid (FAA) vs total hydrolysable amino acid (THAA) racemisation of Elephantidae enamel samples from Stanton Harcourt gravel pit and samples from other UK sites (Dickinson et al. 2019). The other UK sites have been correlated with: MIS 6 (Tattershall Thorpe), MIS 7 (Crayford, Ilford) and MIS 11 (Barnfield pit, Swansombe).

(correlated with MIS 11, Bridgland 1994; Schreve 2001). The racemisation of alanine (Ala) in *M. trogontherii* samples from Stanton Harcourt cluster with Ala racemisation values for enamel samples from other sites correlated with Marine Isotope Stages (MIS) 6 and 7 such as Tattershall Thorpe (Meijer and Preece, 2000; Bridgland *et al.*, 2015), Crayford (Schreve 2001; Bridgland 2014) and Ilford (Bridgland 1994; Coope 2001; Schreve,2001) (Figure 3.1). Phenylalanine (Phe) shows slightly lower racemisation values in the Stanton Harcourt samples than these other MIS 6-7 sites. Therefore, the enamel IcPD values obtained for the Stanton Harcourt samples analysed support a late Middle Pleistocene age, are consistent with correlation for the gravel pit with MIS 7 and can tentatively support correlation with late MIS 7. However, further comparisons to enamel racemisation values from additional UK Middle Pleistocene sites such as the deposits correlated with MIS 5e at St James and Barrington (Schreve 2001) are required for a more comprehensive assessment of the age of the deposits at Stanton Harcourt'.

### *Optically stimulated luminescence*

Optically stimulated luminescence (OSL) dates were attempted on two sand samples from Site 6. One was from sands considered by the authors to be at the base of the MIS 6 sediments. This produced an estimated date of 170 +/-15 ka which would fit with current understanding of this interface. The second sample was associated with some temperate elements and bones believed on other grounds to be of MIS 7 age. The estimated date of the sample was 300+/-50 ka. The proximity of the Oxford Clay made it difficult to interpret the results because of the unknown degree of water saturation of the sediments during burial. Taking that into account, a total range was given of between 200-450ka which, although somewhat inconclusive, at least indicates that these deposits predate the Last Interglacial (Rees-Jones 1995).

### *Electron spin resonance and Uranium series*

Problems with recent uranium uptake proved to be an obstacle when attempting uranium series or electron spin resonance (ESR) dating methods on vertebrate teeth and bones from Stanton Harcourt.

ESR was attempted on eleven bones from Stanton Harcourt but recent uranium uptake caused by water table changes is believed to give an underestimate of age for the outside layers of these bones. However, the inner bone layers of one specimen (bison metatarsal SH8/73) recorded a minimum age of 192.4 ± 19.9 ka. This is consistent with other evidence from the site indicative of a late MIS 7 age (Pike *et al.* 2002; Lewis *et al.* 2006).

Similar difficulties were encountered when attempting to date a mammoth tooth (SH4/4) from the later stages of the channel fill sediments at Site 4 (Zhou *et al.* 1997). ESR measurements were made on enamel from this specimen and mass-spectrometric uranium series measurements were made on both enamel and dentine samples. A minimum age for this tooth of 146.5 ka confirms that the Channel deposits predate the Last Interglacial.

## Biostratigraphy

### *Large vertebrates*

As outlined in Chapter 1, two major climatic events are recorded at Stanton Harcourt: an interglacial channel overlain by cold climate deposits. The only bone recovered from the cold stage deposits was part of an antler of reindeer *Rangifer tarandus.* The following discussion therefore relates only to the interglacial channel.

The large vertebrates from the Stanton Harcourt Channel are summarised in Table 3.2 and Figure 3.2 and detailed in Chapters 4 and 5. Mammoth is the most common species and is of particular biostratigraphic significance. It is represented by approximately 100 complete tusks, a further 100 partial tusks, 188 molar teeth (many still in mandibles and skulls), and more than 500 post-cranial bones. Initially described as the woolly mammoth *Mammuthus primigenius* (Buckingham *et al.* 1996), it was subsequently re-assigned to *Mammuthus trogontherii*, a late form of steppe mammoth now recognised as synonymous with MIS 7 (Lister & Sher 2001; Scott 2007). The teeth from Stanton Harcourt comprise the largest collection in Britain of the MIS 7 mammoth. They have several distinctive characteristics which differentiate them from the ancestral steppe mammoth and from the woolly mammoth (Lister and Scott in press). Particularly noticeable is their small size relative to that of the other species. An analysis of the post-cranial bones suggests shoulder height ranging between 2.1 and 2.9 metres (Scott and Lister in press).

The association of this small steppe mammoth with some of the other species listed in Table 3.2 is unique to MIS 7. Although both horse *Equus ferus* and bison *Bison priscus* are found throughout the British late Middle and Upper Pleistocene, except from the dense woodland habitat of MIS 5e (Currant 1986, 1989; Sutcliffe 1995), the MIS 7 forms of horse and bison are significantly more robust than in the Upper Pleistocene (See Chapter 5). The carnivores at Stanton Harcourt are also distinctive from those in the interglacials before and after MIS 7 in that they are represented by a typically large lion *Panthera spelaea* and by small forms of wolf *Canis lupus* and bear *Ursus arctos* (Chapter 5).

Generalizations about the vertebrates that characterise MIS 7 are made with caution as the climate and environment of this interglacial is more complex than initially believed. During the early years of fieldwork at Stanton Harcourt, some Quaternary specialists disputed the

| | No. of specimens | % of total |
|---|---|---|
| *Canis lupus*, wolf | 1 | 0.08 |
| *Ursus arctos*, brown bear | 10 | 0.1 |
| *Felis spelaea*, lion | 3 | 0.3 |
| *Palaeoloxodon antiquus*, straight-tusked elephant | 57 | 5 |
| *Mammuthus trogontherii*, steppe mammoth | 996 | 81 |
| *Equus ferus*, horse | 34 | 3 |
| *Cervus elaphus*, red deer | 4 | 0.4 |
| *Bison priscus*, bison | 125 | 10 |
| **TOTAL** | **1230** | |

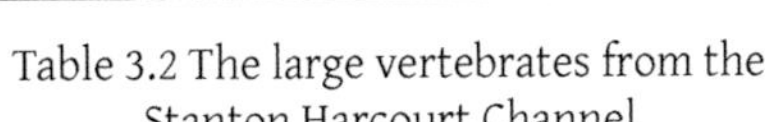

Table 3.2 The large vertebrates from the Stanton Harcourt Channel

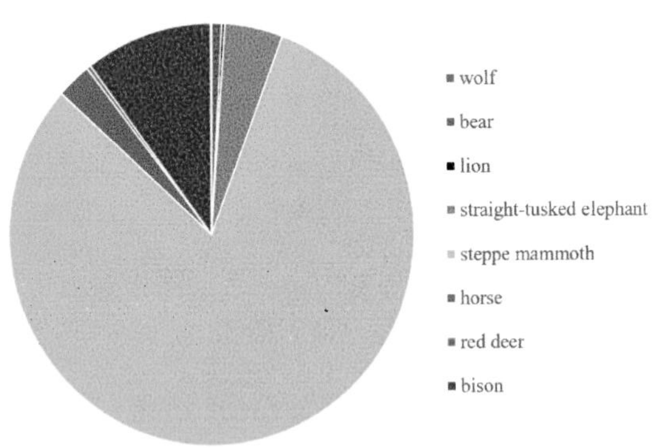


Figure 3.2 Large vertebrate representation at Stanton Harcourt

existence of an interglacial at around 200,000 years. However, MIS 7 not only became firmly established in the terrestrial Pleistocene sequence but was seen to comprise at least two fully temperate climatic episodes separated by cooler intervals (Bridgland 1994). Concurrently, Schreve (1997) re-evaluated the later Middle Pleistocene vertebrates in British museum collections and concluded that some of Britain's most abundant fossil assemblages, many of which had been assigned to the 'Ipswichian' (Last Interglacial), were actually of MIS 7 age. In the majority of the 24 assemblages described by Schreve, mammoth (also referred to as the 'Ilford' mammoth) is the most commonly represented large vertebrate, as at Stanton Harcourt. Schreve suggested that that the presence of the 'Ilford' mammoth might serve as a key indicator for an assemblage of MIS 7 age. In agreement with Bridgland (*op. cit.*), Schreve proposed two fully temperate climatic episodes within MIS7: an earlier forested phase followed by more open grassland conditions. Further studies of the large vertebrate faunas at other British sites supported the argument for the sub-division of this interglacial into at least two distinct warm phases (Murton *et al.* 2001; Schreve 2001).

Emerging details from ice cores, stalagmites, vein calcite and sediment sources indicate that MIS 7 is more complex, with rapid warming at the beginning of the stage and comprised of at least three warm peaks of comparable magnitude (MIS 7 e, c and a). These are separated by cooler intervals, probably involving sea-level fall and connection to the European mainland and are of gradually decreasing magnitude towards the subsequent glacial (Winograd *et al.* 1992; Candy and Schreve 2007). As reviewed by Pettitt and White (2012), opinion varies regarding the extent to which Britain was an island during MIS 7, a factor that would have had a significant effect upon determining the presence or absence of various vertebrates over time. It is agreed that sea-level was sufficiently high during the early part of the interglacial (MIS 7e) to isolate Britain from the Continental mainland. Although sea-levels fell during the two cool intervals (MIS 7b and 7d) of the interglacial, the general consensus is that only during

MIS 7d did sea-levels become sufficiently depressed for Britain to have become a peninsula and thus accessible again from the Continental mainland. As originally proposed by Schreve (1997), the large vertebrates appear broadly to indicate two distinct environments: an earlier phase (equated with MIS 7e) dominated by species with a preference for forest habitat and a later phase with species indicative of more open terrain with woodland in the vicinity. However, the vertebrate assemblages that can be assigned to this later phase are far more numerous than those of MIS 7e (Scott in prep.) and there is considerable species variation in these. This might be indicative of regional habitat variation or might reflect environmental changes through time. However, the absence of absolute dates and reliable stratigraphic data for most sites means that MIS 7c and MIS 7a are indistinguishable on faunal grounds.

Compared with many assemblages, Stanton Harcourt has relatively few species and the assemblage is dominated by open grassland grazers – steppe mammoth, bison and horse. Small numbers of straight-tusked elephant, red deer and bear indicate the proximity of woodland. Forest obligates such as aurochs, forest rhino and giant deer, all characteristic of MIS 7e, are absent. Although the large vertebrates from Stanton Harcourt cannot be used to distinguish between a biostratigraphic age of MIS 7c or MIS 7a, there is sufficient accompanying data to indicate a late MIS 7 age for this material.

### *Molluscs*

Molluscs were found throughout the excavation area. Their abundance and preservation varied enormously across the site depending on their location within the fluvial environment (see Chapter 2). Early in the excavations, a number of bulk sediment samples were taken for the identification of molluscs by D. Keen and C. Gleed-Owen (Table 3.3). Since these identifications were made, there have been a number of changes in the nomenclature (Anderson 2020 and pers. comm.). These changes are noted in Table 3.3 with the updated species names appearing in this chapter. Although it is considered that these Tables broadly represent the species present at the site, much richer deposits found in later years indicate that they are not totally representative of their relative abundance. For example, whole beds of articulated *Corbicula fluminalis* (Müller) and *Potomida littoralis* (Cuvier) were excavated (see Chapter 2). *C. fluminalis* and *P. littoralis* are both molluscs that live in warm water and are not found in periglacial environments. The abundance of these molluscs within the interglacial sediments at Stanton Harcourt is useful in determining the extent of the river channel as they are absent from MIS 6.

Particularly relevant in assigning the deposits at Stanton Harcourt to MIS 7 is *C. fluminalis*, well represented here but absent from Last Interglacial (Ipswichian, MIS 5e) assemblages (Meijer and Preece 2000; Keen 1990, 1995, 2001). The occurrence of *Corbicula* and hippopotamus is mutually exclusive at British sites securely attributable to the Last Interglacial MIS 5e (Keen 2001). This mollusc is not found at Ipswichian sites where hippo is present, unless the shells have been re-worked. Although *C. fluminalis* is also known from earlier interglacials such as at Purfleet, which is believed to date to MIS 9 (Bridgland *et al.* 1995), all other lines of evidence discussed here, point to an MIS 7 age for Stanton Harcourt.

| SPECIES | 1020 | 1021A | 1022 | 1023 | 1024A | 1026 | 1037 | 2002A | 2006 | 2007 | 2008A | 2009 | 2010 | 2011 | 4003 | 6002 |
|---|---|---|---|---|---|---|---|---|---|---|---|---|---|---|---|---|
| *Valvata cristata* (Müller, 1774) | | 1 | | | | | | | | | | | | | | |
| *Valvata piscinalis* (Müller, 1774) | 235 | 188 | 55 | 17 | | 66 | 178 | 170 | 1311 | 357 | 570 | 19 | 209 | 375 | 57 | 212 |
| *Bithynia tentaculata* (L., 1758) | 12 | 40 | 5 | 1 | | 8 | 2 | | | | | | | 2 | 6 | 69 |
| opercula | [8]+ | [67]+ | | [1]+ | | [10]+ | [18] | | | | | | | | | [82] |
| *Lymnaea stagnalis* (L., 1758) | | | | | | | | | 1 | | 1 | | | | | |
| *Galba truncatula* (Müller, 1774) formerly *Lymnaea truncatula* | 6 | 3 | 4 | | | | 2 | 15 | | 8 | 7 | 7 | 1 | 2 | 11 | 1 |
| *Ampullaceana balthica* (L., 1758) formerly *Lymnaea peregra* (Müller, 1774) | 170 | 82 | 29 | 4 | 1 | 31 | 59 | 236 | 964 | 197 | 337 | 14 | 71 | 205 | 25 | 99 |
| *Anisus vortex* (L.,1758) | | | | | | | | 2 | | | | | | | | 1 |
| *Anisus vorticulus* (Troschel,1834) | 1 | | | | | | | | 2 | | | | | | | |
| *Anisus leucostoma* (Millet, 1813) | 1 | 2 | | | | | | 9 | 1 | 1 | | | 2 | | 2 | |
| *Anisus* sp. | | | | | | | | | | 2 | | 1 | 1 | | | |
| *Bathyomphalus contortus* (L., 1758) | | | | | | | | | | 1 | | | | | | |
| *Gyraulus crista* (L., 1758) formerly *Armiger crista* | | | | | | 1 | | | | | | | | | | |
| *Gyraulus laevis* (Alder, 1838) | 21 | 18 | 1 | 1 | | 3 | | | 40 | 24 | 35 | | 2 | 84 | 11 | 10 |
| *Planorbis planorbis* (Müller, 1774) | | | | | | | | 20 | 5 | 4 | 21 | | | 3 | 2 | 16 |
| *Planorbis* sp. | | 2 | 5 | | | | 5 | 5 | 7 | 13 | 25 | | 6 | 61 | 5 | |
| *Ancylus fluviatilis* (Müller, 1774) | 27 | 30 | 9 | 1 | | 11 | 7 | 15 | 17 | 7 | 4 | 1 | 1 | 5 | 12 | 7 |

| SPECIES | 1020 | 1021A | 1022 | 1023 | 1024A | 1026 | 1037 | 2002A | 2006 | 2007 | 2008A | 2009 | 2010 | 2011 | 4003 | 6002 |
|---|---|---|---|---|---|---|---|---|---|---|---|---|---|---|---|---|
| *Acroloxus lacustris* (L., 1758)* | | | | | | | | | | | | | | | | |
| *Potomida littoralis* (Cuvier, 1798) | 7 | | 2 | | 11 | 1 | | ** | | | | | | | 1 | |
| *Anodonta* sp. | 1 | | | | | | | | | | | | 1 | | | |
| Unionidae sp. et gen. undet. | | 1 | | | | | | | | | 1 | 1 | | | 1 | |
| *Corbicula fluminalis* (Müller, 1774) | 84 | 34 | 25 | 7 | 1 | 18 | 21 | 125 | 4 | 28 | 16 | 2 | 16 | 25 | 36 | 18 |
| *Sphaerium corneum* (L., 1758) | 6 | 15 | 2 | 1 | | 1 | 2 | 5 | 3 | 3 | 5 | 1 | 4 | 4 | 5 | 2 |
| *Sphaerium lacustris* (Müller, 1774) | | | | | | | | | | | | | 2 | | | |
| *Pisidium amnicum* (Müller, 1774) | 17 | 8 | 13 | 7 | | 5 | 18 | 7 | 9 | 7 | 11 | 6 | 13 | 14 | 3 | |
| *Euglesa casertana* (Poli, 1791) formerly *Pisidium casertanum* | 2 | 2 | 2 | | | 4 | | 1 | | | | | | | | |
| *Euglesa personata* (Malm, 1855) formerly *Pisidium personatum* | | 1 | | | | 1 | | | | | | | | 1 | | |
| *Euglesa subtruncata* (Malm, 1855) formerly *Pisidium subtruncatum* | 13 | 22 | 6 | 2 | | 10 | | | 5 | 14 | 11 | 5 | 61 | 22 | 10 | |
| *Euglesa supina* (Schmidt,1851) formerly *Pisidium supinum* | | 2 | | | | | 1 | 1 | | 19 | 47 | 2 | 12 | 6 | 6 | |
| *Euglesa henslowana* (Sheppard, 1823) formerly *Pisidium henslowanum* | 88 | 113 | 25 | 11 | | 55 | 38 | 37 | 154 | 174 | 162 | 18 | 104 | 167 | 25 | |
| *f. inappendiculata* (Moquin-Tandon,1855 | | 3 | | | | | | | | | | | | | | |
| *Euglesa hibernica* (Westerland,1894) formerly *Pisidium hibernicum* | | | | | | | | | | 7 | 3 | 1 | 1 | 1 | 4 | |

| SPECIES | 1020 | 1021A | 1022 | 1023 | 1024A | 1026 | 1037 | 2002A | 2006 | 2007 | 2008A | 2009 | 2010 | 2011 | 4003 | 6002 |
|---|---|---|---|---|---|---|---|---|---|---|---|---|---|---|---|---|
| *Euglesa nitida* (Jenyns,1832) formerly *Pisidium nitidum* (Jenyns, 1832) | 10 | 12 | 4 | 1 | | 3 | | 1 | 27 | 33 | 19 | 1 | 17 | 23 | 10 | |
| *Odhneripisidium moitessierianum* (Paladilhe, 1866) formerly *Pisidium moitessierianum* | 61 | 25 | 19 | 2 | | 33 | 3 | 1 | 18 | 65 | 36 | 8 | 83 | 42 | 30 | |
| *Odhneripisidium tenuilineatum* (Stelfax, 1918) formerly *Pisidium tenuilineatum* | | | | | | | | | 2 | 4 | 4 | 1 | 6 | 2 | 7 | |
| *Pisidium* sp. | 32 | 25 | 5 | 3 | | 22 | 18 | 368 | 6 | 4 | 4 | 3 | 24 | 32 | 6 | 56 |
| *Carychium minimum* (Müller,1774) | | | | | | 1 | | | | 2 | 1 | 8 | 2 | | 14 | |
| *Carychium tridentatum* (Risso,1826) | | 2 | 1 | | | | | | | | | | | | | 2 |
| *Carychium* sp. | | 2 | 1 | | | | | | | | | | | | | |
| *Oxyloma elegans* (Risso,1826) formerly *Oxyloma pfeifferi* (Rossmässler, 1835) | | | | | | | | 20 | 1 | 11 | 1 | | | 1 | 2 | 2 |
| *Succinea putris* (L., 1758)* | | | | | | | | | | | | | | | | |
| *Succinella oblonga* (Draparnaud, 1801) formerly *Succinea oblonga* | | 3 | | 1 | | | | | | | | | | | | |
| *Succinea indet.* | 1 | | 1 | | | 3 | 1 | | 3 | | | | | | 2 | |
| *Cochlicopa lubrica* (Müller, 1774) | | 2 | | | | | | 2 | | | | | | | 2 | |
| *Cochlicopa* sp. | | | | | | | | | 1 | 1 | | 1 | | | 1 | 1 |
| *Vertigo antivertigo* (Draparnaud, 1801) | | | | | | | | | | | 1 | | | | 1 | |
| *Vertigo pygmaea* (Draparnaud, 1801) | | 1 | | | | | | 1 | | 1 | | | | | | |

| SPECIES | 1020 | 1021A | 1022 | 1023 | 1024A | 1026 | 1037 | 2002A | 2006 | 2007 | 2008A | 2009 | 2010 | 2011 | 4003 | 6002 |
|---|---|---|---|---|---|---|---|---|---|---|---|---|---|---|---|---|
| *Vertigo sp.* | | | 2 | | | | 1 | 1 | | | | | | | | |
| *Pupilla muscorum* (L., 1758) | 4 | 38 | 5 | | | 3 | 4 | 39 | 16 | 23 | 27 | 6 | 13 | 20 | 15 | 16 |
| *Spermodea lamellata* | | | | | | | | | | | | | | | 2 | |
| *Vallonia costata* (Müller, 1774) | 1 | 7 | 1 | | | | | | 2 | 18 | 11 | 7 | 2 | 12 | 4 | 12 |
| *Vallonia pulchella* (Müller, 1774) | 5 | 13 | 1 | | | 6 | 7 | | 19 | 1 | 2 | | 3 | 11 | 1 | 6 |
| *Vallonia excentrica* (Sterki, 1892) | | | | | | | | | | | 2 | 1 | 1 | 1 | | |
| *Vallonia enniensis* (Gredler,1856) | | 1 | 4 | | | | 2 | | | 9 | 5 | 4 | 4 | 1 | 8 | |
| *Vallonia* sp. | | 13 | 4 | | | 1 | | 52 | 3 | 41 | 14 | 26 | 4 | 8 | 10 | |
| *Ena montana* (Draparnaud, 1801) | | 1 | | | | | | | | | | | | | | |
| *Punctum pygmaeum* (Draparnaud, 1801) | | 1 | 1 | | | | 1 | | | 1 | | 3 | | | 1 | |
| *Vitrina pellucida* (Müller, 1774) | | 4 | | | | | | | | 3 | | | | | 1 | 1 |
| *Vitrea crystallina* (agg.) | | | | | | | 1 | 1 | | | | | | | | |
| *Vitrea* sp. | | 2 | | | | | | | | | 1 | 7 | | | 1 | |
| *Nesovitrea hammonis* (Ström, 1765) | | 1 | 1 | | | | | | | | | | | | 2 | |
| *Aegopinella nitidula* (Draparnaud, 1805) | 1 | 2 | | | | | | | | | | | | | | |
| *Aegopinella pura* (Alder, 1830) | | | | | | | | | 3 | | | | | | | |
| *Limax* sp. | | | | | | | | | | 1 | | | | | 2 | |
| *Zonitoides nitidus* (Müller, 1774) | | 4 | | | | | | | | | | | | | | |
| *Zonitoides* indet. | | | | | | | | | | | | | | | 3 | |

| SPECIES | 1020 | 1021A | 1022 | 1023 | 1024A | 1026 | 1037 | 2002A | 2006 | 2007 | 2008A | 2009 | 2010 | 2011 | 4003 | 6002 |
|---|---|---|---|---|---|---|---|---|---|---|---|---|---|---|---|---|
| *Deroceras* sp. | | 3 | | | | | | | | | | | | | | |
| *Clausilia bidentata* (Ström, 1765) | | 1 | | | | | 1 | | | | | | | | 2 | |
| *Clausilia* sp. | 1 | | | | | | | | | | | | | | | |
| *Trochulus hispidus* (L., 1758) *formerly Trichia hispida* | 4 | 16 | 4 | 1 | | 3 | 10 | 10 | | | | | | | | 11 |
| *Candidula crayfordensis* (Kennard and Woodward, 1915) | 3 | | | | | 1 | 1 | 44 | 60 | 31 | 19 | 11 | 15 | 18 | 9 | 1 |
| *Helicella itala* (L., 1758) | | | | | | | | | 1 | | | | | | | |
| *Arianta arbustorum* (L., 1758) | | | | | | | 1 | 5 | | 2 | 4 | 4 | 3 | 2 | 4 | |
| | **814** | **746** | **237** | **60** | **13** | **291** | **384** | **1193** | **2685** | **1117** | **1407** | **169** | **684** | **1150** | **362** | **543** |

Total taxa identified to species = 54 of which Keen identified 45 at Site 1.
Note: Bivalve counts were divided by 2 to get the number of individuals and single valves have been entered as 1 individual
* indicates additional species identified by Keen from other samples at the site.
** Gleed-Owen noted the presence of fragmentary *Potomida littoralis* but as there were no hinges so they were not counted.

Table 3.3 Molluscs from Site 1(1020-37) Site 2(2002A-11) Site 4(4003) and Site 6(6002) at Stanton Harcourt (Keen 1992; Gleed-Owen 1998).

The presence of the low-spired form of *Valvata piscinalis* also indicates an MIS 7 age (Green *et al.* 1996). The absence of *Belgrandia marginata* may also be significant in eliminating the Ipswichian MIS 5e (Meijer and Preece 2000; Keen 2001).

### *Insects*

Of the many bulk sediment samples taken across the site, some were rejected because the insect remains were too oxidized for species identification. Five bulk samples were thus analysed R.G. Coope (2006). Four were from the excavations and were compared with the slightly larger assemblage recovered earlier from organic sediments at Dix Pit and discussed in Briggs *et al.* (1985). The initial bulk sample weighed 10 kg and the further four bulk samples from the controlled excavations each weighed about 5 kg.

The following results and discussion on the stratigraphical significance of the coleopteran assemblages comes from the following report provided for the authors by Coope (2006):

> 'The insect remains were recovered from the bulk samples in the laboratory by the standard method described by Coope (1986). They were generally well preserved. Most of them were of Coleoptera (beetles) or Trichoptera (caddisflies) but fragments of other insect orders were also recovered including Megaloptera, Hemiptera, Hymenoptera and Diptera. No attempt was made to investigate these other orders in detail.
>
> All the beetle and caddisfly fossils in these assemblages could be identified as species that are still living today. Table 3.4 lists the coleopteran taxa according to the nomenclature and in the taxonomic order given by Lucht (1987). In this table the numbers in each column and opposite each taxon indicate the minimum numbers of individuals present in the sample and is based on the maximum number of an identifiable skeletal element of that taxon. This table augments the list previously published (Briggs *et al.* 1985) by increasing the number of samples and updating the nomenclature to make it compatible with other published beetle assemblages.
>
> Altogether 163 coleopteran taxa were identified of which 118 could be named to species or species group. Of these species, 4 are not now members of the British Fauna.
>
> The Trichoptera from the channel deposits are listed in Table 3.5. These were represented by numerous larval sclerites, but no attempt has been made to give numerical values to their abundance'.

According to Coope there is no significant difference between the faunas of the 5 samples. There are local minor differences between the faunal assemblages from the excavation and the earlier sample but all 'were all deposited in similar interglacial climatic conditions'.

Since the majority of interglacial coleopteran faunas are made up of species that are still living in the British Isles, it is not stratigraphically useful to make crude comparisons of bulk similarity between them. Rather it is the presence of exotic species in these assemblages that have the greatest stratigraphical significance in the sense of providing evidence for

| | 1 | 2 | 3 | 4 | 5 |
|---|---|---|---|---|---|
| **Carabidae** | | | | | |
| Carabus cathratus L. | | | | 1 | |
| Nebria brevicollis (F.) | 1 | | | | 1 |
| Notiophilus palustris (Duft.) | | 1 | | | 1 |
| Dyschirius globosus (Hbst.) | | | | | 1 |
| Dyschirius sp. | | 1 | | | |
| Elaphrus riparius (L.) | | 1 | | | |
| Clivina fossor (L.) | | 1 | | | |
| Trechus secalis (Payk.) | 1 | 1 | | | |
| Bembidion properans (Steph.) | 1 | 1 | | | 2 |
| Bembidion varium (Ol.) | | | | | 1 |
| Bembidion gilvipes Sturm(Panz.) | 1 | | | | 1 |
| Bembidion clarki (Daws.) | | | 1 | | |
| Bembidion quadrimaculatum (L.) | 1 | | | | 1 |
| Bembidion doris (Panz.) | 1 | | 1 | | |
| Bembidion octomaculatum Serv. | 1 | | | | 1 |
| Bembidion articulatum | | | | | 1 |
| Bembidion obtusum Serv. | 2 | 1 | | 1 | 2 |
| Bembidion aeneum Germ. | | | | | 1 |
| Patrobus atrorufus (Ström) | 1 | 2 | | | |
| Harpalus sp. | 1 | 1 | | | 2 |
| Poecilus sp. | | | | | 1 |
| Pterostichus strenuous (Panz.) | | 1 | 1 | | |
| Pterostichus melanarius (L.) | 2 | 2 | | | 1 |
| Calathus fuscipes (Goeze) | | 1 | | 1 | 1 |
| Calathus melanocephalus (L.) | 1 | 1 | | | 1 |
| Agonum sexpuntatum (L.) | | | | | 1 |
| Amara sp. | 1 | 2 | | | 6 |
| Oodes helopioides (F.) | 1 | | | | |
| Syntomus truncatellus (L.) | | 1 | | | |
| Polistichus connexus (Fourcr.) | | | | | 1 |

| | 1 | 2 | 3 | 4 | 5 |
|---|---|---|---|---|---|
| **Haliplidae** | | | | | |
| Haliplus confines Steph. group | 2 | | | | |
| Haliplus sp. | 2? | 1? | | | 1 |
| | | | | | |
| **Dytiscidae** | | | | 1 | |
| Potamonectes depressus (F.) | | | | | 1 |
| Platambus maculatus(L.) | | | | | |
| Agabus nebulosus (Forst.) | | | | | 1 |
| Colymbetes fuscus (L.) | | 1 | | | 1 |
| | | | | | |
| **Glyrinidae** | | | | | |
| Orectochilus villosus (Müll.) | 1 | 1 | | | 3 |
| | | | | | |
| **Hydraenidae** | | | | | |
| Hydraena sp. | 1 | | 1 | 1 | 2 |
| Octhebius minimus (F.) | 3 | 2 | 1 | | 5 |
| Octhebius bicolon Germ. | 2 | 1 | | | |
| Limnebius nitidus (Marsh.) | | | | | 2 |
| Limnebius sp. | 1 | | 1 | | |
| Hydrochus elongatus (Schall.) | | | | | 2 |
| Helophorus nubilus F. | 1 | 2 | | | 1 |
| Helophorus grandis Illiger | | 1 | | | 4 |
| Helophorus 'aquaticus' (L.) = aequalis Thoms. | | 1 | 2 | | 1 |
| Helophorus misc. small spp. | 12 | 3 | 2 | 3 | 11 |
| | | | | | |
| **Hydrophilidae** | | | | | |
| Coelostoma orbiculare (F.) | | | | | 2 |
| Sphaeridium bipustulatum F. | 1 | 1 | | | 1 |
| Cercyon ustulatus (Preyssl.) | | 1 | | | 2 |
| Cercyon melanocephalus (L.) | | | | | 1 |
| Cercyon pygmaeus Illiger | 1 | | | | 1 |
| Hydrobius fuscipes (L.) | | 1 | | | 1 |
| Laccobius sp. | 1 | | | | 1 |

| | 1 | 2 | 3 | 4 | 5 |
|---|---|---|---|---|---|
| **Histeridae** | | | | | |
| Hister quadrimaculatus L. | | | | | 1 |
| Gen. et sp. indet. | 1 | 2 | | | 2 |
| | | | | | |
| **Silphidae** | | | | | |
| Phosphuga atrata L. | | | | | 1 |
| | | | | | |
| **Clambidae** | | | | | |
| Clambus sp. | | | | | 1 |
| | | | | | |
| **Orthoperidae** | | | | | |
| Corlophus cassidoides (Marsh.) | 2 | | | 1 | |
| | | | | | |
| **Ptiliidae** | | | | | |
| Acrotrichis sp. | 1 | | | 1 | |
| | | | | | |
| **Staphylinidae** | | | | | |
| Olophrum piceum (Gyll.) | | | | | 1 |
| Trogophloeus spp. | 2 | | 1 | | |
| Aploderus sp. | 2 | | | 1 | |
| Oxytelus rugosus (F.) | 1 | 1 | | | |
| Oxytelus nitidulus Grav. | 1 | 1 | | | 2 |
| *Oxytelus gibbulus Epp. | 3 | 3 | | | 5 |
| Oxytelus tetracarinatus (Block.) | | | | 1 | |
| Oxytelus sp. | 2 | | | 1 | |
| Platystethus arenarius (Fourcr.) | 1 | | | | |
| Platystethus comutus Grav. or deneger Muls. | 1 | 1 | | | 4 |
| Platystethus nitens (Sahib.) | 3 | 1 | 1 | 2 | 4 |
| *Bledius cribricollis Heer | | | | | 1 |
| Stenus spp. | 1 | 1 | 1 | | 4 |
| Paederus riparius (L.) | | | | | 1 |
| Lathrobium angusticolle Boisd. Lacord. | | | | | 1 |
| Aechenium depressum (Grav.) | | | | | 1 |
| Xantholinus sp. | | | 1 | | |

| | 1 | 2 | 3 | 4 | 5 |
|---|---|---|---|---|---|
| Staphylinus sp. | | | | | 1 |
| Tachyporus chrysomelinus (L.) | | 1 | | | 2 |
| Tachyporus sp. | 1 | 1 | | | 2 |
| Tachinus lignorum (L.) | 1 | | | | |
| Tachinus subterraneus (L.) | | | | | 1 |
| Tachinus corticinus Grav. | | | | 1 | |
| Alaeocharinae Gen. et sp. indet. | 2 | 3 | 1 | 1 | 2 |
| | | | | | |
| **Pselaphidae** | | | | | |
| Bryaxis sp. | | | | 1 | 1 |
| | | | | | |
| **Cantharidae** | | | | | |
| Cantharis nigricans (Müll.) | | | | | 1 |
| | | | | | |
| **Elateridae** | | | | | |
| Agriotes sp. | | | | | 3 |
| Adelocera murina (L.) | | | | | 1 |
| Athous cf. haemorrhoidalis (F.) | 1 | 1 | 1 | | |
| | | | | | |
| **Throscidae** | | | | | |
| Throscus sp. | | | | | 1 |
| | | | | | |
| **Dascillidae** | | | | | |
| Dacillus cervinus (L.) | 1 | | | | |
| | | | | | |
| **Dryopidae** | | | | | |
| Helicus substriatus (Müll.) | 1 | | | | |
| Dryopus sp. | | 1 | | | 2 |
| Esolus parallepipidus (Müll.) | 3 | 1 | | 1 | 9 |
| Oulimnius tuberculatus (Müll.) or troglodytes (Gyll.) | | | | 1 | |
| Limnius volckman (Panz.) | | 1 | | | 3 |
| Normandia nitens (Müll.) | | | 1 | | |
| Machronychus quadrituberculatus Müll. | 2 | | | | 2 |

| | 1 | 2 | 3 | 4 | 5 |
|---|---|---|---|---|---|
| **Georissidae** | | | | | |
| Georissus crenulatus (Rossi) | | | | | 1 |
| **Heteroceridae** | | | | | |
| Heterocerus sp. | | | | | 1 |
| | | | | | |
| **Dermestidae** | | | | | |
| Dermestes murinus L. | | 1 | | | 1 |
| | | | | | |
| **Byrrhidae** | | | | | |
| Limnichus pygmaeus (Sturm) | | 1 | | | |
| Byrrhus sp. | 1 | | | | 1 |
| Porcinolus murinus (F.) | | 1 | | | |
| | | | | | |
| **Nitidulidae** | | | | | |
| Meligethes aeneus (F.) | | | | | 1 |
| | | | | | |
| **Cryptophagidae** | | | | | |
| Atomaria mesomelaena (Hbst.) | | 1 | | | 2 |
| | | | | | |
| **Lathridiidae** | | | | | |
| Enicmus sp. | | | | | 2 |
| Corticaria sp. or Melanophthalma | | | | 1 | |
| Corticarina sp. | 2 | | | 1 | |
| | | | | | |
| **Coccinellidae** | | | | | |
| Anisosticta novemdecimpunctata (L.) | | | | 1 | |
| | | | | | |
| **Anobiidae** | | | | | |
| Dorcatoma chrysomelina Sturm | | | | | 1 |
| | | | | | |
| **Anthicidae** | | | | | |
| Anthicus antherinus (L.) | | | | 1 | 1 |
| Anthicus sp. | 2 | | | | 1 |
| | | | | | |
| **Tenebrionidae** | | | | | |
| Opatrum sabulosum (L.) | | | | | 1 |
| | | | | | |
| **Scarabaeidae** | | | | | |
| Aphodius sabuleti (Panz.) | | | | | 1 |

| | 1 | 2 | 3 | 4 | 5 |
|---|---|---|---|---|---|
| Aphodius erraticus (L.) | | 1 | | | 1 |
| Aphodius fossor (L.) | 1 | 1 | | | 2 |
| *Aphodius cf carpetanus Grav. | 2 | 1 | | | |
| Aphodius depressus (Kug.) | | | | | 1 |
| Aphodius fasciatus (Ol.) | | | | | 2 |
| Aphodius spp. | 5 | 5 | 1 | 2 | 7 |
| Heptaulacus villosus (Gyll.) | 1 | | | | |
| Heptaulacus sus (Hbst.) | | | | 1 | 2 |
| Ryssemus germanus (L.) | 1 | | | | |
| | | | | | |
| **Chrysomelidae** | | | | | |
| Macroplea appendiculate (Panz.) | 1 | | | 1 | |
| Donacia dentata Hoppe. | | | 1 | | 1 |
| Donacia semicuprea Panz. | 1 | 2 | | 1 | 1 |
| Donacia bicolor Zschach | 1 | | | | |
| Donacia sp. | 2 | | | | 1 |
| Plateumaris affinis (Kunz.) | | | | | 1 |
| Lerna sp. | | | | | 1 |
| Hydrothassa marginella (L.) | | | | | 1 |
| Phyllotreta atra (F.) | | | | | 1 |
| Phyllotreta sp. | | | | | 2 |
| Longitarsus sp. | | | | | 1 |
| Haltica sp. | | | | 1 | 1 |
| Chaetocnema spp. | 2 | 1 | 2 | 2 | 9 |
| | | | | | |
| **Bruchidae** | | | | | |
| Bruchus rufimanus Boh. | | | | | 1 |
| | | | | | |
| **Scolytidae** | | | | | |
| Scolytus sp. | 1 | | | | |
| Xyleborus dryographus (Ratz.) | | | | | 1 |
| Gen. et sp. indet | 1 | | | | |
| | | | | | |
| **Curculionidae** | | | | | |
| Rhynchites sp. | 1 | | | | |

| | 1 | 2 | 3 | 4 | 5 |
|---|---|---|---|---|---|
| Apion spp. | 6 | 7 | 1 | 2 | 10 |
| Cathormiocerus sp. | 1 | | | | |
| Otiorhynchus ligneus (Ol.) | 1 | 3 | | | 10 |
| *Stomodes gyrosicollis (Boh.) | | 1 | | | |
| Trachyphloeeus sp. | | | | | 1 |
| Strophosoma sp. | | | | | 1 |
| Barynotus obscurus (F.) | | 1 | | 1 | 1 |
| Sitona Lepidus Gyll. | | | | | 4 |
| Sitona sp. | 1 | 1 | | | |
| Coniocleonus nebulosus (L.) | | 2 | | | 2 |
| Cleonus piger (Scop.) | | | | | 1 |
| Notaris scirpi (F.) | | 1 | | | 2 |
| Notaris acridulus (L.) | 1 | 1 | 1 | | 1 |
| Orthchaetes setiger (Beck) | | | | | 1 |
| Liparis germanus (L.) | | 2 | 1 | | 1 |
| Alophus triguttatis (F.) | 1 | 5 | | | 2 |
| Hypera postica (Gyll.) | | 1 | | | 1 |
| Limnobaris pilistriata (Steph.) | | 1 | | | |
| Ceutorhynchus erysimi (F.) | | | | | 2 |
| Ceutorhynchus sp. | | | | | 1 |
| Rhynchaenus quercus (L.) | 7 | | 2 | 1 | 4 |
| Rhynchaenus rufitarsus (Germ.) | | | | | 2 |
| | 116 | 93 | 26 | 36 | 230 |

* indicates species not now living in the British Isles.
Sample 1 - 1064B Site 1: Sq J16
Sample 2 - 2012 Site 2: Sq I32
Sample 3 - Site 2
Sample 4 - Site 4 Sq C42 E0.80; N0.80; D63.71-63.78
Sample 5 – summary of initial sample taken in Dix Pit and published in Briggs *et al.* 1985

Table 3.4 Coleoptera from the Stanton Harcourt channel deposits identified by G.R. Coope. The nomenclature and taxonomic order follow that of Lucht 1987. The numbers in each column and opposite each species indicate the minimum number of individuals of that species in the sample.

correlation. Nevertheless, such correlations must always be on probation; their reliability always subject to the revisions that subsequent discoveries dictate.

The Stanton Harcourt assemblage has two species which Coope considered to be of stratigraphical significance. *Oxytelus gibbulus* has a sporadic fossil history in the British Isles. It occurs rarely in Middle Devensian (Weichselian) interstadials (Coope *et al.* 1961) and during the latter half of the Ipswichian (Eemian) Interglacial (e.g. at Itteringham and Coston; Coope unpublished data). However, it is extraordinarily abundant in deposits that have been attributed to Marine Isotope Stage (MIS) 7 at Upper Strensham, Worcestershire (de Rouffignac *et al.* 1995), at Stoke Goldington, Buckinghamshire (Green *et al.* 1996), and at Marsworth, Buckinghamshire (Murton *et al.* 2001). At Stanton Harcourt, *Oxytelus gibbulus* is the most abundant recognisable staphylinid species. The second species that may be of stratigraphical importance is *Stomoides gyrosicollis* which has, so far, only been found in the deposits attributed to MIS 7; i.e. at Stoke Goldington (Green *et al.* 1996) and at Marsworth (Murton *et al.* 2001). On the basis of the occurrences of these two exotic beetle species it is likely that the Stanton Harcourt material is also of MIS 7 age.

| Hydropsychidae | *Hydropsyche angustipennis* Curtis |
|---|---|
| | *Hydropsyche contubernalis* McL. |
| | * *Hydropsyche cf. bulbifer* McL. |
| | |
| **Polycentropidae** | *Cymus trimaculatus* Curtis |
| | |
| **Limnephilidae** | *Apatania* spp. |
| | *Anabolia nervosa* Curtis |
| | *Halesus* sp. |
| | |
| **Lepidostomatidae** | *Lepidostoma hirta* F. |

Table 3.5 List of Trichoptera obtained from the original sedimentary sample at Dix Pit (Coope in Briggs *et al.* 1985). All specimens were disarticulated larval sclerites and no attempt has been made to indicate their relative abundance.

*indicates species not now living in the British Isles

Coope reiterated this conclusion by maintaining that the Stanton Harcourt insect assemblage was similar to that from Aveley (MIS 7) and different to those from other interglacials. The insect fossils from known Hoxnian Interglacial sites in the neighbourhood are very different and none contain any of the critical species discussed above. These sites include Hoxne itself, the Woodstone beds near Peterborough, Nechelles near Birmingham and Quinton near Birmingham (Coope 2001). The same was deemed to be true for Barling (MIS 9).

### *Vegetation*

This is an unusual site in that, apart from the usual sediment samples containing various seeds and pollen, a great many logs, branches, twigs, leaves and *in situ* roots were excavated (see, for example, Figure 2.24 in Chapter 2). Apart from the 250 items of wood that were identified, abundant pieces of wood of all sizes were excavated and recorded. There were also many hazel nuts and several acorns (Table 3.6).

The vegetation from Stanton Harcourt is indicative of fully interglacial conditions at the time of deposition. Some of the wood samples are those of mature deciduous trees indicative of stable climatic conditions over a long period, although there is nothing distinctive in terms of assigning the vegetation to a particular interglacial episode. It is now generally accepted that there have been five major interglacials since the Anglian Glaciation, at MIS 11, 9, 7, 5e and the Holocene, and many more short term warm climatic events. The presence of deciduous woodland at Stanton Harcourt supports one of the more major warm periods.

Although from a different interglacial, many of the floral species from the Stanton Harcourt Channel are the same as those found in sub-stage III of the classic Hoxnian, which show a more maritime influence than some of the more continental species seen during the Ipswichian (Godwin, 1975). This, together with the presence of oak, beech and hornbeam woodland at Stanton Harcourt, supports an environment during the early part of the second half of an interglacial. During the time period of this research, a number of other sites in Britain have been discovered, or have been re-evaluated, that appear to have characteristic faunal and floral assemblages indicating a distinct interglacial interval equivalent to MIS 7. These include Ailstone (Bridgland *et al.* 1989), Marsworth (Murton *et al.* 2001), Stoke Goldington (Green *et al.* 1996), Strensham, (Coope 2001) and West Thurrock (Bridgland 1994). Of these, Marsworth in

Buckinghamshire is particularly important because MIS 7 interglacial material is separated from overlying MIS 5e (Ipswichian or Last Interglacial) temperate deposits by periglacial sediments.

The climatic and environmental implications of the species identified are discussed in Chapter 6.

| | Item identified | Number |
|---|---|---|
| **Trees & shrubs** | | |
| *Abies sp.* (Fir) | seeds, pollen and spores Site 1 (JC) | 80 |
| *Acer sp.* (Maple) | seeds (MR) | 2 |
| *Alnus* sp. (Alder) | seeds Site 2; seeds, pollen and spores Site 1 (JC) | 2 at Site 2; 10 at Site 1 |
| *Betula* sp. (Birch) | wood (JH); wood at Site 2 (JC) | 5 (JH); 1 (JC) |
| *Carpinus sp.* (Hornbeam) | wood (JH) | 78 |
| *Cornus sanguinea* L. (Dogwood) | seeds at Site 2 (JC); wood at Site 2 (RG); seeds (MR) | 4 (JC); 8 (MR) |
| *Corylus sp.* (Hazel) | wood (IG)(JH); seeds, pollen and spores at Site 1 (JC) | 2 (IG); 1 (JH); 5 (JC) |
| *Corylus avellana L.* (Hazel / cobnut) | seed (MR) | 1 |
| *Crataegus sp.* (Hawthorn) | seed (MR) | 1 |
| *Fagus silvatica* (Beech) | wood (JH) | 107 |
| *Fraxinus excelsior* (Ash) | wood (IG) | 1 |
| *Juniperus* sp. (Juniper) | wood (JH) | 2 |
| *Pinus* sp. (Pine) | seeds at Site 2 (JC) | 1 |
| *Prunus* sp. | seeds at Site 2 (JC) | 6 |
| *Prunus spinosa* L. (Blackthorn, sloe) | seeds at Site 2 (JC); seeds (MR) | 3 (JC); 1 (MR) |
| *Quercus sp.* (Oak) | wood (IG) (JH) (RG); 3 bud scales and 1 leaf abscission pad (MR); seeds, pollen and spores (JC) | 17 + ?10 (IG); 60 (JH); 4 (MR); 25 (JC) |
| *Rubus sp.* (Blackberry) | seed (MR) | 1 |
| *Sambucus nigra* L. (Elder) | seeds Site 2 (JC) | 18 |
| *Taxacaeae* (Yew) | wood (IG) | 4 |
| *Taxacaeae* ? (Conifer probably yew) | wood (IG) | 5 |
| *Salix sp.* (Willow) | wood (IG) | 1 +?4 (many twigs) |
| **Herbs and grasses of dry land** | | |
| *Atriplex* sp. (Orache) | seeds Site 2 (JC) | 3 |
| *Brassica* sp. | seeds Site 2 (JC) | 3 |
| *Calluna* (Heather) | seeds, pollen and spores at Site 1 (JC) | 20 |
| *Chenopodium* sp. (Goosefoot) | seeds Site 2 (JC) | 8 |
| *Cirsium palustre* L. Scop. (Marsh thistle) | seeds Site 2 (JC) | 12 |
| *Filipendula ulmaria L.* (Meadowsweet) | seed (MR) | 1 |

| | Item identified | Number |
|---|---|---|
| *Gramineae* (grasses) | pollen, seeds and spores Site 1 (JC) | 100 |
| *Leontodon autumnalis* L. (Autumnal hawkbit) | seeds Site 2 (JC) | 7 |
| *Lychnis flos-cuculi* L. (Ragged robin) | seeds Site 2 (JC) | 16 |
| *Polygonum* sp. (Knotweed or knotgrass) | seeds Site 2 (JC) | 2 |
| *Polygonum cf. lapathifolium* (Knotweed) | seeds (MR) | 3 |
| *Polygonum aviculare* (Knotgrass) | seed (MR) | 1 |
| *Ranunculus acris* L. (Meadow buttercup), *repens* L. (Creeping buttercup) and *bulbosus* L. (Bulbous buttercup) | seeds (JC) | 57 |
| *Ranunculus ranunculus sp.* (Buttercup) | seed only to subspecies (MR) | 1 |
| *Scabiosa columbaria* L. (Small scabious) | seeds Site 2 (JC) | 10 |
| *Scleranthus annuus* L. (Annual knawel) | seeds Site 2 (JC); leaf abscission pad only to genera (MR) | 6 (JC); 5 (MR) |
| *Silene* sp. | seeds Site 2 (JC) | 1 |
| *Solanum dulcamara* L. (Bittersweet, Woody nightshade) | seeds Site 2 (JC) | 13 |
| *Stellaria cf. nemoram* (Stitchwort) | seeds Site 1 (MR) | 3 |
| **Heliophytes** | | |
| *Bidens cernua* L. (Nodding bur-marigold) | seeds Site 2 (JC) | 3 |
| *Cladium mariscus*(L.) Pohl. (Sedge) | seeds Site 2 (JC) | 38 |
| *Lycopus europaeus* L. (Gipsy-wort) | seeds Site 2 (JC) | 1 |
| *Schoenoplectus lacustris* (L.) Palla. (Bullrush) | seeds Site 2 (JC) | 15 |
| **Aquatics** | | |
| *Alisma plantago-aquatica* L. (Water plantain) | seeds Site 2 (JC) | 6 |
| *Chara* sp. (Stonewort) | seeds Site 2 (JC) | 18 |
| *Hippuris vulgaris* (Mare's-tail) | seeds Site 2 (JC) | 1 |
| *Potomogeton* sp. (Pondweed) possibly includes *P. compressus* L. - the Grass wrack pondweed; Broadleaf pondweed *P. natans* L. or Loddon pondweed -*P.nodosus* Poiret - latter used to occur in the Thames in slow flowing calcareous water | seeds (MR (JC) | 4 (MR); 246 (JC) |
| *Ranunculus* (Batrachium) sp. (common water crowfoot/water buttercup) | seed (MR); seeds Site 2 (JC) | 20 (MR); 6 (JC) |
| *Sagittaria sagittifolia* L. (Arrowhead) | seed Site 2 (JC) | 1 |
| *Sparganium* erectum L. (Branched bur-reed) | seeds (JC) | 10 |
| *Zanichellia palustris* L. (Horned pondweed) | seeds (MR) (JC) | 3 (MR); 18 (JC) |

| | Item identified | Number |
|---|---|---|
| **Unclassified** | | |
| *Carex* sp. (Sedge) | seed Site 2 (JC); seed (MR) check | 17 (JC); 1 (MR) |
| *Cyperaceae indet.* (Sedge) | seed (MR); pollen, seeds, spores Site 1 (JC) | 1 (MR); 60 (JC) |
| *Hieracium* sp. | seeds Site 2 (JC) | 2 |
| *Hipericum* sp. | seed Site 2 (JC) | ?1 |
| *Mentha* sp. | seeds Site 2 (JC) | 6 |
| Moss fragments | fragments (MR) | |
| Pteridium *(algal cyst)* | pollen, seeds and spores Site 1 (JC) | 15 |
| Rosaceae | pollen, seeds and spores Site 1 (JC) | 5 |
| *Viola* sp. | seeds Site 2 (JC) | 3 |

**Identifications by**
Jim Campbell and Mike Field - plant macro-fossils, seeds, spores etc. (JC)
Rowena Gale (RG) – wood at Site 2
John Hather (JH) - wood
Ian Gourlay (IG) - wood
Mark Robinson (MR) - seeds from sample 1069

Table 3.6 Vegetation identified at Stanton Harcourt

## *Artefacts*

A total of 36 artefacts came from the Channel excavations: 11 flint handaxes, four quartzite handaxes, 11 flint flakes, 4 flint cores, 5 pieces of flint debitage and 1 quartzite flake. They are described in Chapter 7. As regards their usefulness as a dating tool, as an assemblage, these artefacts are comparable to other Lower Palaeolithic material of the British late Middle Pleistocene.

# Chapter 4

# The Mammoths

The Stanton Harcourt mammoths represent an exceptional sample of well documented *Mammuthus trogontherii* from the MIS 7 interglacial. More than 140 tusks were excavated, 188 molars (many in skulls and mandibles), and nearly 500 post-cranial bones (Tables 4.1-4.5). Consequently, they have proved invaluable in the reassessment of mammoth evolution in the Late Middle Pleistocene of Europe (Lister and Scott in press). When first discovered at Stanton Harcourt, the mammoths were described as the woolly mammoth *Mammuthus primigenius* (Buckingham *et al.* 1996). It was later recognised that these are the remains of a small form of steppe mammoth *Mammuthus trogontherii.* Their identification became increasingly significant to the interpretation of the Stanton Harcourt material, as outlined below.

Mammoth remains have long been known from British Pleistocene sites. The majority, with their distinctive, high-crowned molars, are of the woolly mammoth *Mammuthus primigenius*, frequently associated with indicators for cool or cold climate and the open steppe-like habitat of late Middle and Late Pleistocene glacial periods. Less common is another mammoth, the steppe mammoth *Mammuthus trogontherii.* Occurring in Europe at sites of late Early and Middle Pleistocene age (although relatively rarely in Britain), these mammoths are generally, but not exclusively, from interglacial context and associated with evidence for open as well as woodland habitats. Although it was originally proposed that *M. trogontherii* in Europe evolved into *M. primigenius* in the late Middle Pleistocene, earlier fossils referable to *trogontherii* and *primigenius* have since been recorded in eastern Asia, suggesting that their appearance in Europe was the result, at least in part, of immigration events rather than purely *in situ* evolution (Lister and Sher 2001, 2015).

From the early days of the excavation at Stanton Harcourt, it was evident that the adult mammoth teeth were small relative to the many mammoth molars being collected by the author and colleagues from gravel pits in the vicinity. Early 20$^{th}$C publications had recorded the existence in Britain of mammoth molars smaller than was regarded as typical for the woolly mammoth. Of particular importance is the large mid-19$^{th}$C collection from Ilford, Essex, where the small size of the mammoth teeth by comparison with woolly mammoth teeth from other locations was recorded by Hinton (1909-10). Small mammoth teeth were also reported from gravels of the Upper Thames by Sandford (1925) who commented on thicker enamel and a reduced number of lamellae (the plates of which each tooth is comprised). The small mammoth was thereafter variously referred to as of 'Ilford' type or as an 'early' or 'primitive' form of *M. primigenius.* Accordingly, in the initial publication on the Stanton Harcourt excavation, the mammoths were described as *M. primigenius* (Buckingham *et al.* 1996). However, a curious anomaly was noted at the time: that although the large vertebrate fauna from the site was dominated by woolly mammoth, all other biological material from the site indicated fully interglacial conditions.

The excavations were underway at a particularly significant time in palaeoclimatic studies. As discussed in Chapter 1, a revision of the Pleistocene glacial-interglacial sequence in the late 1980s/early 1990s resulted in the recognition of a 'new' interglacial equated with Marine

Isotope Stage 7 (MIS 7). A combination of data on Thames terrace stratigraphy, AAR dates, and abundant interglacial faunal and floral remains from the excavation, placed the Stanton Harcourt Channel deposits within MIS 7 (Chapters 3 and 6). Concurrently, the evidence for an additional interglacial, enabled Quaternary specialists to reassess the stratigraphic relationship of many British fossil vertebrate assemblages. It was of particular relevance to the interpretation of Stanton Harcourt when it transpired that the small 'Ilford' mammoth tended to characterise many assemblages also attributed to MIS 7 and that it might even be considered as a 'marker' species for the interglacial (Schreve 1997). Of even greater import was the conclusion reached by Lister & Sher (2001) that, in terms of the plate number of the last molars, the mammoths from Ilford were statistically indistinguishable from typical *M. trogontherii* of the early Middle Pleistocene. The 'Ilford' mammoth was thus considered by Lister and Sher to be of 'trogontherioid' molar form, even if very much smaller in size than early Middle Pleistocene *M. trogontherii*. The re-assignment of the Stanton Harcourt mammoths from *M. primigenius* to *M. trogontherii* was compatible with the rest of the biological data which described a fully interglacial environment at the site. Thus, the anomaly of woolly mammoth in this habitat could be explained (Scott 2001).

Mammoth is present at 23 of the 28 British sites now referred to MIS 7 (Scott in prep.). Although their dentition is trogontherioid and they are characteristically small by comparison with typical *M. primigenius* or with *M. trogontherii* of earlier interglacials, recent analysis indicates that the MIS 7 mammoths are not homogenous and that the term 'Ilford mammoth' is therefore no longer appropriate (Lister and Scott in press). Of particular interest with regard to Stanton Harcourt, is that the Ilford mammoths are considered to be of mid-MIS 7 age and less advanced dentally than those from Stanton Harcourt which, on independent evidence, is believed to date to the end of MIS 7. All discussion on the evolutionary status of the small MIS 7 mammoths has thus far focussed on the dentition. No estimates have been made of their overall size (shoulder height) on the basis of post-cranial remains. These are far less numerous and less well preserved than molars but a study of the post-cranial remains from Stanton Harcourt and other British MIS 7 sites serves to show that, indeed, the late Middle Pleistocene steppe mammoth was smaller than the woolly mammoth and considerably smaller than the ancestral steppe mammoth (Scott and Lister in press). These findings are discussed further in this chapter.

| SITE | REF | ELEMENT | TOOTH | L/R | PART | LAWS | AEY | COMMENT |
|---|---|---|---|---|---|---|---|---|
| 1 | 2 | DENT | M3 | R | CO | | | |
| 1 | 11 | DENT | M3 | R | CO | XIX | 32 | |
| 1 | 12 | DENT | dP3 | L | CO | V | 3 | dp3 (with dp4 SH1-13); associated maxillary fragments |
| 1 | 13 | DENT | dP4 | L | CO | V | 3 | dp4 (with dp3 SH1-12); associated maxillary fragments |
| 1 | 55 | DENT | M2 | R | CO | XV | 24 | |
| 1 | 70 | DENT | | | FR | | | 4 posterior lamellae |
| 1 | 87 | DENT | dP4 | R | CO | VII | 6 | |
| 1 | 91 | DENT | M1 | L | CO | XII | 18 | |
| 1 | 108 | DENT | M3 | R | CO | XXVI | 49 | |

| SITE | REF | ELEMENT | TOOTH | L/R | PART | LAWS | AEY | COMMENT |
|---|---|---|---|---|---|---|---|---|
| 1 | 109 | DENT | M3 | L | CO | XXII | 39 | |
| 1 | 111 | DENT | | | FR | | | 1 lamella |
| 1 | 119 | DENT | M3 | R | CO | XIX.5 | 32 | |
| 1 | 151 | DENT | M3 | L | CO | XXVIII | 55 | |
| 1 | 158 | DENT | M2 | L | CO | XV | 24 | |
| 1 | 171 | DENT | M3 | R | CO | XIX | 57 | |
| 1 | 189 | DENT | M3 | L | CO | XX | 34 | |
| 1 | 196 | DENT | M3 | L | CO | XXVIII | 55 | |
| 1 | 222 | DENT | M3 | R | CO | XXVI | 49 | |
| 1 | 223 | DENT | M3 | L | CO | XX | 34 | |
| 1 | 229 | DENT | M3 | R | CO | XXII | 39 | tooth in skull SH1-241 |
| 1 | 230 | DENT | M3 | L | CO | XXII | 39 | tooth in skull SH1-241 |
| 1 | 292 | DENT | M3 | R | CO | XXI | 36 | |
| 1 | 310 | DENT | M3 | L | CO | XXVI | 49 | |
| 1 | 335 | DENT | M1 | L | ANT | | | 3 lamellae |
| 1 | 378 | DENT | M3 | L | POST | XXIII | 43 | |
| 1 | 403 | DENT | M2 | L | CO | XVII | 28 | |
| 1 | 241 | MXD | M3 | L+R | CO | XXII | 39 | M3s: SH1-229 and SH1-230 |
| 1 | 242 | MXD | M3 | L | CO | XXVIII | 55 | tooth in collapsed skull |
| 2 | 33 | DENT | M2 | R | CO | XVII | 28 | |
| 3 | 9 | DENT | M2/M3 | L | PART | | | |
| 3 | 16 | DENT | M3 | R | CO | XXVI | 49 | |
| 4 | 4 | DENT | M2 | L | CO | XVII | 28 | |
| 4 | 7 | DENT | dP4/M1 | L | POST | | | |
| 4 | 19 | DENT | M3 | R | HALF | XXII | 39 | |
| 4 | 26 | DENT | M3 | R | CO | XXIV | 45 | |
| 4 | 27 | DENT | | ? | HALF | | | |
| 4 | 67 | DENT | M3 | R | CO | XXII | 39 | |
| 4 | 72 | DENT | M3 | R | CO | XXIX | 57 | deformed |
| 4 | 89 | DENT | M3 | L | CO | XXII | 39 | |
| 4 | 94 | DENT | M3 | L | CO | XVIII | 30 | |
| 4 | 99 | DENT | M3 | R | POST | XXIV | 45 | |
| 4 | 106 | DENT | M3 | L | CO | XXX | 60 | |
| 4 | 124 | MXD | ? | R | CO | | | molar+skull in fibreglass |
| 4 | 141 | DENT | | | FR | | | |
| 5 | 5 | DENT | ?M2 | ? | PART | | | |
| 5 | 13 | DENT | M3 | L | CO | XXIV | 45 | |
| 5 | 21 | DENT | | L | CO | | | |
| 5 | 58 | DENT | ?M2 | L | CO | | | |
| 5 | 59 | DENT | M2 | L | CO | XIX | 32 | |
| 5 | 62 | DENT | M3 | L | CO | XXVI | 49 | |
| 5 | 69 | DENT | | | FR | | | |
| 5 | 70 | DENT | M2 | R | CO | XVII | 28 | |
| 5 | 71 | DENT | M3 | R | CO | XXIV | 45 | |
| 6 | 21 | DENT | M2 | L | CO | XVII | 28 | |
| 6 | 32 | DENT | dP4 | L | CO | V | 3 | 8 plates in wear |

| SITE | REF | ELEMENT | TOOTH | L/R | PART | LAWS | AEY | COMMENT |
|---|---|---|---|---|---|---|---|---|
| 6 | 37 | DENT | | | POST | | | |
| 6 | 59 | DENT | | | PART | | | 1 lamella |
| 6 | 66 | DENT | M3 | L | CO | XVII | 28 | |
| 6 | 80 | DENT | | | | | | no data; maybe not retrieved |
| 6 | 90 | DENT | M3 | L | CO | XXII | 39 | |
| 6 | 91 | DENT | dP4 | R | CO | V | 2 | 12 lamellae and talon; 4 lamellae in wear |
| 6 | 100 | DENT | M3 | L | CO | XXIX | 57 | |
| 6 | 119 | DENT | M3 | R | PART | XXIII | 43 | |
| 6 | 126 | DENT | M3 | L | CO | XIX | 32 | |
| 6 | 131 | DENT | M3 | L | CO | XIX | 32 | |
| 6 | 134 | DENT | | | FRS | | | 4 posterior lamellae |
| 6 | 138 | DENT | ?M3 | L | CO | ? | | fragmentary |
| 6 | 142 | DENT | M1/M2 | R | CO | XVIII/XIV | 30 or 22 | |
| 6 | 154 | DENT | M3 | L | CO | XXIII | 43 | |
| 6 | 155 | DENT | M3 | L | CO | XXI | 36 | |
| 6 | 194 | DENT | M2/M3 | L | HALF | XII/XIX | 18 or 32 | |
| 6 | 205 | DENT | M1/M2 | L | HALF | VII/XII | 28 or 18 | |
| 6 | 211 | DENT | M3 | R | CO | XXIX | 57 | |
| 6 | 239 | DENT | M1/M2 | L | ANT | | | |
| 6 | 241 | DENT | M3 | R | ANT | XX | 34 | |
| 6 | 262 | DENT | M2 | L | CO | XVI | 26 | |
| 6 | 267 | DENT | M3 | L | CO | XXII | 39 | |
| 6 | 279 | DENT | M1/M2 | L | PART | mw | | plate formula -9X |
| 6 | 290 | DENT | M3 | R | CO | XXIII | 43 | |
| 6 | 302 | DENT | M3 | L | CO | XXIII | 43 | |
| 6 | 174 | MXD | dP4 | L | PART | IX/X | 10 or 13 | |
| 7 | 27 | DENT | M3 | L | CO | XIX | 32 | 20 lamellae + talon; 7 lamellae in wear |
| 7 | 30 | DENT | M3 | L | PART | XXVII/XVIII | | |
| 7 | 45 | DENT | M3 | L | CO | XXII | 39 | |
| 7 | 103 | DENT | M3 | L | CO | XXV | 47 | |
| 7 | 122 | DENT | M3 | R | CO | XXIII | 43 | |
| 7 | 176 | DENT | M3 | L | POST | XXIII | 43 | posterior section |
| 7 | 177 | DENT | M3 | L | CO | XXIII | 43 | |
| 7 | 185 | DENT | M1/M2 | R | ANT | | | 7 lamellae |
| 7 | 214 | DENT | M1/M2 | ? | ANT | +++ | | 6 lamellae, very fragmentary |
| 7 | 220 | DENT | M3 | L | CO | XXV | 47 | |
| 7 | 235 | DENT | M3 | R | POST | | | 8 lamellae |
| 7 | 238 | DENT | M3 | R | CO | XXVII | 53 | |
| 7 | 240 | DENT | M1/M2 | L | CO | XII/XVIII | 18 or 30 | |

| SITE | REF | ELEMENT | TOOTH | L/R | PART | LAWS | AEY | COMMENT |
|---|---|---|---|---|---|---|---|---|
| 7 | 249 | DENT | M3 | L | CO | XXIII | 43 | |
| 7 | 253 | DENT | M2 | L | CO | XV | 24 | |
| 7 | 284 | DENT | M3 | R | CO | XXVII | 53 | |
| 7 | 286 | DENT | M3 | L | CO | XXVIII | 55 | |
| 7 | 7 | MXD | M2+M3 | L+R | CO | XIII | 20 | |
| 7 | 37 | MXD | dP4 | L+R | CO | VIII | 8 | skull crushed and collapsed |
| 8 | 22 | DENT | M3 | R | CO | XXIII | 43 | |
| 14 | 43 | DENT | M3 | R | CO | XXIII | 43 | |
| 14 | 44 | DENT | M1 | L | CO | IX | 10 | |
| 14 | 45 | DENT | M1 | R | CO | XII | 18 | |
| 15 | 36 | MXD | M2+M3 | L+R | CO | XX | 34 | excellent condition |
| 14 | 47 | DENT | M3 | R | CO | XXII | 39 | |
| 14 | 51 | DENT | dP4 | R | CO | VI | 4 | |
| 15 | 2 | DENT | M3 | R | CO | XXI | 36 | |
| 15 | 13 | DENT | M3 | L | CO | XXIII | 43 | |
| 1 | 397 | DENT | | | CO | | | no data; stolen |
| 1 | 136 | DENT | | | FRS | | | 6 lamellae, fragmentary |
| 1 | 353 | DENT | | | FRS | | | |
| 1 | 405 | DENT | | | FR | | | |
| 1 | 411 | DENT | | | FRS | | | |
| 1 | 412 | DENT | | | FRS | | | 2 lamellae |
| 2 | 36 | DENT | | | FR | | | |
| 4 | 71 | DENT | | | FR | | | |
| 4 | 79 | DENT | | | FR | | | |
| 4 | 81 | DENT | | | PART | | | |
| 4 | 122 | DENT | | | FRS | | | |
| 5 | 2 | DENT | | | CO | | | |
| 5 | 53 | DENT | | | FRS | | | |
| 5 | 77 | DENT | | | | | | |
| 6 | 157 | DENT | | | FR | | | |
| 6 | 160 | DENT | | | FRS | | | |
| 7 | 173 | DENT | | | FR | | | 2 lamellae |
| 7 | 189 | DENT | | | FRS | | | 2 tooth fragments |
| 8 | 12 | DENT | | | FR | | | |

Key:

| | |
|---|---|
| DENT | dentition/tooth |
| MXD | maxilla with dentition |
| CO | complete |
| FR/S | fragment(s) |
| ANT | anterior section of tooth |
| POST | posterior section of tooth |

Table 4.1 All mammoth upper dentition from Stanton Harcourt. The Laws categories and the estimated age at death of the individual represented (AEY) are discussed in the Chapter 4.

| SITE | REF | ELEMENT | PART | L/R | | SITE | REF | ELEMENT | PART | L/R |
|---|---|---|---|---|---|---|---|---|---|---|
| 1 | 6 | CRA | FR | | | 1 | 168 | LAC | CO | R |
| 1 | 86 | CRA | FR | | | 7 | 43 | NAS | FR | |
| 1 | 113 | CRA | CH | | | 1 | 249 | OCC | CO | |
| 1 | 162 | CRA+PMX | SECT | | | 1 | 278 | OCC | FR | L |
| 1 | 188 | CRA | FR | | | 6 | 27 | OCC | FR | L |
| 1 | 198 | CRA | FR | | | 1 | 379 | OCC | FR | L+R |
| 1 | 242 | CRA | CO | | | 6 | 235 | OCC | PART | L+R |
| 1 | 287 | CRA | FRS | | | 8 | 70 | OCC | CH | |
| 1 | 330 | CRA | CH | | | 1 | 64 | OCC | FR | |
| 1 | 366 | CRA | FRS | | | 1 | 178 | PMX | FR | |
| 1 | 368 | CRA | FR | | | 1 | 190 | PMX | FR | |
| 2 | 14 | CRA | FR | | | 2 | 32 | PMX | FRS | |
| 2 | 60 | CRA | FR | | | 4 | 69 | PMX | FR | |
| 2 | 62 | CRA | FR | | | 4 | 90 | PMX | FR | |
| 4 | 59 | CRA | FRS | | | 7 | 81 | PMX | FR | |
| 4 | 98 | CRA | CH | | | 7 | 86 | PMX | CH | |
| 4 | 103 | CRA | FR | | | 7 | 104 | PMX | FRS | |
| 4 | 115 | CRA | FR | | | 7 | 218 | PMX | SECT | |
| 5 | 54 | CRA | FR | | | 7 | 294 | PMX | CH | |
| 6 | 1 | CRA | FR | | | 14 | 25 | PMX | CHS | |
| 6 | 44 | CRA | FR | | | 4 | 111 | PMX | PART | |
| 6 | 255 | CRA | FRS | | | 6 | 189 | PMX | FR | |
| 7 | 35 | CRA | FR | | | 7 | 57 | PMX | FR | |
| 7 | 65 | CRA | FR | | | 5 | 37 | ZYG | FR | |
| 7 | 78 | CRA | FR | | | 1 | 149 | ZYG | CO | |
| 7 | 201 | CRA | FR | | | 4 | 97 | ZYG | CO | |
| 7 | 297 | CRA | FR | | | 6 | 93 | ZYG | CO | |
| 1 | 346 | CRA | FR | | | 7 | 272 | ZYG | CH | |
| 4 | 124 | CRA | CO | | | | | | | |

KEY:

| | |
|---|---|
| CRA - unspecified cranial | ZYG - zygomatic |
| LAC - lacrymal | FR/S - fragment(s) |
| NAS - nasal | OCC - occipital |
| PMX - premaxilla | CH - chunk/large part |
| CO - complete or nearly so | SECT – section/large part |

Table 4.2 Mammoth cranial remains without dentition

| SITE | REF | ELEMENT | TOOTH | L/R | PART | LAWS | AEY | COMMENT |
|---|---|---|---|---|---|---|---|---|
| 1 | 80 | DENT | dp4 | R | CO | VIII | 8 | |
| 1 | 232 | DENT | m3 | L | CO | XXIII | 43 | |
| 1 | 268 | DENT | dp4/m1 | L | ANT | worn | | anterior small tooth; plate formula X4 |
| 1 | 269 | DENT | m1 | L | ANT | unworn | < 6 | plate formula X3- |
| 1 | 365 | DENT | m3 | L | CO | XXVI | 49 | |
| 1 | 367 | DENT | dp4/m1 | - | PART | | | plate formula -4- |
| 2 | 34 | DENT | m3 | R | CO | XIX | 32 | |
| 2 | 50 | DENT | m3 | R | CO | XXV | 47 | |
| 4 | 18 | DENT | m3 | L | CO | XXV | 47 | |
| 4 | 52 | DENT | m1 | L | ANT | unworn | <6 | plate formula X5- |
| 4 | 140 | DENT | m3 | R | CO | XIX | 32 | |
| 5 | 26 | DENT | m1/m2 | R | ANT | unworn | | |
| 5 | 75 | DENT | m3 | L | CO | XXI | 36 | |
| 6 | 4 | DENT | m3 | R | CO | XXV | 47 | |
| 6 | 101 | DENT | m3 | R | CO | XXVII | 53 | |
| 6 | 152 | DENT | m2 | R | CO | XV | 24 | |
| 6 | 153 | DENT | m3 | R | CO | XVIII | 30 | |
| 6 | 186 | DENT | m3 | R | CO | XX | 34 | |
| 6 | 238 | DENT | m2 | L | CO | XIII | 20 | |
| 6 | 240 | DENT | m3 | R | HALF | XXI | 36 | |
| 6 | 242 | DENT | m2 | L | CO | XIX | 32 | |
| 6 | 268 | DENT | m2/m3 | R | HALF | not possible | | |
| 7 | 10 | DENT | m3 | R | CO | XXVII | 53 | |
| 7 | 23 | DENT | m1/m2 | L | ANT | unworn | | plate formula X7- |
| 7 | 96 | DENT | m3 | R | CO | XXIII | 43 | |
| 7 | 128 | DENT | m1/m2 | L | CO | X | 13 | |
| 7 | 244 | DENT | m3 | R | CO | XXIII | 43 | |
| 7 | 257 | DENT | m3 | R | POST | XXV | 47 | |
| 7 | 303 | DENT | m3 | R | PART | not possible | | |
| 12 | 1 | DENT | m3 | R | CO | XXIII | 43 | |
| 14 | 29 | DENT | m3 | L | CO | XIX | 32 | |
| 14 | 30 | DENT | | | HALF | | | |
| 15 | 6 | DENT | m3 | L | CO | XXVI/XXVII | 49/53 | |
| 15 | 26 | DENT | m3 | L | CO | XXVIII | 55 | |
| 15 | 27 | DENT | m2 | L | PART | worn | | plate formula -7- |
| 15 | 37 | DENT | m3 | R | CO | XXVI | 49 | |
| 15 | 41 | DENT | m2/m3 | L | PART | unworn | | plate formula -5- |
| 15 | 53 | DENT | dp4/m1 | R | POST | | | |
| 15 | 54 | DENT | m3 | R | PART | mw | | |
| 1 | 22 | MND | m3 | L | CO | XXVI | 49 | |
| 1 | 180 | MND | m3 | L+R | CO | XXV | 47 | |
| 1 | 237 | MND | m3 | L+R | CO | XXIII | 43 | |

| SITE | REF | ELEMENT | TOOTH | L/R | PART | LAWS | AEY | COMMENT |
|---|---|---|---|---|---|---|---|---|
| 1 | 239 | MND | m3 | L+R | CO | XXVII | 53 | complete mandible, L+R m6 |
| 2 | 12 | MND | m1 | L+R | CO | | | symphysis unfused; sockets for dp4; m1s in situ |
| 3 | 20 | MND | m1+ m2 | L | CO | XV | 24 | |
| 4 | 1 | MND | m3 | L+R | CO | XXVI | 49 | needs repair |
| 4 | 37 | MND | m3 | L+R | CO | | | mandible with L+R m6; stolen |
| 4 | 95 | MND | m3 | L+R | CO | XXVI.5 | | |
| 6 | 22 | MND | m3 | L | CO | XXIV | 45 | |
| 6 | 123 | MND | m3 | L+R | CO | XXIII | 43 | |
| 6 | 168 | MND | m3 | L | CO | XXIII | 43 | |
| 6 | 233 | MND | dp3 | L+R | CO | III/IV | 1 - 2 | R dp2+dp4 erupting; L dp3(s), dp4 erupting |
| 6 | 253 | MND | m2+ m3 | L+R | CO | XXII | 39 | |
| 6 | 281 | MND | m3 | L | PART | XXIII | 43 | |
| 7 | 1 | MND | m3 | L+R | CO | XXII +HALF | 41 | |
| 7 | 54 | MND | m3 | L | CO | XXVII | 53 | |
| 7 | 203 | MND | m3 | L+R | CO | XXVI | 49 | |
| 7 | 256 | MND | m3 | R | PART | XXV | 47 | |
| 7 | 259 | MND | m3 | L | PART | XXII | 39 | |
| 7 | 289 | MND | m3 | R | CO | XIX | 32 | mandible very fragmented |
| 7 | 310 | MND | | | CH | | | |
| 8 | 45 | MND | dp4 | R | CO | IV/V | 2 - 3 | |
| 1 | 24 | MNO | | | SYM | | | symphysis, weathered |
| 1 | 101 | MNO | | L+R | CO | | | fair condition; also part R. asc. ramus |
| 1 | 103 | MNO | | L+R | SYM | | | |
| 1 | 199 | MNO | | L | CH | | | large part |
| 1 | 207 | MNO | | | SYM | | | symphysis, very eroded |
| 1 | 256 | MNO | | L | CO | | | symphysis and mandible ?deformed |
| 1 | 298 | MNO | | | CHS | | | two weathered chunks |
| 1 | 325 | MNO | | L+R | PART | | | fair condition; possibly male |
| 1 | 350 | MNO | | L+R | PART | | | part left side, complete right. Poor condition |
| 2 | 35 | MNO | | R | CH | | | chunk mandible, no dentition |
| 4 | 16 | MNO | | R | CO | | | Some of left side, most of right |
| 4 | 20 | MNO | | | FR | | | fragment juvenile symphysis |
| 5 | 4 | MNO | | L | FR | | | ascending ramus |
| 5 | 22 | MNO | | L | FR | | | part of SH5-21 |
| 5 | 35 | MNO | | | FRS | | | |
| 5 | 38 | MNO | | L+R | CO | | | needs repair; weathered; male |
| 5 | 78 | MNO | | | FRS | | | |
| 6 | 17 | MNO | | | SYM | | | symphysis + part L. mandible; machine damage |

| SITE | REF | ELEMENT | TOOTH | L/R | PART | LAWS | AEY | COMMENT |
|---|---|---|---|---|---|---|---|---|
| 6 | 60 | MNO | | | SYM | | | juvenile symphysis and part right mandible |
| 6 | 139 | MNO | | R | PART | | | very juvenile symphysis, probably neonatal |
| 6 | 147 | MNO | | | CH | | | |
| 6 | 149 | MNO | | | SYM | | | symphysis, sockets for dentition;?male |
| 6 | 150 | MNO | | | SYM | | | symphysis |
| 6 | 180 | MNO | | L+R | CO | | | symphysis: asc. rami broken |
| 6 | 224 | MNO | | | FRS | | | |
| 7 | 29 | MNO | | | SYM | | | |
| 7 | 49 | MNO | | | SYM | | | |
| 7 | 51 | MNO | | | SYM | | | part of SH7-52 and SH7-54 (now MND SH7-54) |
| 7 | 52 | MNO | | L | CH | | | part of SH7-51 and SH7-54 (now MND SH7-54) |
| 7 | 94 | MNO | | | FR | | | |
| 7 | 147 | MNO | | | SYM | | | symphysis; mandible very weathered |
| 7 | 171 | MNO | | | SYM | | | very weathered and machine damage |
| 7 | 231 | MNO | | R | CH | | | lingual ramus |
| 7 | 253 | MNO | | | FRS | | | |
| 7 | 258 | MNO | | | CH | | | |
| 7 | 289 | MNO | | | SYM | | | weathered, machine damage |
| 7 | 309 | MNO | | | CO | | | |
| 7 | 33 | MNO | | | CH | | | mandible fragment; ? part of SH7-1 |
| 7 | 132 | MNO | | | CH | | | mandible fragment |
| 8 | 75 | MNO | | | CH | | | |
| 14 | 49 | MNO | | | CH | | | |
| 15 | 8 | MNO | | | FRS | | | |
| 15 | 28 | MNO | | L | PART | | | broken; 2 chunks |
| 15 | 35 | MNO | | | FR | | | part ramus |
| 15 | 43 | MNO | | | SYM | | | symphysis looks like male |
| 15 | 52 | MNO | | | CHS | | | |

Key:

| | |
|---|---|
| DENT | dentition/tooth |
| MND | mandible with dentition |
| MNO | mandible without dentition |
| SYM | mandibular symphysis |
| CO | complete |
| CH/S | chunk/ large part(s) |
| FR/S | fragment(s) |
| ANT | anterior section of tooth |
| POST | posterior section of tooth |

Table 4.3 All mammoth isolated lower dentition and mandibles (with or without dentition). The Laws categories and the estimated age at death of the individual represented (AEY) are discussed in the Chapter 4.

| | |
|---|---|
| Complete or virtually complete | 85 |
| Estimated to have been complete but damaged by machines | 15 |
| Estimated to have been complete but one end destroyed by overlying MIS 6 gravels | 3 |
| Large section, one or both ends missing | 42 |
| Many fragments of one disintegrated tusk | 15 |
| Isolated tusk fragments | 26 |

Table 4.4 Mammoth tusks at Stanton Harcourt

| SITE | NO | Original state | Outside curvature | Inside curvature | Proximal diameter | COMMENT |
|---|---|---|---|---|---|---|
| 1 | 16 | CO | 80 (e.90) | 70 (e80) | 6 | poor condition, tip damaged (c.10 cm missing) |
| 1 | 18 | CO | >110 | | 18 | very poor condition; destroyed |
| 1 | 19 | CO | >50 | | >12 | very poor condition; visible as a trace; destroyed |
| 1 | 25 | CO | 174 | 150 | 16 | fair condition; distal damaged |
| 1 | 37 | CO | 202 | 167 | 17 | distal damaged |
| 1 | 38 | CO | >203 | >180 | 13 | very near surface; prox. end scraped away by machines |
| 1 | 39 | CO | 185 | 174 | 7 | complete but very fragile |
| 1 | 43 | CO | >180 | | 11,5 | destroyed in flood |
| 1 | 52 | CO | 220 | 197 | 13 | good condition; destroyed in flood |
| 1 | 60 | CO | 270 | 225 | 15 | excellent condition |
| 1 | 88 | CO | 200 | 150 | e 17 | fragile; destroyed |
| 1 | 105 | CO | 236 | 155 | 19 | |
| 1 | 125 | CO | 130 (e140) | e93 | 10 | excellent condition; distal damaged |
| 1 | 138 | CO | >28 | | 5 | slender tusk, part of SH1/139: split by ice wedge cast |
| 1 | 139 | CO | | | | central and distal end of SH1/138 |
| 1 | 159 | CO | >65 | | 15 | complete but crushed |
| 1 | 170 | CO | 185 | 115 | 10 | excellent condition; accidentally destroyed |
| 1 | 172 | CO | 200 | 130 | e15 | complete; damaged distal end |
| 1 | 179 | CO | 147 | 105 | 7 | excellent condition; juvenile |
| 1 | 201 | CO | e180 | e160 | 17 | very fragile; destroyed |
| 1 | 226 | CO | 138 | 94 | 12 | |
| 1 | 244 | CO | 81 | 68 | 6 | complete; juvenile |
| 1 | 247 | CO | 125 | 100 | 12 | fair condition; distal damaged |
| 1 | 248 | CO | 210 | 170 | 11 | fair condition; distal damaged |
| 1 | 251 | CO | 265 | 75 | 18 | very curved large tusk; destroyed |
| 1 | 264 | CO | 95 | | 9 | most of tusk; distal missing |
| 1 | 285 | CO | >130 | | e14 | poor condition: crushed, distal missing |
| 1 | 316 | CO | e250 | e140 | e19 | large tusk, distal crushed |

| SITE | NO | Original state | Outside curvature | Inside curvature | Proximal diameter | COMMENT |
|---|---|---|---|---|---|---|
| 1 | 331 | CO | >110 | | 17 | hit in trench; estimated complete; destroyed |
| 1 | 351 | CO | 250 | 170 | 16 | fragile; crushed; destroyed |
| 1 | 17 | CO | 165 | 155 | 15 | prox. end cut through by ice wedge cast |
| 1 | 34 | FR | | | | |
| 1 | 45 | FR | | | | |
| 1 | 46 | FR | | | | |
| 1 | 69 | FR | | | | distal end |
| 1 | 205 | FR | | | | crushed section |
| 1 | 220 | FR | | | | |
| 1 | 329 | FR | | | | |
| 1 | 363 | FR | | | | |
| 1 | 414 | FRGS | | | | shattered tusk section |
| 1 | 228 | FRS | | | | |
| 1 | 361 | FRS | | | | |
| 1 | 1 | SECT | >35 | | 12 | hit by quarry machine; fair condition |
| 1 | 120 | SECT | | | >12 | crushed section |
| 1 | 133 | SECT | | | | very crushed |
| 1 | 135 | SECT | | | | shattered section; seems damaged by MIS6 gravel |
| 1 | 202 | SECT | >73 | | 14 | fragile; destroyed |
| 1 | 225 | SECT | >55 | | 12 | crushed section; destroyed |
| 1 | 309 | SECT | >30 | | 12.5 | |
| 1 | 364 | SECT | >20 | | 6 | juvenile; distal end |
| 2 | 13 | CO | e130 | | 15 | complete but destroyed in trench excavation |
| 2 | 22 | CO | e115 | | 13 | under tusk SH2-13; destroyed |
| 2 | 25 | CO | e63 | | 4 | juvenile |
| 3 | 7 | FRS | | | | |
| 4 | 6 | CO | >88 | | 12 | complete but destroyed in trench excavation |
| 4 | 13 | CO | e220 | e80 | 16 | good condition; damaged distal |
| 4 | 35 | CO | 235 | | 16 | |
| 4 | 46 | CO | >86 | | 11 | fragile; destroyed |
| 4 | 47 | CO | >100 | | e14 | |
| 4 | 53 | CO | | | | juvenile; damaged |
| 4 | 64 | CO | | | | shattered trace of complete tusk |
| 4 | 139 | CO | | | e14 | estimate complete; destroyed in trench cutting |
| 4 | 34 | FR | | | | |
| 4 | 61 | FR | | | | |
| 4 | 145 | FR | | | | |
| 4 | 100 | FRS | | | | |
| 4 | 25 | SECT | | | | |
| 4 | 40 | SECT | >10 | | >3 | |

| SITE | NO | Original state | Outside curvature | Inside curvature | Proximal diameter | COMMENT |
|---|---|---|---|---|---|---|
| 4 | 41 | SECT | | | | |
| 4 | 87 | SECT | >50 | | | partial tusk damaged in trench excavation |
| 4 | 88 | SECT | >50 | | 9 | |
| 4 | 130 | SECT | | | | |
| 5 | 10 | CO | >170 | | c.15 | |
| 5 | 12 | CO | 64 | | 7 | juvenile |
| 5 | 24 | CO | | | | fragile; destroyed in trench cutting |
| 5 | 34 | CO | | | | |
| 5 | 39 | CO | 165 | 132 | 14 | complete but destroyed in trench excavation |
| 5 | 42 | CO | e85 | e75 | 7 | juvenile; fragile; distal missing |
| 5 | 43 | CO | 125 | 90 | 8 | virtually complete; damaged by machine |
| 5 | 82 | CO | 70 | | 7 | juvenile |
| 5 | 72 | FR | | | | |
| 5 | 23 | FRS | | | | |
| 5 | 3 | SECT | >90 | | | |
| 6 | 7 | CO | 130 | 107 | 8 | |
| 6 | 13 | CO | 180 (e200) | e135 | 16 | complete; damaged distal |
| 6 | 114 | CO | 170 | 115 | 16 | |
| 6 | 118 | CO | | | | complete; crushed |
| 6 | 128 | CO | e180 | e155 | 16 | complete; damaged distal |
| 6 | 144 | CO | 250 | | 18 | excellent condition |
| 6 | 167 | CO | 235 | | 12 | |
| 6 | 181 | CO | >100 | | | |
| 6 | 199 | CO | e140 | e80 | e12 | |
| 6 | 246 | CO | >55 | 42 | 6.5 | juvenile; distal end destroyed by digger; ancient break at proximal end |
| 6 | 260 | CO | >>80 | | e10 | complete tusk cut through by trench |
| 6 | 266 | CO | 130 | 85 | 12 | destroyed in trench cutting |
| 6 | 269 | CO | 190 (e200) | e97 | 15 | crushed in antiquity, tip damaged; destroyed |
| 6 | 270 | CO | >110 | | e10 | slender tusk, fragile and both ends damaged; destroyed |
| 6 | 273 | CO | 245 (e255) | 150 | 17 | damaged tip; tusk very weathered in antiquity |
| 6 | 277 | CO | 138 (e150) | 98 | 17 | poor condition; destroyed |
| 6 | 45 | FR | | | | |
| 6 | 64 | FR | | | | |
| 6 | 95 | FR | | | | |
| 6 | 104 | FR | | | | |
| 6 | 107 | FR | | | | |
| 6 | 109 | FR | | | | |
| 6 | 120 | FR | | | | |
| 6 | 193 | FR | | | | |

| SITE | NO | Original state | Outside curvature | Inside curvature | Proximal diameter | COMMENT |
|---|---|---|---|---|---|---|
| 6 | 258 | FR | | | | |
| 6 | 54 | FRS | | | | |
| 6 | 203 | FRS | | | | |
| 6 | 210 | FRS | | | | |
| 6 | 108 | SECT | >40 | | >5 | destroyed in trench excavation |
| 6 | 110 | SECT | | | | |
| 6 | 115 | SECT | | | | |
| 6 | 117 | SECT | | | | |
| 6 | 135 | SECT | | | | destroyed in trench cutting |
| 6 | 207 | SECT | >48 | | 10 | |
| 6 | 256 | SECT | >25 | | 7 | juvenile; fragile section; destroyed |
| 6 | 265 | SECT | | | | |
| 6 | 296 | SECT | | | | destroyed in trench cutting |
| 6 | 304 | SECT | 17 | | 3.5 | poor condition |
| 7 | 15 | CO | >85 | >74 | 11 | juvenile; fragile |
| 7 | 22 | CO | 158 (e168) | | 13 | poor condition |
| 7 | 28 | CO | >140 | | 12 | poor condition |
| 7 | 46 | CO | 140 | | 8 | complete slender tusk; destroyed by machine |
| 7 | 75 | CO | 230 | | >15 | |
| 7 | 80 | CO | 155 (e160) | | e17 | |
| 7 | 97 | CO | 160 (e175) | | distorted | poor condition |
| 7 | 99 | CO | e170 | | e16 | slightly flattened |
| 7 | 102 | CO | 85 (e95) | | 9 | juvenile; damaged |
| 7 | 142 | CO | >150 | | e15 | estimate complete; destroyed in trench cutting |
| 7 | 237 | CO | 130 | 115 | 17 | proximal missing |
| 7 | 241 | CO | 210 (e235) | 77 | 24 (e17) | very circular tusk; crushed prox. end. ? paired with 7/242 |
| 7 | 242 | CO | >>58 | | 16 | estimate complete; destroyed in trench cutting |
| 7 | 250 | CO | 128 | 95 | 9 | fragile; distorted; destroyed |
| 7 | 266 | CO | 235 (e245) | 105 | 16 | distal end crushed and missing |
| 7 | 267 | CO | 175 | 145 | 13 | excellent condition |
| 7 | 268 | CO | 210 | 170 | 14 | excellent condition; distal crushed |
| 7 | 269 | CO | 120 | | 13 | complete; both ends destroyed in trench cutting |
| 7 | 270 | CO | 197 | 170 | 13 | |
| 7 | 271 | CO | e120 | e105 | 15,5 | |
| 7 | 276 | CO | 275 | 250 | 18 | |
| 7 | 277 | CO | ? | 226 | e16 | |
| 7 | 280 | CO | 155 (e175) | e120 | 15 | |
| 7 | 282 | CO | 122 (e132) | 105 | 10 | |
| 7 | 307 | CO | 195 | 100 | 15 | |
| 7 | 36 | FR | | | | |
| 7 | 91 | FR | | | | |
| 7 | 92 | FR | | | | |

| SITE | NO | Original state | Outside curvature | Inside curvature | Proximal diameter | COMMENT |
|---|---|---|---|---|---|---|
| 7 | 93 | FR | | | | |
| 7 | 154 | FR | | | | |
| 7 | 300 | FR | | | | |
| 7 | 120 | FRS | | | | |
| 7 | 175 | FRS | | | | |
| 7 | 179 | FRS | | | | |
| 7 | 58 | SECT | >>40 | | e12 | distal section; destroyed |
| 7 | 63 | SECT | | | 12 | |
| 7 | 111 | SECT | >40 | | 8 | |
| 7 | 149 | SECT | >57 | | 10 | juvenile; section of tusk |
| 7 | 153 | SECT | >85 | | 15 | |
| 7 | 243 | SECT | >31 | | 6 | very young tusk, cut off by trench |
| 7 | 265 | SECT | >>60 | | 12 | both ends removed by overlying gravel or machine |
| 7 | 278 | SECT | >>55 | | 15 | |
| 7 | 279 | SECT | >>25 | | 10 | |
| 8 | 9 | CO | 128 | 122 | * | juvenile; fragile |
| 8 | 35 | CO | 165* | 145* | 15 | both ends damaged |
| 8 | 61 | CO | 140* | 120* | 8 | poor condition |
| 8 | 63 | CO | 140 (e160) | e110 | e14 | very fragmentary and east end removed by MIS6 gravel |
| 8 | 65 | CO | 125 (e150) | e110 | e14 | Prox. end slightly damaged, distal end missing c. 15cm |
| 8 | 37 | SECT | >10 | | 1 | |
| 11 | 1 | FR | | | 11 | |
| 14 | 14 | CO | 215 | 170 | e16 | distorted and crushed |
| 14 | 34 | CO | e170 | | 15 | tip missing in antiquity; estimate length = 170cm |
| 14 | 48 | FRGS | | | | |
| 14 | 7 | FRS | | | | |
| 14 | 1 | SECT | >>60 | | 12 | |
| 14 | 4 | SECT | >>80 | | 17 | |
| 14 | 20 | SECT | | | | |
| 14 | ? | SECT | | | | |
| 15 | 1 | CO | 215 | 113 | 16 | large tusk; excellent condition |
| 15 | 16 | CO | 190 | 106 | 15 | fair condition |
| 15 | 5 | FR | | | | juvenile |
| 15 | 10 | FR | | | | |
| 15 | 12 | FR | | | | |
| 15 | 4 | SECT | >>25 | | 8 | distal end |
| 15 | 11 | SECT | >60 | | e12 | |
| 15 | 14 | SECT | | | | poor condition; destroyed |
| 15 | 40 | SECT | >>120 | | 14 | crushed distal section |
| 15 | 24 | SECT | | | | poor condition; destroyed in trench cutting |

Table 4.5 Tusks and fragments recorded at Stanton Harcourt (measurements in cm)

## The compostion of the mammoth assemblage

The fact that such a large number of skulls, tusks, teeth and bones were recovered from fluvial deposits raises a number of questions.

- To what extent are they contemporaneous with one another and with the remains of the other faunal and floral material excavated?
- What factors might have affected the differential preservation of cranial *vs* post-cranial elements?
- How closely might the composition of the assemblage resemble the population(s) from which they derived?
- How did they come to be incorporated into the river sediments?

The question of the relative contemporaneity of the excavated fauna and flora is discussed in Chapter 2. It is concluded that the deposits show no evidence for a single flood event bringing all the items together, but that they represent a gradual accumulation under fully interglacial conditions. While there is some evidence for variation in the rate of water flow, perhaps indicative of seasonal rainfall, there is no evidence for significant climatic or environmental change during the timespan of the channel deposits.

The Stanton Harcourt mammoths may thus be regarded as a fairly homogenous assemblage for the purpose of discussing how they occur in these river deposits and what they signify in terms of mammoth populations in the Upper Thames region in the late Middle Pleistocene. In the analysis of fossil material, the use of the terms 'sample', 'assemblage' and 'population' can have different connotations. This is due to the fact that, in most instances, the fossil assemblage is the end result of a series of processes and is unlikely to resemble the original animal population(s) from which it was derived. Klein and Cruz-Uribe (1984) summarise the stages through which a fossil fauna is likely to have passed:

1. The life assemblage – the community of live animals in their 'natural' proportions
2. The death assemblage – the carcasses of animals available to predators or any other agent of bone accumulation
3. The deposited assemblage – the carcasses or portions of carcasses that come to rest at a site
4. The fossil assemblage – the animal remains that survive until excavation or collection many millennia after deposition
5. The sample collection – that part of the fossil assemblage that is excavated or collected

Except in rare instances, at each stage from the death assemblage to the fossil assemblage there is a successive loss of elements. The nature and degree of this loss will be determined by one or more of a variety of post-depositional factors: destruction of bones at the death site or their transport away from it by predators, damage to surface bones caused by exposure to the elements, the speed with which some bones become buried and preserved, and the effect of water. The greater the time that elapses between the deposited assemblage and the fossil assemblage, the greater the likelihood that one or more of the post-depositional processes will be repeated.

Before considering the processes that might have been involved in the transition from mammoths on the landscape to excavated remains in the channel, we discuss the representation of the various skeletal elements in Stanton Harcourt mammoth assemblage.

The total assemblage of all mammoth remains from the excavation is summarised in Figure 4.1. It is clear that there are far more teeth and tusks than parts of the post-cranial skeleton, the reasons for which are discussed below.

### *Dental and cranial remains*

In mammoths, as in elephants, six molariform teeth form successively in each jaw during the life of the animal. They are recognised as three 'deciduous' molars followed by three 'permanent' molars. The upper molars are referred to as dP2-4 and M1-3 and the lower molars as dp2-dp4 and m1-3.

All cranial and dental remains are presented in Tables 4.1- 4.5 as follows:

> Table 4.1 includes all upper dentition whether isolated or in skulls.
> Table 4.2 lists additional cranial material without dentition
> Table 4.3 includes all mandibles with or without dentition
> Table 4.4 summarises the representation of complete *vs.* partial tusks and tusk fragments
> Table 4.5 lists all tusks, their measurements and other data

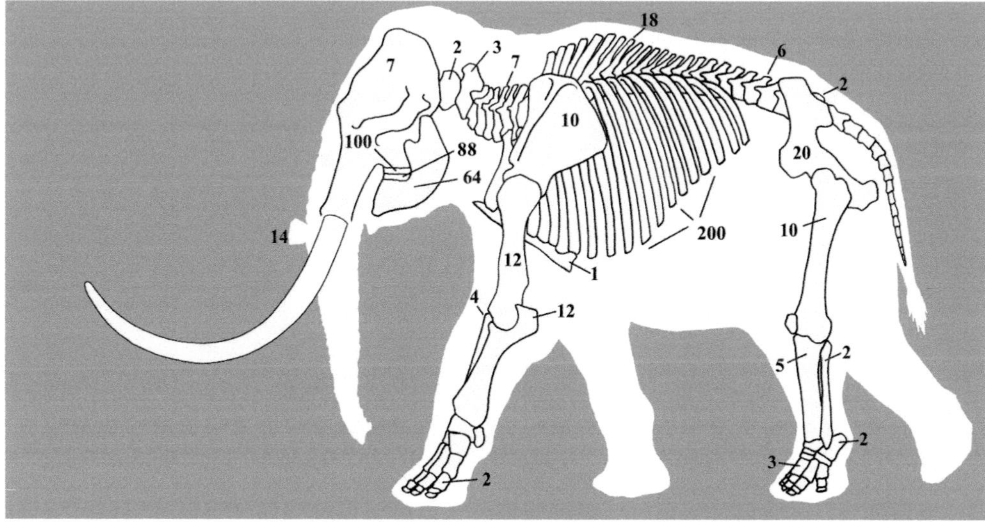


Figure 4.1 Skeletal remains of mammoth from Stanton Harcourt. Numbers represent individual specimens (NISP) rather than the number of mammoths (MNI)

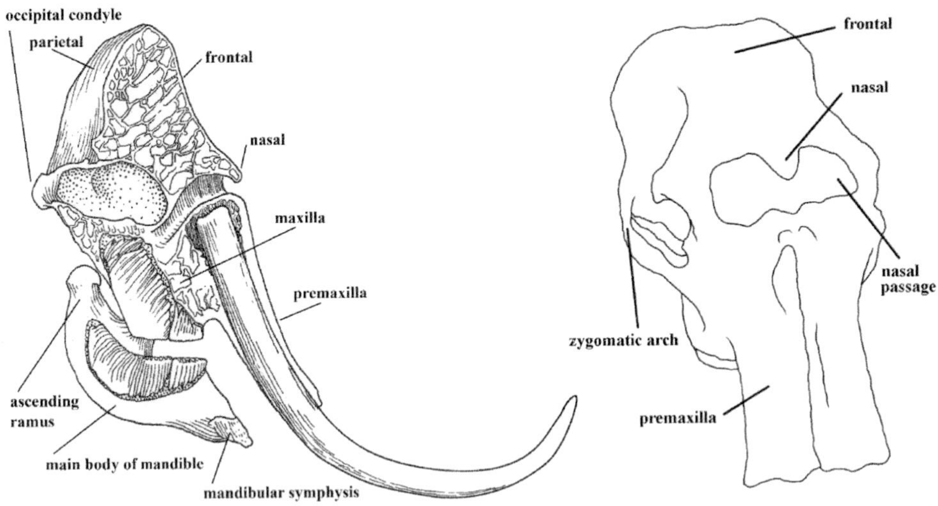


Figure 4.2 Diagrams of a mammoth skull and mandible from the side and a skull from the front

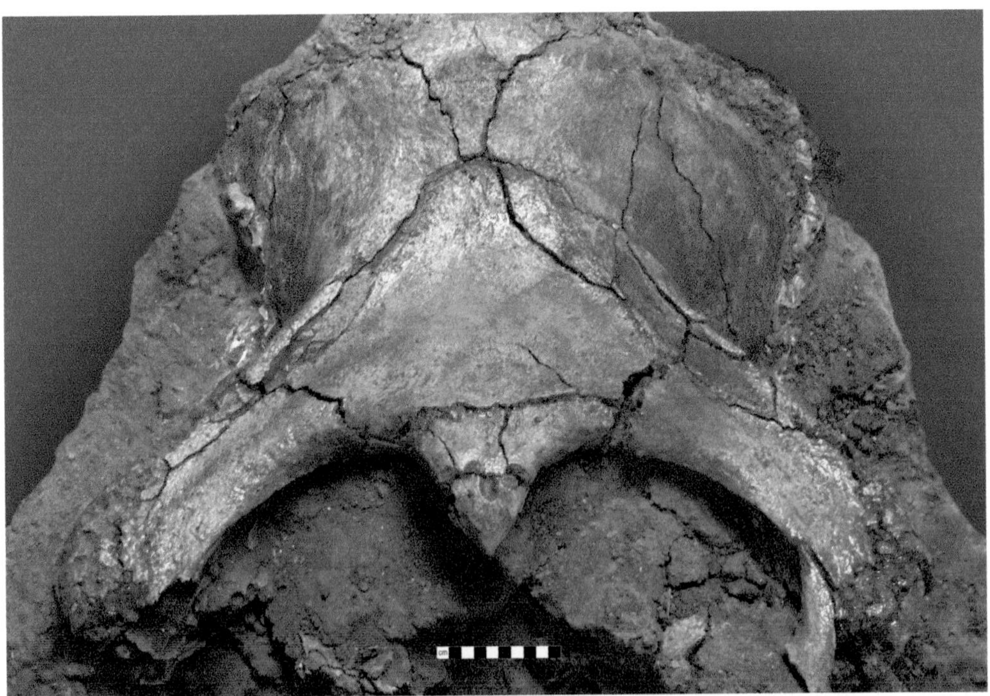

Figure 4.3 Frontal and nasal of partial mammoth skull SH1-242. Although the occipital region of this skull was missing, this skull has one premaxilla, the palate, and part of the left maxilla with M3.

*Skulls*

There were no complete skulls, although there were seven partial skulls. The frontal/parietal area of a mammoth is comprised of a relatively thin bone cortex encasing a network of delicate bone tissue creating air filled sinus cavities (Figure 4.2). This region is relatively easily damaged but, in one specimen, it is more or less complete (Figure 4.3).

Most of the partial skulls comprise the maxillae with dentition (Figure 4.4). The premaxillae – the fairly thick tube-like structures that form the sockets for the tusks - are incomplete in most specimens. These would probably have broken away from the skulls as the tusks became detached. A similar pattern of preservation was observed for the many mammoth skulls at La Cotte de St Brelade, Jersey (Scott 1986).

*Upper dentition*

Apart from the dentition in skulls, there are 100 isolated upper teeth, the majority of which are complete. Being heavy, they probably became embedded in the sediments reasonably quickly and were thus protected. Most are quite well preserved, and a number have intact roots suggesting that they were still in skulls when they became part of the river deposit and these disintegrated later (Figure 4.5). Others have suffered damage as the result of weathering or fluvial action. This is most evident as complete teeth with the roots broken off (Figure 4.6).

Figure 4.4 Mammoth skull SH15-36 with palate facing upwards showing left and right M2s and M3s *in situ*

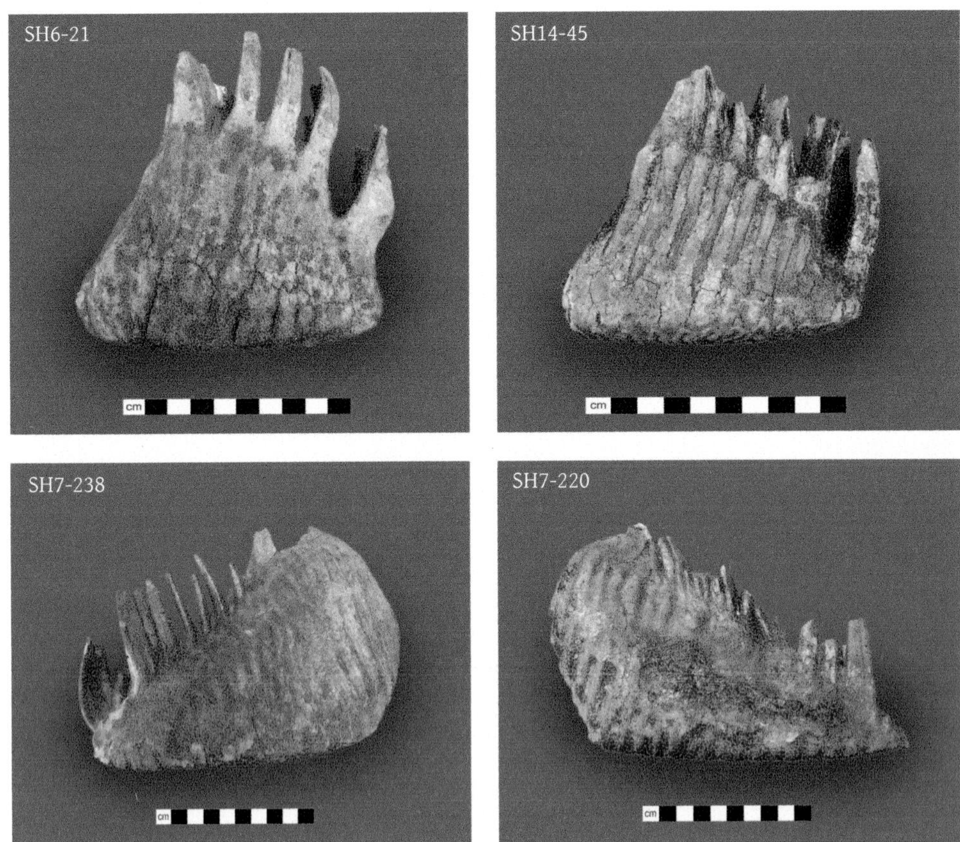


Figure 4.5 Mammoth upper molars showing little damage to the fragile roots. SH6-21: L M2; SH14-45: R M1; SH7-238: R M3; SH7-220: L M3

*Mandibles and lower dentition*

There are 64 complete and partial mandibles (Table 4.3). Mammoth molars are heavy and the mandibular bone each side of them is surprisingly thin. Thus, the occurrence of so many mandibles at Stanton Harcourt with dentition suggests relatively little post-depositional movement of these specimens.

A mammoth mandible comprises three principal parts: the ascending ramus, the main body of the mandible (corpus) and the mandibular symphysis (Figure 4.7). At Stanton Harcourt, the majority of mandibles were found with the occlusal surface of the molars uppermost (as in Figure 4.8). In such a position, in a fluvial environment, the ascending ramus is likely to be the most susceptible to damage and is missing in almost all the Stanton Harcourt specimens. Some of the otherwise complete mandibles with teeth were sufficiently weathered (cracked and exfoliated surfaces) when excavated as to suggest that they may have lain on the riverbank for some time before becoming incorporated into the river sediments. However, the surface condition of many is good indicating that they entered the water fairly soon after the death of the individual mammoths.

Figure 4.6 Mammoth upper molars in fair condition but having lost their roots. SH1-223: L M3; SH4-89: L M3; SH7-27: L M3; SH6-126: L M3

There are 22 mandibles with dentition, 13 of which have both right and left teeth *in situ*. These tend to be in a good state of preservation. Other mandibles, however, have evidently undergone more turbulence and incurred more damage. There are 38 isolated lower molars that have broken out of the body of the mandibles. The majority of the teeth, being heavy, seem to have been buried relatively soon after the break-up of the mandible and have preserved well with very little damage to the delicate roots (Figure 4.9). Once the mandible had broken and the teeth were lost, the surviving section of the mandible (mainly the mandibular symphysis) would have been more vulnerable to damage by weathering, trampling, and water transport. There are 18 of these, a few of which have survived reasonably intact, but the majority are damaged, weathered and/or water rolled (Figure 4.10) and there are many mandibular fragments.

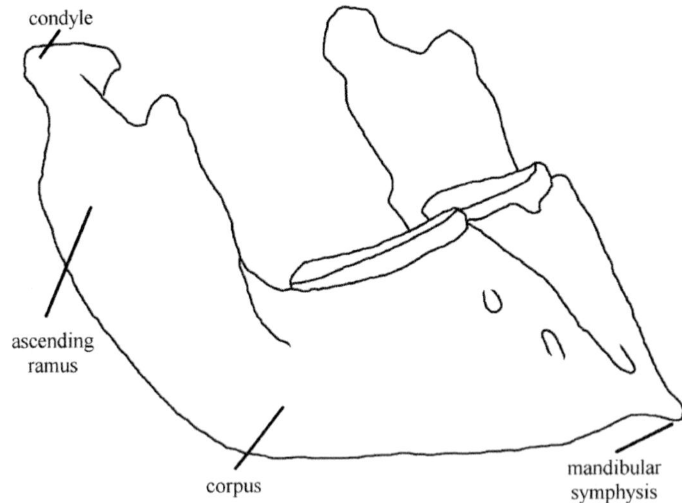


Figure 4.7 Diagram showing parts of mammoth mandible

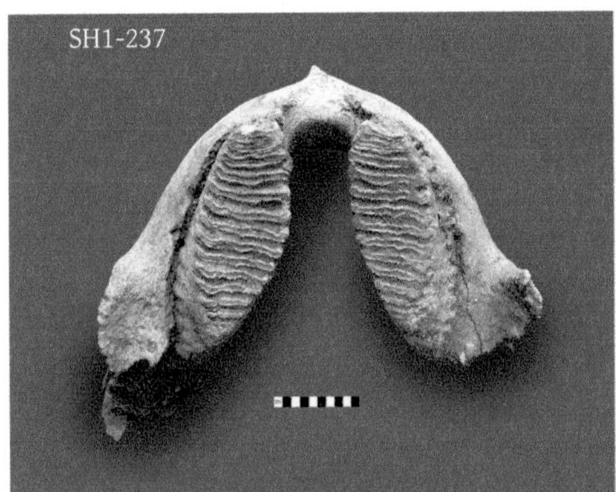

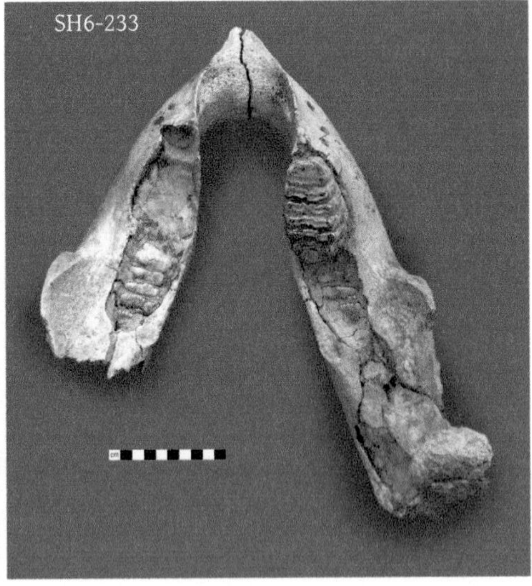


Figure 4.8 Three mammoth mandibles with dentition in different stages of eruption and wear. For estimated age at death (AEY) see Chapter 4. SH6-233 with Rdp3 in medium wear and L+R dp4 in crypt – AEY 1-2 years. SH6-253 with L+R m2 (worn) and L+R m3 in medium wear - AEY 39 years. SH1-237 with L+R m3 (worn) – AEY 43 years

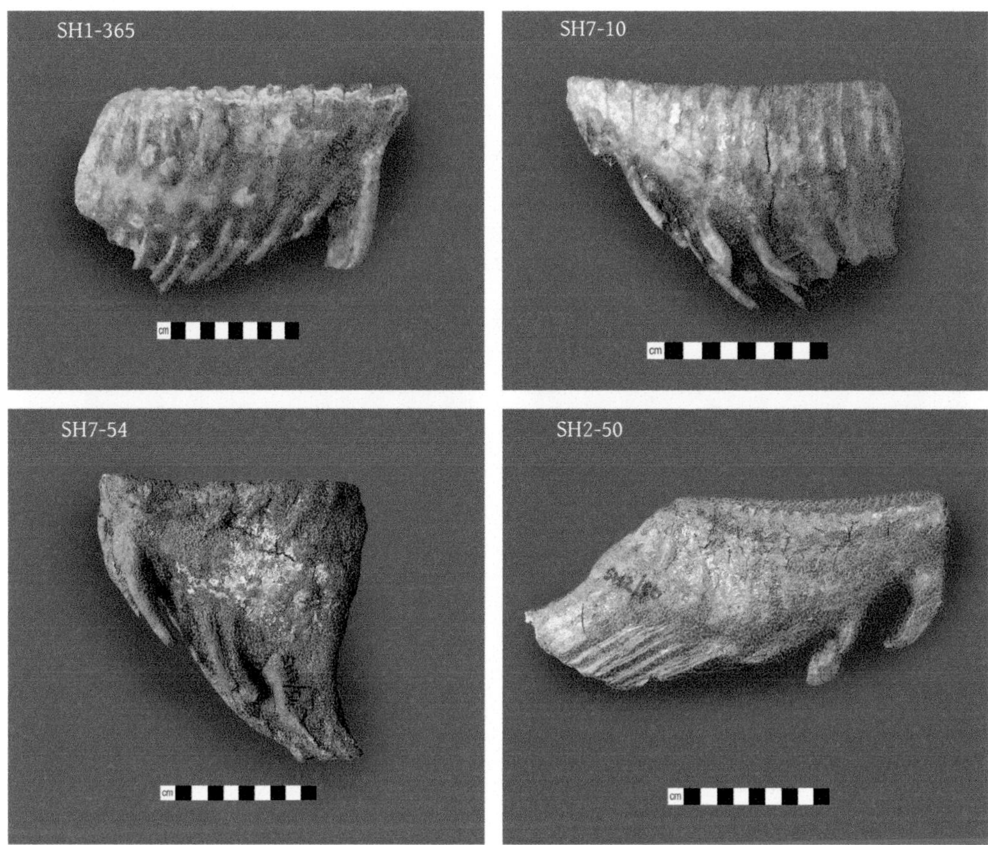


Figure 4.9 Mammoth lower teeth showing relatively undamaged roots. SH1/365: L m3; SH7/10: R m3; SH7/54: L m3; SH2-50: Rm3

*Tusks*

The excavations revealed an extraordinary accumulation of more than 100 complete or virtually complete tusks. Additionally, there were more than 40 sections of tusk ranging in length from 20 – 200cm (Figure 4.11; Tables 4.4 and 4.5). The density with which they occurred across the site may be seen for example in Chapter 2 Figure 2.

An elephant or mammoth tusk is comprised of a series of layers reminiscent of a sliced onion when seen in section. These layers of ivory (dentine) exfoliate rapidly and the tusks disintegrate more quickly than the postcranial bones (Haynes 2006). Following the death of an elephant, once the tusks are exposed to the elements, they are vulnerable to two agents of destruction: weathering and trampling. Animals frequently return to easily accessible banks of rivers to drink and thus the bones of animals that died at these locations will become trampled. Tusks are noted to be particularly vulnerable. For example, of 200 elephant bones and tusks around a water source in Zimbabwe Haynes (*op. cit.*) records that all the tusks (26) were broken but only a third of the long bones.

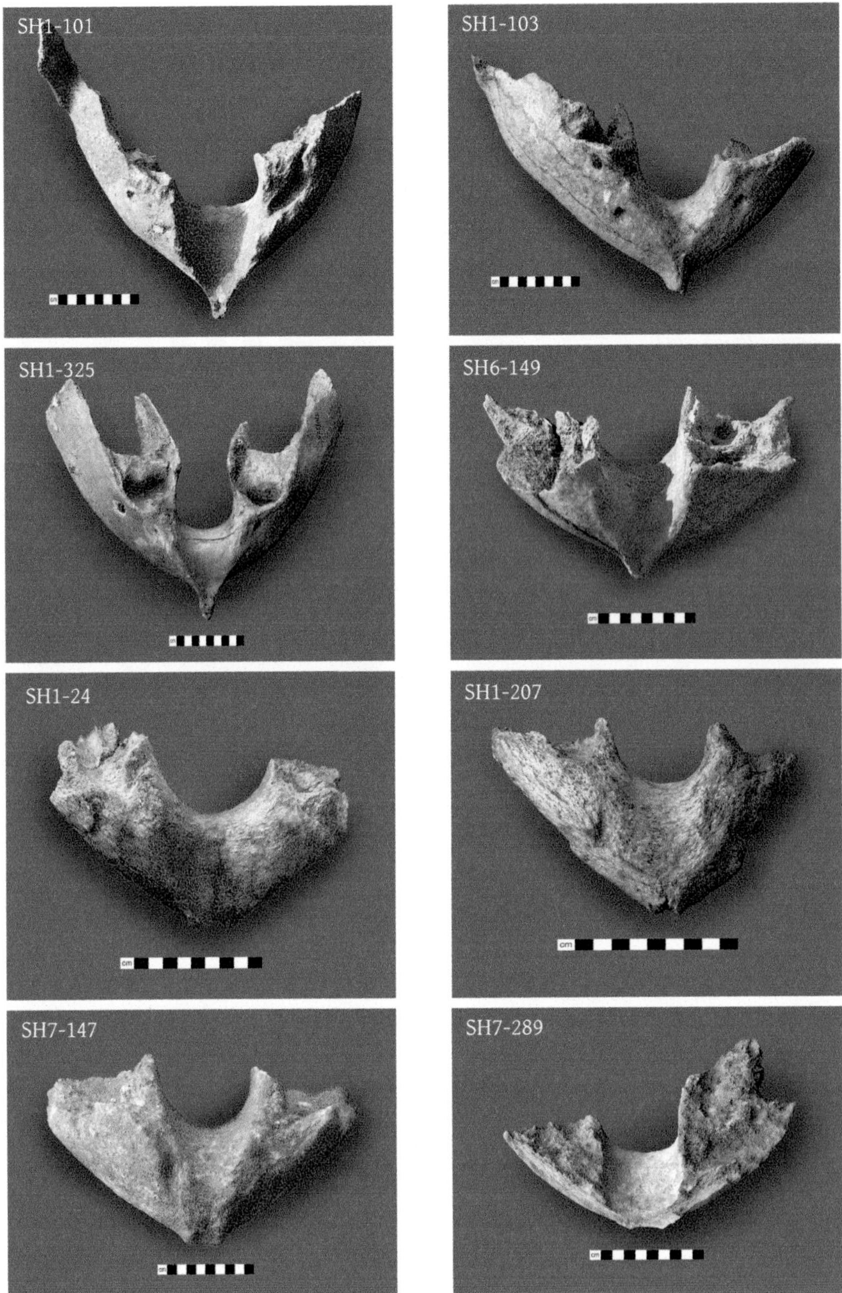


Figure 4.10 Mammoth mandibles without dentition illustrating two types of post-depositional preservation. In the top two rows SH1-101, SH1-103, SH1-325 and SH6-149 have lost molars but have distinct morphological features and sharp-edged breaks. These probably underwent little movement in the water after deposition. In the lower two rows SH1-24, SH1/207, SH7-147 and SH7/289 have rounded contours and show significant weathering and/or fluvial damage

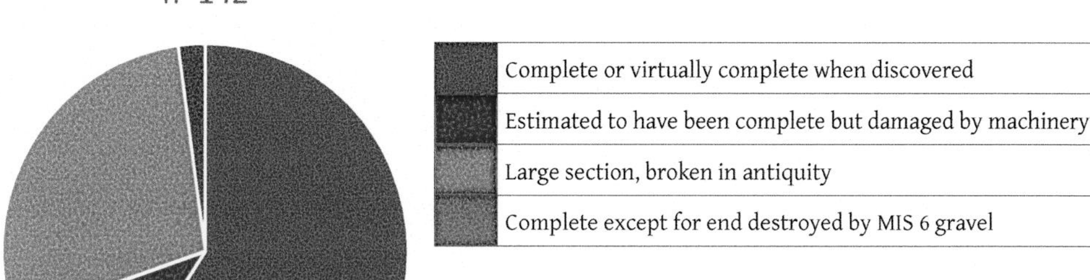


Figure 4.11 Relative frequencies of complete and partial tusks at Stanton Harcourt

The good state of preservation of so many complete tusks at Stanton Harcourt (Figures 4.12 and 4.13) suggests that they became part of the river sediments relatively soon after the death of the mammoths. Several tusks (and bones) in the uppermost levels had evidently been damaged by the influx of the succeeding 'cold' gravels which had cut off and removed one or both ends. Although their condition generally ranged from fair to excellent, they were not heavily mineralised and all required to be plastered or fibre glassed before they could be lifted (Figure 4.14). Some however, even though complete when discovered, could not be retrieved for a variety of reasons. Those near the quarry surface were vulnerable to frost damage and were shattered or had a putty-like consistency. Many were intersected by exploratory excavation trenches. Others had been crushed or damaged during the quarrying of the overlying gravel deposits leaving a mere trace of the original tusk (Figure 4.15).

Field observations of the process of decay of elephant skeletons (Crader 1983, Haynes 1988) show that tusks exfoliate and split relatively rapidly if they lie around on the land surface for more than a few years (Figure 4.16). In the main therefore, the fair condition of the majority of complete tusks at Stanton Harcourt, resembling the elephant tusks that had not been exposed for any length of time, indicate that they became buried relatively soon after the death of the mammoth. The partial tusks are characterised by being a metre or more in length and had lost one or both the ends. As most of these had surfaces in generally fair condition it is suggested that they did not get broken by trampling or weathering while on land but after the complete tusk had entered the water. Here, their weight and curvature possibly caused them to get caught up among rocks and logs and one or both ends to get broken off.

*The age range of the mammoth tusks at Stanton Harcourt*

Both male and female elephants (and mammoths) have tusks. From about 6 months to 1 year of age they carry a pair of milk tusks (tushes). These are only a few centimetres in length and never visible externally. These are replaced by a pair of permanent tusks which become visible at about 30 months (Hanks 1979). They grow, by the addition of layers of dentine, in width and length throughout most of the individual's life. In African elephants, female tusks grow lengthwise through life but circumference growth ceases around the age of sexual maturity. In males the tusks grow in length and circumference throughout life. After puberty the pulp

4.12 a
1 m

4.12 b

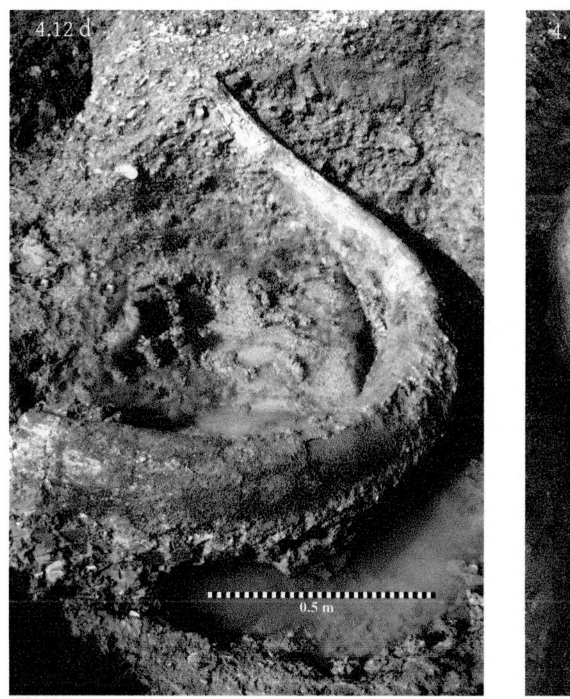

Figure 4.12 Complete tusks in the process of excavation showing relatively little surface damage. Note that the tusk (e) is still encased in the premaxilla

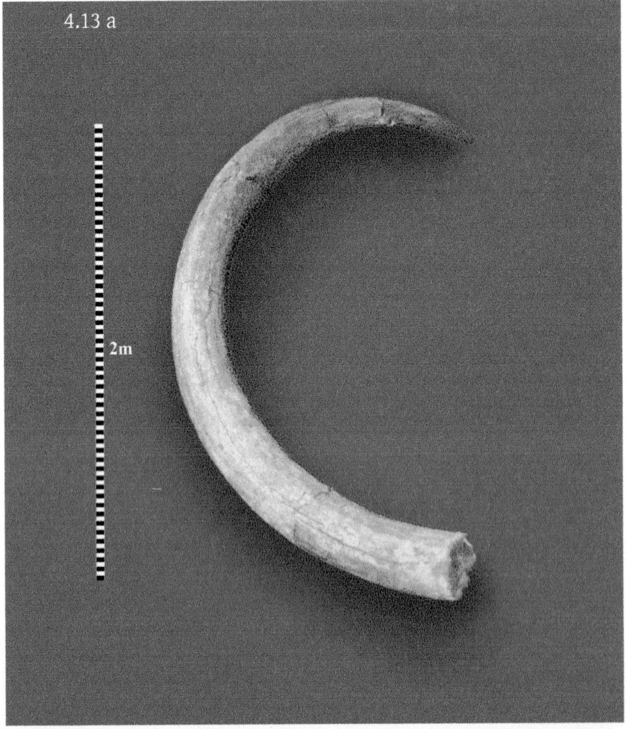


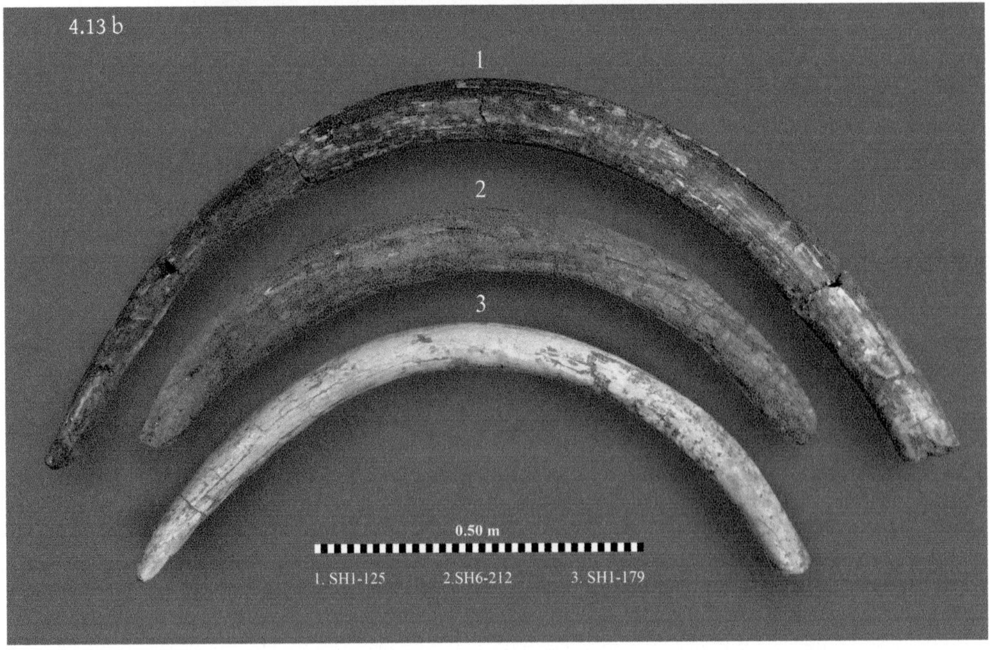


Figure 4.13 Some of the complete tusks from Stanton Harcourt: (a) SH6/144 (b) SH1/125, SH6/212 and SH1/179

Figure 4.14 Tusks removed from the field in fibreglass and Plaster of Paris

4.15 a
SH7-243
SH7-242
SH7-241

4.15 b
SITE 7
Sq G6
19 May '97

Figure 4.15 Examples of post- depositional damage to tusks: (a) Three tusks: SH7/241 has both ends removed by quarrying to the interface of the MIS 6 (quarried) gravel and the underlying channel gravel. SH7/243 has had the proximal end cut off during quarrying and the distal end destroyed by the excavation of a trench. The small tusk SH7/242 has also been intersected by the trench. (b) the proximal end of SH7/80 was evidently sheared off by the incoming MIS 6 gravel and the distal end was damaged during trench excavation. (c) Distal end of tusk SH4/125 quarried away (d) Tusk SH7-280 damaged by quarrying/surface scraping leaving only traces of the complete tusk

Figure 4.16 African elephant and mammoth tusks in various stages of weathering. (a) the larger grey tusks of African elephants had been lying about for less than 10 years and are similar in surface condition to many complete tusks from Stanton Harcourt. (b) Africa elephant tusk that had lain on the surface for 10-15 years. (c) detail of weathered tusk from Stanton Harcourt (elephant tusk photos with kind permission of Gary Haynes)

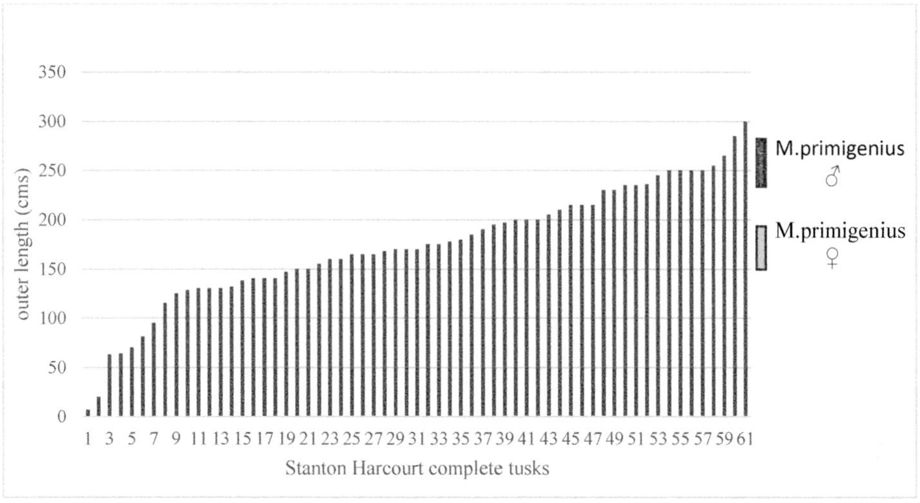


Figure 4.17 Stanton Harcourt mammoth tusks on which it was possible to take complete outside length measurements (Table 4.5). On the right hand are shown the average range of tusk lengths for male and female European woolly mammoth *M. primigenius* (data from Lister and Bahn 2015)

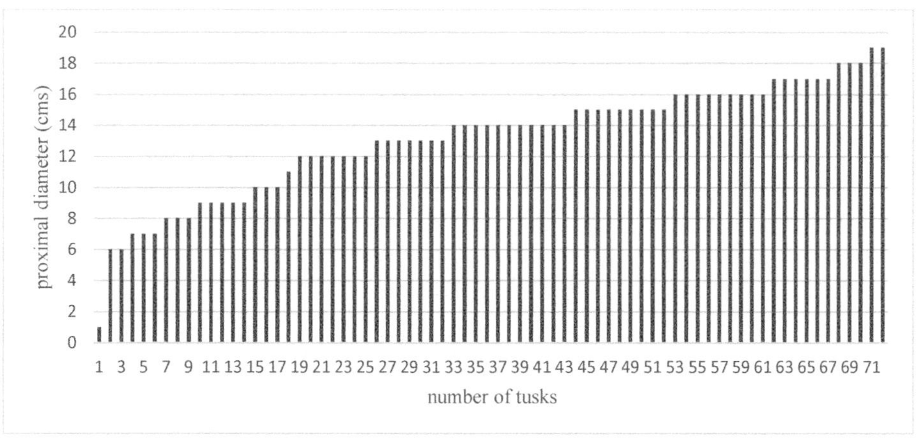


Figure 4.18 Proximal diameter of Stanton Harcourt mammoth tusks (for measurements see Table 4.5)

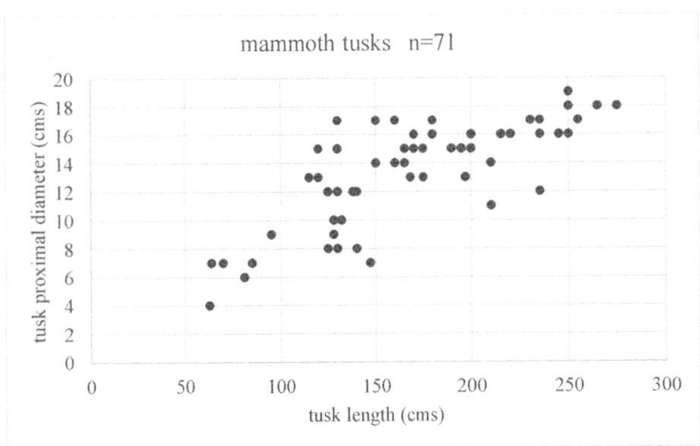


Figure 4.19 Proximal width vs. length of complete tusks from Stanton Harcourt

cavity of female tusks begins to 'fill in' decreasing in depth with time. In males pulp cavity continues to increase in depth with age (Perry 1954). Siberian woolly mammoths appear to corroborate this (Haynes 1991). In mammoths, males develop larger, more distinctly spiralled tusks than females. Typically, the tusks of the male Eurasian woolly mammoths achieved a length (measured along the outer curve) of between 2.4 and 2.7m and weighed in the region of 45kg. Female tusks were smaller with a length between 1.5 and 1.8m and weighing only 9-11kg (Lister and Bahn 2007, 2015).

A wide size/age range of tusks is represented at Stanton Harcourt (Fig 4.17). Approximately 10% are less than 1m in length and denote young individuals. The remaining tusks range in length from 1m - 2.5m showing all age categories to be evenly represented, with a few large ones reaching 2.8 – 3m. The measurement of the diameter of the proximal end of the tusk is another means of illustrating the range of sizes of the individuals represented (Figure 4.18). Here the sample is increased by including tusks with broken distal ends. Another way of looking at the range of sizes of the tusks is the ratio of the proximal width to the length in complete tusks (Figure 4.19).

The fact that several of the Stanton Harcourt tusks are as large than those of the typical British woolly mammoths is intriguing. As discussed earlier, the teeth are ascribed to a small form of steppe mammoth *Mammuthus trogontherii*, now recognised as characteristic of MIS 7. As described later in this chapter, their post-cranial bones indicate a shoulder height of between 2.2 and 2.9m. Shoulder height in the ancestral early Pleistocene *M. trogontherii* was approximately 3.9m (Lister and Stuart 2010) and in the woolly mammoth *Mammuthus primigenius* shoulder height ranged between 2.75m and 3.4m (Lister and Bahn 20007, Lister 2014).

Although the MIS 7 mammoths were significantly smaller than the woolly mammoths (with the exception of some very Late Pleistocene and Holocene examples), their tusks are large relative to their body size. Approximately a third of the tusks at Stanton Harcourt are within the size range of adult female woolly mammoths, 40% exceed the range for adult female woolly mammoths, and several are within the size range for adult male woolly mammoths (Figure 4.17).

### *Post-cranial remains*

By comparison with the combined number of more than 400 teeth and tusks representing more than 80 individuals, the identifiable post cranial remains are relatively few (Figure 4.1; Table 4.6).

#### *Forelimbs*

There are five scapulae, more or less complete except for damage to the distal margin (Table 4.7 and Figures 4.20 and 4.21). The long bones of the limbs generally comprise a shaft (diaphysis) with a separate bone at each end (the proximal and distal epiphyses). The epiphyses fuse to the shaft as the mammoth matures. Fusion in various skeletal elements in elephantids takes place in sequence well into maturity (Table 4.8). Consequently, the earlier the fusion takes place of the component parts of a bone, the better its chance of preservation. Apart from one damaged complete specimen with fused epiphyses, humeri at Stanton Harcourt are represented almost exclusively by diaphyses Figure 4.22. As can be seen from Table 4.9 and Figure 4.23 very few measurements were possible on these specimens. Of the 16 ulnae, only one is complete (Table 4.10; Figures 4.24 and 4.25) and most of the rest are represented by the proximal end. There are five incomplete radii, being either fragmentary or diaphyses lacking (unfused) epiphyses (Table 4.11; Figure 4.26).

#### *Hindlimbs*

Pelvic bones are the best represented skeletal elements. There are 57 complete or partial pelvic bones, of which 21 are reasonably intact: 15 right and 6 left. A selection is illustrated in Figure 4.27 and their measurements are given in Table 4.12 and Figure 4.28. Pelvic remains that could not be measured are listed in Table 4.13.

There are six complete (or almost complete) femora representing animals of approximately 20+ years of age and several limb shafts and unfused epiphyses of smaller, younger individuals (Table 4.14 and Figures 4.29 and 4.30). Only one of the five tibiae is complete (Table 4.15 Figures 4.31 and 4.32). The fibula is a relatively slender bone compared with the tibia and is represented by only two specimens, both lacking (unfused) epiphyses (Figure 4.33).

#### *Ribs, vertebrae and feet*

Here were many ribs (complete and broken) but vertebrae are few and bones of the feet virtually absent. The disproportionate body-part representation of the mammoths at Stanton Harcourt is discussed in detail later in this chapter. In summary, it is considered to be due to a combination of post-mortem processes on the land and the effect of being in the river. After the death of individual mammoths, carnivore activity and the length of time a carcass lay on the land surface will have begun a process of selective preservation in favour of larger elements. Once in the river, differential preservation of the mammoth skeletal elements appears to have been determined by the state of fusion of the epiphyses, the size and shape of the bones, and their weight.

| Skeletal element | NISP | L/R |
|---|---|---|
| Vertebrae - atlas | 2 | |
| axis | 3 | |
| cervical, complete | 2 | |
| partial | 5 | |
| thoracic, complete | 6 | |
| partial | 12 | |
| lumbar | 6 | |
| sacrum | 2 | |
| fragments | 13 | |
| Scapula - complete | 5 | L=2; R=3 |
| proximal | 5 | L=1; R=4 |
| fragments | 13 | |
| Humerus - complete | 2 | R=2 |
| proximal fragments | 3 | |
| diaphysis | 10 | L=3; R=2 |
| distal | 2 | L=1 |
| Radius - complete | 1 | R=1 |
| proximal | 2 | R=2 |
| diaphysis | 1 | R=1 |
| distal | 1 | |
| Ulna - complete | 3 | L=1; R=2 |
| proximal +diaphysis | 9 | L=1; R=8 |
| proximal fragments | 3 | L=2; R=1 |
| distal | 1 | L=1 |
| Pelvis - complete | 14 | L=5; R=9 |
| ilium | 4 | L=1; R=3 |
| acetabulum/ilium | 4 | L=1; R=3 |
| acetabulum/ischium | 2 | |
| pubis | 2 | |
| fragments | 24 | |
| Femur - complete | 7 | L=3; R=4 |
| proximal | 3 | L=2; R=1 |
| prox. epiph. fragments | 5 | |
| diaphysis | 10 | L=6; R=4 |
| diaphyseal fragments | 9 | |
| distal | | |
| Patella | | |
| Tibia - complete | 1 | L=1 |
| proximal | | |
| diaphysis | 5 | L=4 |
| distal | | |
| Fibula | 2 | L=1; R=1 |
| Tarsals - calcaneum | 2 | R=2 |
| astragalus | | |
| naviculo-cuboid | | |
| Other carpal/tarsal | 1 | |
| Metacarpal - complete | | |
| proximal | | |
| distal | | |
| Metatarsal - complete | 2 | |
| proximal | 1 | |
| distal | | |
| Phalanges - 1st | 2 | |
| Indet. fragments | 93 | |
| Sternum | 2 | |
| Ribs and fragments | c.250 | |
| Epiphyseal frags., not identified | 8 | |
| Diaphyseal frags., not identified | c.130 | |

Table 4.6 Post-cranial remains of mammoth from Stanton Harcourt

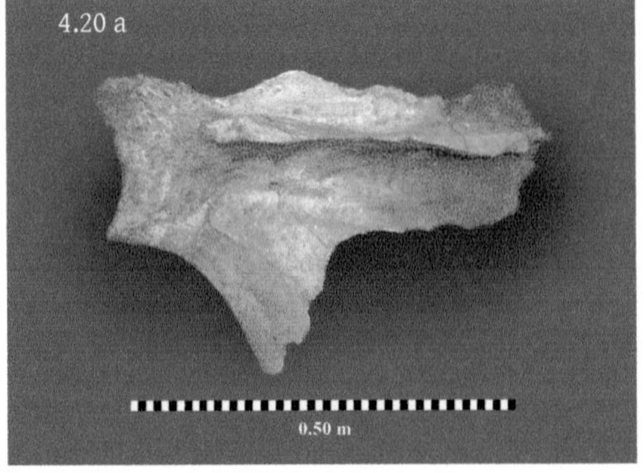

Figure 4.20 Examples of mammoth scapulae from Stanton Harcourt showing typically good preservation of proximal ends and damage to distal margins: (a) SH1/261 and (b) SH7/174

| SITE | REF | L/R | 1. Articular width of glenoid cavity | 2. Max. width of glenoid epiphysis | 3. Depth of glenoid epiphysis | 4. Max. length of scapula | 5. Max. dorsal width | Comments |
|---|---|---|---|---|---|---|---|---|
| SH6 | 112 | R | 180 | 240 | 107 | 721 | | virtually complete; distal margin fused |
| SH15 | 44 | R | 120 | 150 | 96 | 590 | ? | complete, small amount distal damage |
| SH1 | 154 | R | 143 | | 76 | >435 | BRO | prox. and blade; distal damaged |
| SH4 | 102 | R | 160 | 205 | 95 | 615 | BRO | almost complete; distal margin unfused |
| SH1 | 294 | R | 152 | 200 | 98 | >450 | BRO | poor condition |
| SH1 | 261 | L | 164 | 190 | 100 | >540 | BRO | fairly complete, distal damaged |
| SH5 | 31 | R | 70 | | | >210 | | proximal and part blade; damaged |
| SH6 | 214 | R | 145 | 185 | 89 | >230 | BRO | proximal and part blade |
| SH6 | 276 | L | 161 | 175 | 87 | 590 | ? | virtually complete, distal margin damaged |
| SH6 | 275 | L | 175 | | 82 | 730 | | virtually complete, distal margin damaged |
| SH6 | 282 | L | | c.150 | | >550 | | almost complete; proximal and distal damaged |

Table 4.7 Measurements (mm) of mammoth scapulae (see Figure 4.21)

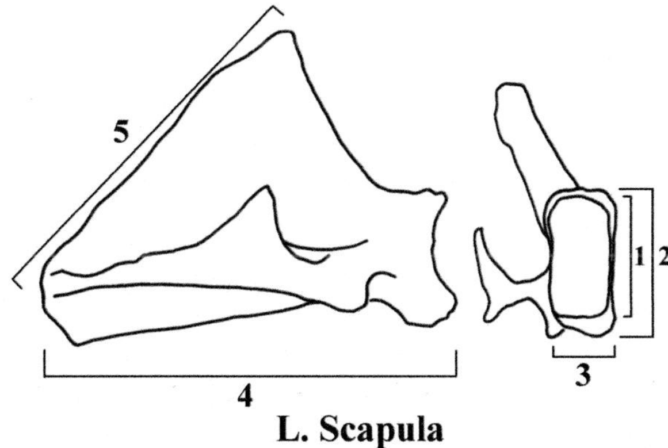


Figure 4.21 Measurements taken on mammoth scapulae

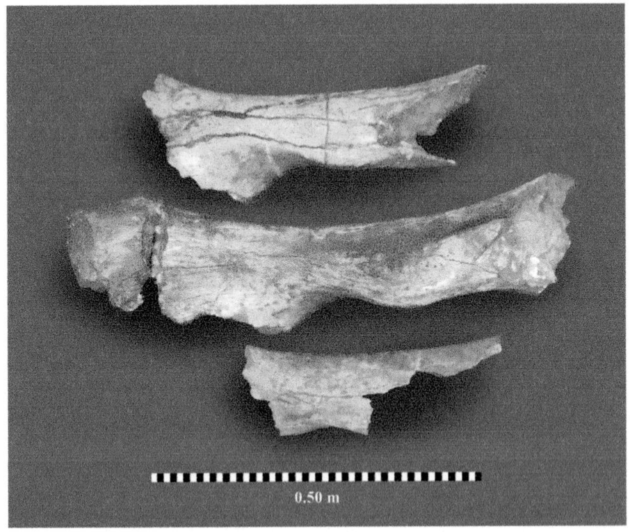


Figure 4.22 Selection of mammoth humeri from Stanton Harcourt: SH4/73, SH1/4 and SH3/4

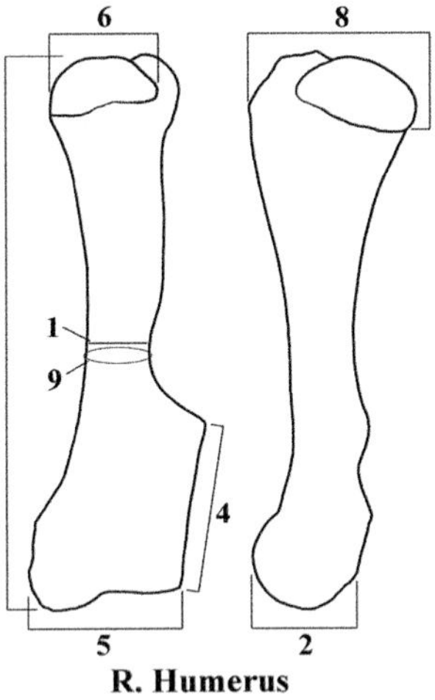

R. Humerus

Figure 4.23 Measurements taken on mammoth humeri

| APPROX. AGE IN YEARS | FEMALES<br>**Element and state of fusion** | MALES<br>**Element and state of fusion** |
|---|---|---|
| 8 | 3 bones of innominate fusing | 3 bones of innominate fusing |
| 18-19 | Humerus – distal fused<br>Humerus – proximal fusing<br>Radius/ulna – proximal fusing<br>Tibia – distal fused<br>Innominate – fused with sacrum<br>Vertebrae – some centra fusing | Humerus – distal fused |
| 18-23 | Tibia – proximal fuses<br>Vertebrae – sacrum fused<br>Femur – distal fuses | |
| 28-32 | Femur – proximal fused/fusing<br>Radius/ulna – distal fuses | Femur – distal fuses<br>Femur – proximal fuses<br>Radius/ulna – proximal fuses<br>Tibia – proximal fuses<br>Tibia – distal fuses<br>Vertebrae – sacrum fuses |
| 36+ | Scapula – vertebral border fuses<br>Innominate – iliac crest fuses | Scapula – vertebral border fuses<br>Humerus – proximal fuses |
| 40+ | | Innominate – fuses with sacrum |
| 50+ | | Innominate – iliac crest fuses<br>Radius/ulna – distal fuses<br>Ribs – articular ends fuse |

Table 4.8 Epiphyseal fusion rates in the African elephant *Loxodonta africana* (data from Haynes 1987, 1993 and Haynes *pers. comm.*)

| SITE | REF | L/R | 1. Minimum diameter diaphysis | 4. Length supra-condyloid ridge to base lateral epicondyle | 7. Total length including epiphyses | 9. Minimum circum-ference | Comments |
|---|---|---|---|---|---|---|---|
| 1 | 4 | L | 104 | e300 | >750 | 345 | Prox. epiphysis missing; dist. epiphysis broken |
| 3 | 4 | L | 106 | | >380 | | part diaphysis, no epiphyses |
| 4 | 73 | L | 103 | | >520 | 320 | diaphysis only |
| 4 | 65 | L | 60 | | >95 | 192 | very young, diaphysis only |
| 5 | 40 | R | 107 | | | e360 | very poor condition; epiphyses crushed |
| 7 | 139 | R | 108 | | >530 | >>340 | diaphysis only |
| 7 | 245 | R | 97 | | >400 | 310 | diaphysis only |
| 8 | 58 | R | 96 | 250 | 680 | 295 | complete with both epiphyses fused |

Table 4.9 Measurements (mm) of mammoth humeri (see Figure 4.23)

| SITE | REF | L/R | 1. Depth olecranon process | 2. Maximum antero-posterior width | 3. Width proximal articular surface | 4. Length | 5. Distal antero-posterior width | Comments |
|---|---|---|---|---|---|---|---|---|
| 1 | 260 | R | 185 | 225 | 186 | >646 | e150 | distal epiphysis unfused and missing |
| 1 | 301 | L | 145 | 250 | 188 | >510 | e100 | proximal and distal damaged |
| 7 | 266 | R | 152 | 180 | BRO | >485 | e100 | broken distal end |
| 7 | 239 | R | | 190 | BRO | >470 | e100 | unfused epiphyses missing |
| 6 | 162 | R | 179 | 210 | 165 | BRO | BRO | distal crushed and missing |
| 15 | 7 | R | 180 | | | | | very crushed and distal missing |
| 8 | 46 | L | 173 | 225 | 184 | 747 | 160 | excellent condition |
| 1 | 344 | R | 165 | 240 | 165 | BRO | BRO | distal epiphysis broken |
| 1 | 127 | R | | 195 | | >500 | e113 | both epiphyses missing |

Table 4.10 Measurements (mm) taken on mammoth ulnae (see Figure 4.25)

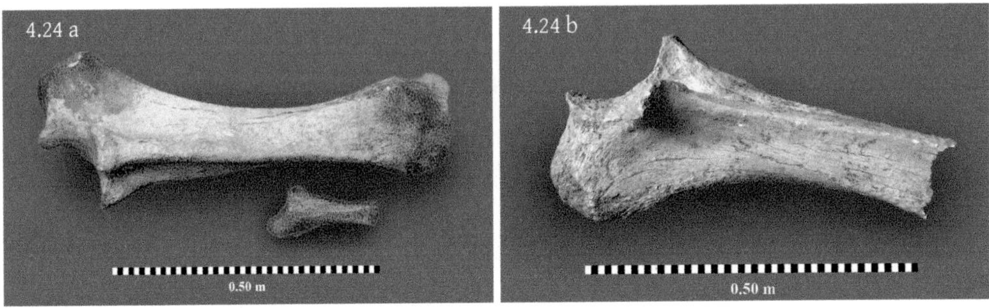


Figure 4.24 (a) complete mammoth ulnae SH8/46 and juvenile SH6/45 (b) typical preservation of ulnae at the site - SH1/284

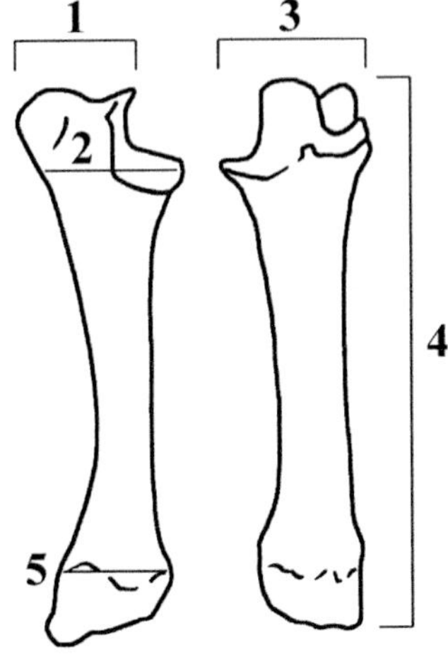


Figure 4.25 Measurements taken on mammoth ulnae

Figure 4.26 Fragmentary mammoth radii from Stanton Harcourt - SH6/88, SH8/36 and SH6/218

| SITE | REF | L/R | 1. Proximal epiphysis medio-lateral | 2. Proximal epiphysis antero-posterior | 3. Total length | 4. Distal medio-lateral | 5. Distal antero-posterior | Comments |
|---|---|---|---|---|---|---|---|---|
| 4 | 91 | R | 100 | | >566 | | | distal missing |
| 6 | 218 | R | | | | 100 | 140 | distal epiphysis |
| 8 | 36 | R | | | | 85* | 44* | *distal diaphysis without epiphysis; measurement of diaphysis at point of fusion |

Table 4.11 Measurements (mm) of mammoth radii

| SITE | REF | L/R | PART | 1. Medio-lateral width acetabulum | 3. Max. width ilium | 4. Min. width base ilium | 8. Antero-posterior length acetabulum |
|---|---|---|---|---|---|---|---|
| 1 | 14 | L | ILI/ACE | | | e170 | 150 |
| 1 | 51 | R | CO | 148 | | | |
| 1 | 128 | R | ILI/ACE | | | 163 | |
| 1 | 184 | R | CO | 120 | e530 | 130 | 131 |
| 1 | 233 | L | CO | 135 | >495 | 150 | |
| 2 | 49a | R | CO | 124 | >520 | 138 | 123 |
| 2 | 49b | L | CO | 123 | >520 | 130 | 121 |
| 3 | 13 | L | CO | 130 | | 135 | 130 |
| 5 | 41 | R | CO | | >470 | 180 | 145 |
| 6 | 67 | R | CO | 145 | >500 (e520) | 175 | 160 |
| 6 | 147 | L | CO | 126 | | 139 | 113 |
| 6 | 163 | R | CO | 145 | e570 | 185 | e130 |
| 7 | 69 | L | CO | e120 | >460 | 160 | 140 |
| 7 | 119 | R | CO | | e840 | | 156 |
| 7 | 292 | R | CO | | | e160 | |
| 8 | 44 | R | ACE/ILI/ISC | 118 | >490 (e540) | 128 | 130 |
| 8 | 72 | R | CO | e105 | >450 | 124 | e130 |
| 6 | 308 | R | CO | 128 | e570 | 130 | 134 |

KEY:

CO – complete;
ILI - ilium;
ISC – ischium;
ACE – acetabulum;
PUB - pubis

Table 4.12 Mammoth pelvic bones on which it was possible to take measurements (mm). See Figure 4.28

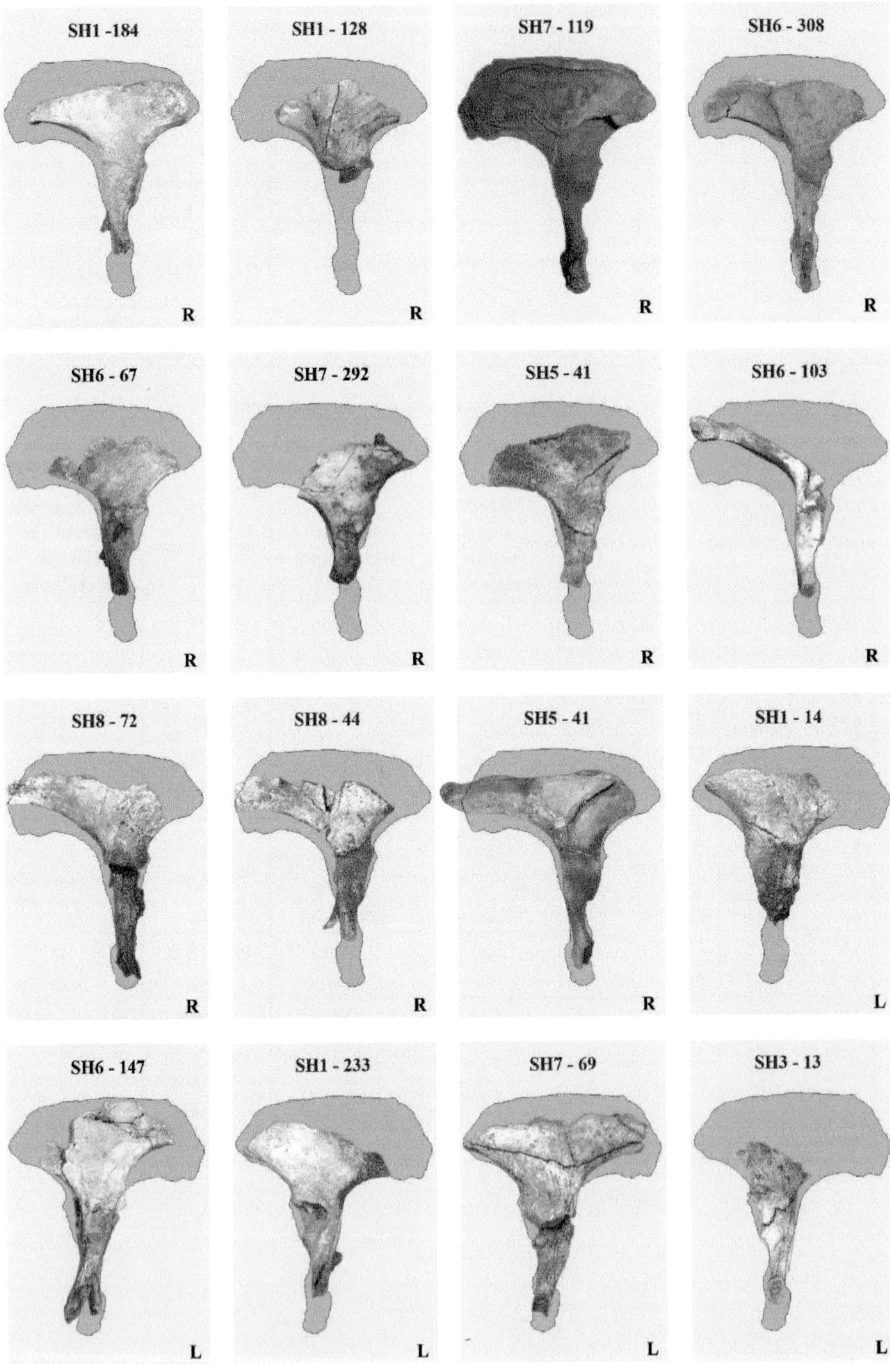


Figure 4.27 Typical preservation of mammoth innominates from Stanton Harcourt

| SITE | NO | ELEMENT | L/R | PART |
|---|---|---|---|---|
| 1 | 293 | ILI | R | CH |
| 1 | 337 | ILI | | CHS |
| 1 | 8 | INN | L | CO |
| 1 | 48 | ILI | | CH |
| 1 | 140 | ILI | | PART |
| 1 | 173 | ILI | L | PART |
| 1 | 175 | ILI | R | PART |
| 1 | 258 | INN | | FR |
| 1 | 373 | ILI | | FR |
| 1 | 42 | ISC | | CH |
| 2 | 47 | INN | | ILI/ACE |
| 2 | 63 | INN | | FRS |
| 4 | 10 | ILI | | CHS |
| 4 | 48 | ILI | | CH |
| 5 | 32 | ILI | | CH |
| 6 | 62 | ILI | | FRS |
| 6 | 78 | ILI | | FR |
| 6 | 84 | ILI | R | FR |
| 6 | 6 | INN | R | CO |
| 6 | 82 | INN | | CHS |
| 6 | 103 | ILI/ACE/ISC | R | PART |
| 6 | 130 | PUB/ILI | | PART |
| 6 | 191 | ILI | | PART |
| 6 | 206 | ILI | | CHS |
| 6 | 261 | ILI/ACE | | PART |
| 6 | 286 | INN | | CHS |
| 7 | 25 | ILI | | PART |
| 7 | 143 | ILI | | FRS |
| 7 | 158 | ILI | | CH |
| 7 | 200 | ILI | | CH |
| 7 | 248 | ILI/ACE | R | PART |
| 7 | 285 | INN | | PART |
| 7 | 288 | INN | | CH |
| 7 | 287 | ISC | | CH |
| 7 | 166 | ISC/ACE | | PART |
| 7 | 306 | ISC/ACE | | PART |
| 8 | 32 | ISC/ACE/ PUB | | PART |
| 15 | 23 | CHECK | | PART |

KEY:

| | |
|---|---|
| INN | innominate, significant part of |
| ILI | ilium |
| ISC | ischium |
| PUB | pubis |
| ACE | acetabulum |
| CO | complete or almost |
| PART | large part of whole |
| CH/S | chunk(s) |
| FR/S | fragment(s) |

Table 4.13 Mammoth pelvis remains on which no measurements could be taken

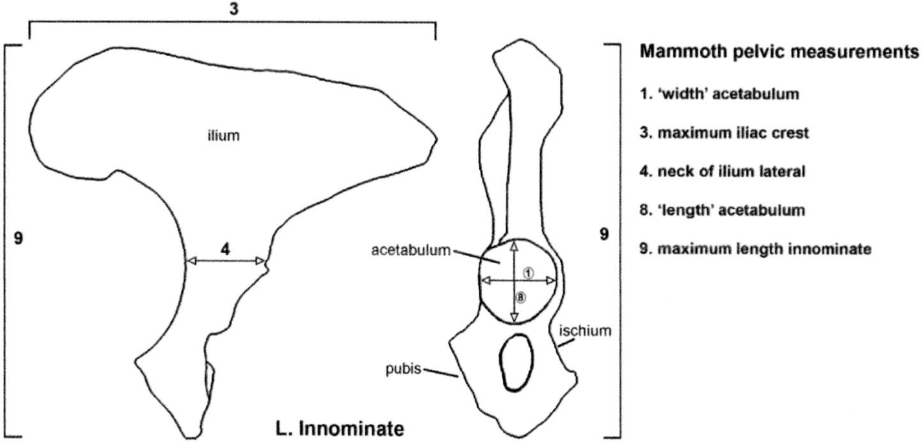


Figure 4.28 Measurements taken on mammoth pelvic bones (innominates)

| SITE | REF | L/R | 1. Minimum diameter diaphysis | 2. Antero-posterior width proximal epiphysis | 3. Length anterior diaphysis | 4. Max. width distal diaphysis | 5. Circum-ference diaphysis | 6. Medio-lateral width proximal epiphysis | 7. Total length femur | 8. Width distal epiphysis | Comment |
|---|---|---|---|---|---|---|---|---|---|---|---|
| 1 | 5 | R | 83 | | | | 235 | | >360 | | Diaphysis, ? juvenile |
| 1 | 163 | L | 97 | | | | 250 | | >510 | | Diaphysis |
| 1 | 308 | R | 135 | | | 210 | 345 | | >740 | 185 | Complete except for prox. epiphysis; dist. epiphysis fused |
| 4 | 9 | R | 105 | | | | 295 | 105 | >750 | | Diaphysis with unfused proximal end |
| 5 | 52 | R | 125 | | | | 340 | | >30 | | Diaphysis, destroyed by machine |
| 6 | 137 | L | | 134 | | | | 137 | | | Proximal epiphysis; unfused |
| 6 | 177 | R | | | | | e 345 | | >74 | | Diaphysis |
| 6 | 148 | L | 121 | >120 | 800 | 180 | 330 | >110 | 1000 | >190 | Complete but crushed/damaged |
| 6 | 122 | L | 126 | | | | 330 | | >770 | | Diaphysis with unfused prox. end |
| 6 | 179 | L | 128 | | | 195 | 345 | | | 175 | Diaphysis and fused broken distal |
| 6 | 8 | L | 97 | | | | 280 | | | | Diaphysis, crushed ? juvenile |
| 7 | 167 | R | 129 | 132 | 830 | | 356 | 136 | 1020 | 180 | Complete; epiphyses fused; proximal ? gnawed |
| 7 | 236 | L | 139 | | | | 380 | | >850 | | Diaphysis |
| 7 | 108 | R | 129 | 152 | 850 | 195 | 350 | 145 | 1020 | 183 | Complete; epiphyses fused |
| 8 | 71 | L | 112 | | | | 300 | | >470 | | Diaphysis |
| 8 | 67 | ? | | | | | | 141 | | | Proximal epiphysis; fused |
| 8 | 25 | L | 102 | 141 | 850 | e 200 | 325 | 141 | 1050 | 178 | Complete; proximal and distal fused |
| 8 | 5 | R | e130 | | | | | | >900 | | Complete except for prox. epiphysis; dist. epiphysis fused |
| 15 | 22 | R | | 132 | | | | e 155 | | | Proximal; damaged |

Table 4.14 Measurements (mm) of mammoth femora (see Figure 4.30)

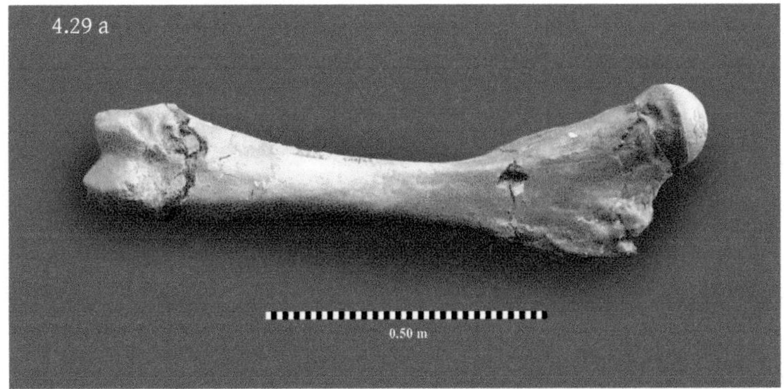


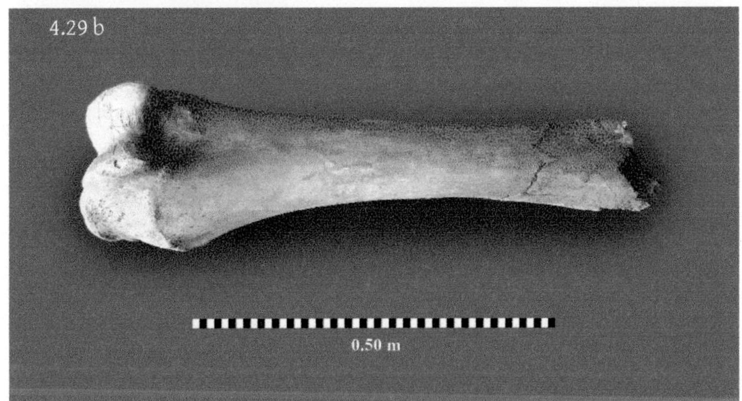


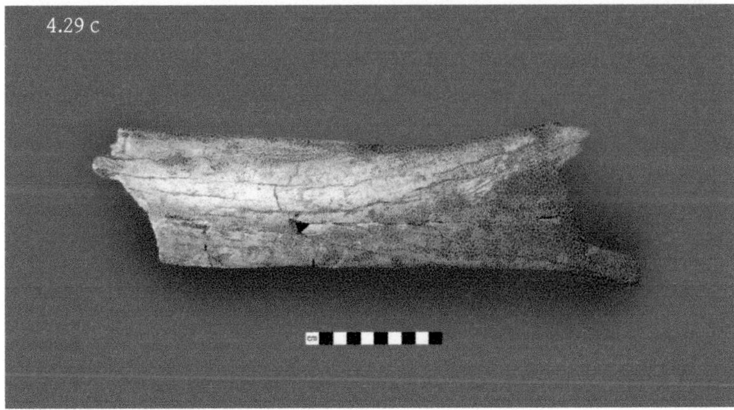


Figure 4.29 Mammoth femora from Stanton Harcourt. (a) complete femur (SH8/25) with both proximal and distal epiphyses fused. (b) femur SH1/308 with fused distal epiphysis. (c) diaphysis of young mammoth (SH1/5)

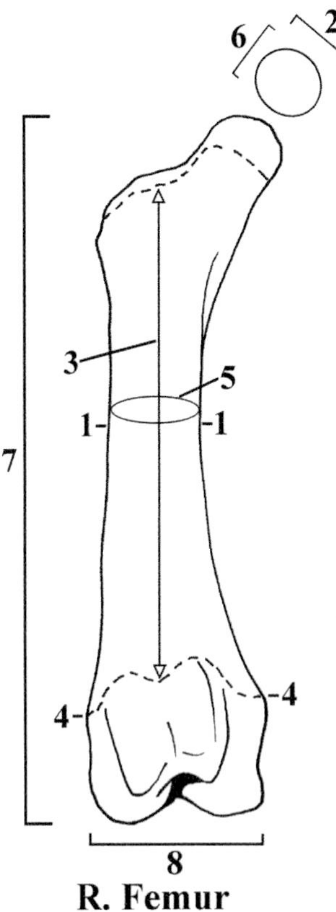


Figure 4.30 Measurements taken on mammoth femora

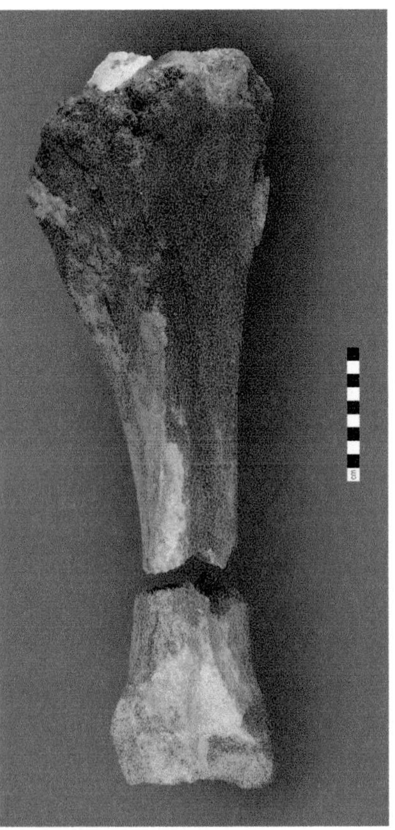

Figure 4.31 Tibia SH2/58 (modern break)

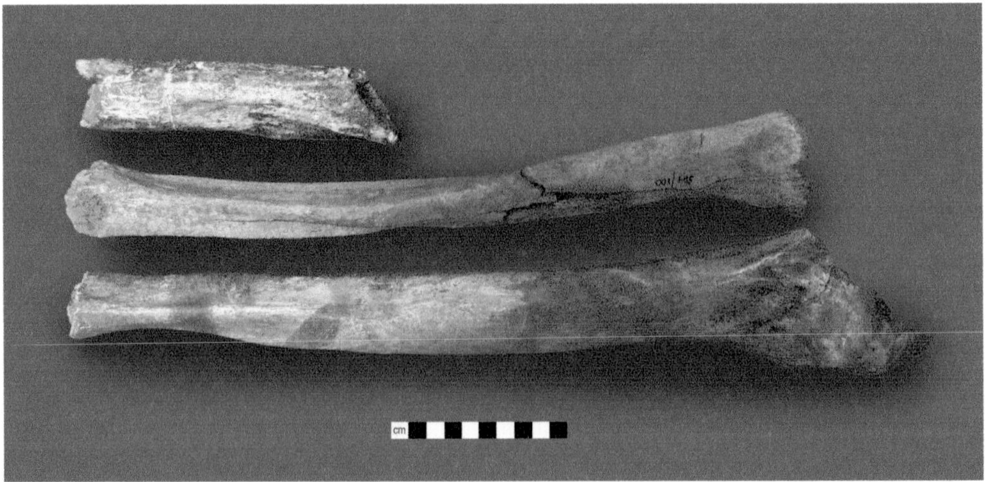

Figure 4.33 Almost complete mammoth fibulae SH7/100 (centre) and SH7/230 (below) with small section of elephant fibula SH1/15 above

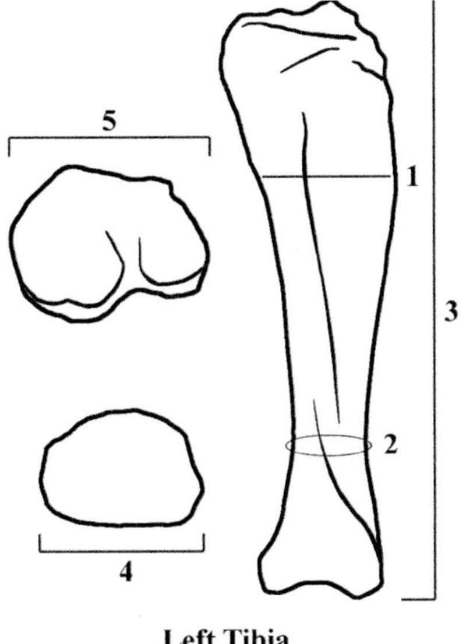


**Left Tibia**

1. Maximum antero-posterior width
2. Minimum circumference of distal diaphysis
3. Length
4. Width distal epiphysis
5. Width proximal epiphysis

Figure 4.32 Measurements taken on mammoth tibiae (Table 4.15)

| SITE | REF | L/R | 1. Maximum anterior-posterior width of diaphysis | 2. Circumference of narrowest point of distal tibia | 3. Maximum length including epiphyses | 4. Maximum width distal epiphysis | 5. Maximum width proximal epiphysis | 6.Medial condyle anterior-posterior | 7.Width of proximal diaphysis at point of fusion with epiphysis | 8.Width of distal diaphysis at point of fusion with epiphysis | Comment |
|---|---|---|---|---|---|---|---|---|---|---|---|
| 1 | 246 | L | | | >610 | | | | | | part diaphysis |
| 1 | 322 | L | | | >380 | | | | | | part diaphysis |
| 2 | 58 | L | | 260 | 530 | 130 | 180 | | 180 | 140 | complete, excellent condition |
| 4 | 85 | L | e170 | 288 | >>530 | | | | 156 | 180 | diaphysis; missing unfused epiphyses |
| 6 | 176 | L | | 260 | | | | | | | part diaphysis |

Table 4.15 Measurements (mm) of mammoth tibiae (see Figure 4.32)

## The sex of the Stanton Harcourt mammoths

From the point of view of understanding the taphonomy of the site it would be of interest to be able to determine the sex of the mammoths. Any interpretation of resulting sex ratio could be due to biological or taphonomic factors, bearing in mind that the assemblage does not represent a single event such as a flood or a mass culling but rather the deaths of many mammoths over time. At the least, if sex can be determined from the mammoth remains, it might be interesting to ascertain whether there is any bias in the preservation and representation of males *vs.* females.

There are various criteria by which males might be distinguished from females. Averianov (1996 maintains that the shape of the mandible, in particular the mandibular symphysis, is different in male and female mammoths. Three other criteria discussed by Lister and Agenbroad (1994) are the size and robusticity of the skeleton, dimensions of the tusks and their alveoli, and the morphology of the pelvis. These four criteria are discussed below with reference to the Stanton Harcourt material.

### *Mandible shape*

The mandible is a very robust bone and is frequently found in relatively complete condition in fossil assemblages which allows the age of death of the specimens with dentition to be estimated based on tooth development. The size and shape of the mandible changes during the life of an elephant and similar changes are seen in mammoths with adult males being generally significantly larger than females. However, examination of a series of mandibles of several elephantid species makes it clear that the considerable individual variation in form within a species makes it impossible to define a set of criteria that would allow certain differentiation between males and females (Lister 2017). Averianov (1996) states that the length of the symphysis is sexually dimorphic and that mandibles with the presence of a long and narrow symphysial process most likely belong to male individuals. However, other authors doubt the reliability of this trait. Álvarez-Lao and Méndez (2011) measured a large number of morphometric variables on a sample of 45 well-preserved mammoth mandibles from the North Sea and concluded that determining sexual dimorphism from the mandible is uncertain. Moreover, mandible shape is exceptionally variable with individual age and traits such as the symphysial process follow a quantitative, unimodal variation and are not diagnostic for sexing purposes and some very robust specimens (almost undoubtedly males) lacked a long or distinctive symphysis. Even if the shape of the symphysial process had been shown to be a reliable indicator of sex, as this process is especially vulnerable to damage (Figure 4.10), such a distinction would be obscured in most of Stanton Harcourt mandibles.

### *Size and robusticity of the skeleton*

Although there is strong sexual dimorphism in mammoths, the prolonged growth period and staggered epiphyseal fusion of their bones (summarised in Table 4.8) means that, with such a small number of complete limb bones, it is not possible to differentiate males from females at Stanton Harcourt on the basis of individual bones.

### *Tusks and alveoli*

As male mammoths have relatively more robust tusks than females one might hope to determine the relative proportion of males to females in a large sample. At Stanton Harcourt there are innumerable tusks with varying degrees of length, proximal diameter and curvature (Table 4.5; Figures 4.18 and 4.19). Several tusks are evidently those of juveniles, many larger ones are notably gracile and are probably those of females, and there are several very large strongly twisted tusks, at least as large as those of *M. primigenius*, which are very likely those of males. The range of ages and sizes represented thus precludes any clear separation of males from females (Figure 4.19).

### *Pelvic morphology*

Although the relatively high number of innominates suggest that it might be possible to distinguish males from females, the damage to most of the bones hindered certain identification in many cases. Measurements taken on all the Stanton Harcourt innominates are given in Table 4.12 and the more completely preserved are illustrated in Figure 4.27. Of the 58 innominates (or parts of these) listed in Table 4.12 only one is virtually complete: SH7/119.

Lister (1996) found that the most reliable index for sexual differentiation among mammoths was the ratio between the diameter of the pelvic aperture and the width of the ilium (Lister *op. cit.* Figure 25.1). Females give a higher ratio than males, the biological basis for this being i) the larger pelvic aperture for birthing in the female and ii) increased robusticity of the ilium for muscle attachment in the male. Although too few of the Stanton Harcourt pelves are sufficiently complete for this method of determination, Lister (pers. comm.) considers the ratio of the width of the base of the ilium against the diameter of the acetabulum also to be a good index of animal size. Based on measurements given in Table 4.12, Figure 4.34 shows clear bimodality and indicates 8 females and 5 males. Another male might be added to this group in the form of the large, most complete pelvis SH7/119 on which only one of the two relevant measurements could be taken because the bone is still encased in Plaster of Paris.

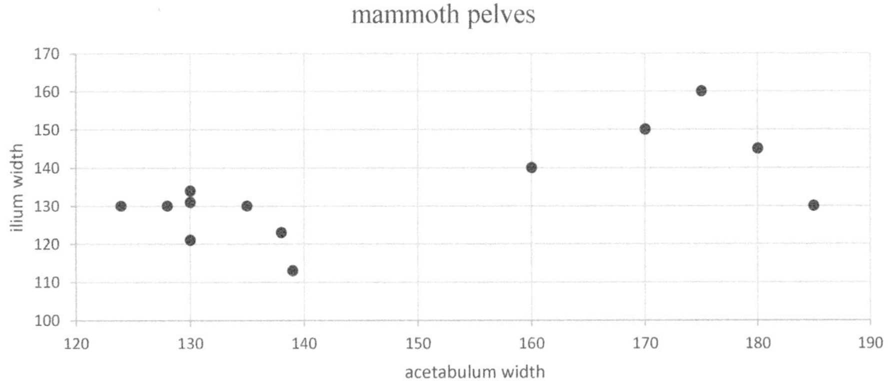


Figure 4.34 The ratio of the diameter of the acetabulum against the width of the base of the ilium is a good index of animal size and appears to show clear bimodality of the pelves of the Stanton Harcourt mammoths. (Measurements 1 and 4 in Figure 4.28, Table 4.12)

## Interpreting the mammoth remains: death, carcass dispersal and the effect of the river

Stone artefacts in the same horizons as some of the mammoth remains indicate the certain presence of hominins when mammoths were in the area. Although there are very few indisputable cut-marks on bones, occasional hunting or scavenging on the part of hominins is likely. There is one skull that could be interpreted as having been levered into the position in which it was discovered with the aid of a length of oak (see Chapter 8). In the following section we consider the possible causes of death of the mammoths and the taphonomic processes that might have had a role in transforming the complete carcasses of a large number of mammoths into this unbalanced assemblage of skeletal remains in the riverbed.

### *Causes of death among elephants*

An African elephant has a potential lifespan of around 70-75 years (Lee *et al.* 2012). Of the individuals that do not achieve this, the cause of death may be predation by lions and hyaenas, accidents, disease, killing by humans, or adverse climatic conditions.

Hanks's (1979) study of elephant mortality in African game reserves shows predation on adults by carnivores to be insignificant and then generally when the elephant is weakened by disease or drought. Predation on sub-adult elephants by lions or hyaenas is usually when young males have left the protection of the herd (at or near the age of sexual maturity). Very young elephants, however, are more vulnerable. Over a fifteen-year period in Etosha National Park, Lindeque (1988) attributed 32% of all deaths of elephants under 3 years old to predation, mainly by lions.

Accidents claim a high number of young elephants every year. Carr (1950) records young elephants in Zambia regularly being swept down rivers. Bere (1966) estimates that one out of three calves fail to survive their first year through drowning, getting stuck in mud, or falling down embankments. Apart from such mishaps, disease introduced by domestic animals, and the effect of human activities (hunting, poaching, culling), drought is one of the most important factors controlling elephant populations in Africa. For example

1. Hanks (1979) recorded 31 elephant carcasses within a 2km radius of one remaining pool during a drought in western Zambia in 1970. A very high percentage of young animals died.
2. During the 1968 drought in Wankie National Park, Zimbabwe, Williamson (1975) recorded 96 carcasses within a 1km radius of a water hole. The highest mortality was among the young.
3. In the Tsavo Park, Kenya, it was estimated that 5,900 elephants died in the drought of 1970/1971 (Corfield 1973). The majority were juveniles and females.

The high incidence of death among young animals and females during drought conditions is the result of social behaviour (Haynes 1991). Elephants are organised as a matriarchal group of cows and calves. The simplest group is a cow with one or more calves but two or three such groups may be associated. At puberty, males are driven out and join or form bachelor herds. At times of water scarcity, the elephants overgraze the area in the proximity of the water source and have to travel further and further each day to eat. The young die of thirst or exhaustion and their mothers won't abandon them, so they die too. Bands of males are less affected by harsh conditions.

Given the high incidence of mishaps among young elephants, it is likely that some young mammoths at Stanton Harcourt mammoths will have met with accidents or drowned in the river. While it is possible that some of the bones represent animals hunted by lions or wolves (both represented by fossil remains), or hominins (represented by artefacts), there is no significant evidence for the activities of any of these. A few bones seem to have been gnawed (possibly by lion or wolf) and a few have possible cut-marks but there is nothing to signify that these predators played a role in the deaths of the mammoths. As regards major climatic episodes causing large scale deaths, there is neither evidence for drought nor for major flood conditions at Stanton Harcourt. The river, however, is undoubtedly highly significant in two other respects: in the accumulation of the mammoth remains in the first instance, and in the subsequent differential preservation of the bones.

Proximity to water is essential to elephants. They consistently use the same pathways to the same water sources and, when injured or old, stay near water. Thus skeletons might begin to build up around a particular water source over several decades as the result of individual mortality due to old age, disease, or accident (Haynes 1988, 1991). When they die in the vicinity of a river, if the river should breach its banks periodically, parts of carcasses on the bank will become part of the fluvial deposit. Any bones that become embedded in the river sediments will have a better chance of preservation, with the likely differential preservation of heavier skeletal elements while smaller, lighter bones are carried downstream.

The Stanton Harcourt mammoths appear to represent just such a scenario: sporadic deaths of animals over time at or in the vicinity of the river, after which parts of carcasses became buried and preserved by fluvial sediments while others were transported away in the current. Several lines of supporting evidence are discussed here: the age profiles of the mammoths, the differential preservation of the bones, and the nature of the deposits in which they occur. Once again, these are reviewed against a background of observations on what happens at African elephant death sites.

### *The dispersal of elephant carcasses after death*

What survives into the fossil record is a matter of chance. Very few sites form in a few moments or are preserved at once by rapid burial. Far more often, an extended amount of time is involved in site formation and fossilization. Haynes's (1988) field studies around African water sources indicate that elephant bone sites are dynamically undergoing several different processes of site formation and modification such as recurring death events, or different degrees of scavenging and trampling over time. The sites change, and attributes of the bone assemblages change as old bones are trampled, weathered, and/or destroyed and new bones are added. The sediments containing the bones may be reworked or redeposited. Bones that become buried in the subsoil might preserve into the fossil record while others (from the same skeleton) become trampled or decay over time. Those that become buried, particularly in a silty anaerobic deposit such as at Stanton Harcourt, may preserve virtually intact. Additional bones from other animals may become incorporated relatively soon after or over a long period of time.

Some of the Stanton Harcourt mammoths may have drowned but the majority of deaths probably occurred (as with modern elephants) on land. After death, two destructive processes begin almost simultaneously: the carcass begins to decay and scavengers come to feed. The speed at which a carcass decays and becomes disarticulated will vary depending on such factors as carnivore activity, weather, and vegetation cover but under temperate or warm conditions Haynes (1988) notes a fairly consistent sequence of elephant skeletal disarticulation (Table 4.16).

As the various body parts come adrift, carnivores may access them. In sub-Saharan Africa, the spotted hyaena makes the greatest modifications to the carcass of an elephant, dispersing bones as far away as 150m from the death site (Crader 1983). Hyaena activity is not relevant at Stanton Harcourt but wolf and lion are present in the assemblage. The modern counterparts of both species break and sometimes consume bones while feeding and a few of the mammoth bones have possible tooth marks. When an elephant dies near a water source, in addition to the processes outlined above, some bones might become buried by elephants or other herbivores returning to drink. Others might be trampled; trampling of elephant bones by other elephants is widely recorded (Coe 1978; Douglas-Hamilton and Douglas-Hamilton 1975; Haynes 1991) and frequent returns can cause considerable damage to even the largest bones.

### *Surface condition: weathering and gnawing*

The Stanton Harcourt mammoth bones and teeth show evidence of a variety of post-depositional histories. Some are in near perfect condition; some are very fragmentary and there are many variants in between. Interestingly, although many post-cranial bones have suffered weathering and fracturing indicating that they lay exposed on the bank with ensuing damage, relatively few could be classified as 'rolled' suggesting that once bones entered the water and became embedded in the river sediments, they moved very little.

| | |
|---|---|
| **1st week** | The thick skin pads on the soles of the feet detach after several days even if carnivores do not feed on them. Foot bones usually detach before any other bones of the carcass |
| **1st month** | **Wet and dry decay (no carnivore interference):**<br>1. Invertebrates and microbial decay consume much of soft tissue; skin becomes hard and dries over bones<br>2. Scapula slides away<br>3. Tusks become loose in their sockets (although some remain rooted); worn teeth with shallow roots in alveoli may fall out of jaws<br>4. Foot bones, carpals and tarsals detach<br>5. Individual long bones may be moved up to 100m by hyaenas, lions or elephants |
| **2nd month** | **Dry/ bleaching stage:**<br>1. Dry skin may remain over articulated bones while invertebrates consume from the inside.<br>2. Skull and mandible detached by carnivores or other elephants<br>3. Long bones moved apart by carnivores or other elephants<br>4. Some but not all ribs scattered and trampled<br>5. Lower leg bones, phalanges, tarsals, carpals and tail bones scattered or removed by scavengers, elephants or other herbivores, or moving water |

Table 4.16 Sequence of disarticulation observed among African elephant carcasses (data from Haynes 1988).

It is difficult to standardise the recording of degrees of weathering on bone surfaces particularly as some bones appear to have been subjected to more than one post-depositional process: they have lain on the land surface, then become incorporated into the river and, in some instances, been moved by the current. Consequently, there were few instances where one could certainly distinguish tooth marks from other damage. Furthermore, bone surfaces often lost their original patina during the process of plastering or fibre glassing them. It is possible that there were few tooth marks to observe in any case as none of the three carnivores represented (lion, wolf and brown bear) is noted for damaging bones at kill/death sites in the extensive manner which hyaenas gnaw bones taken to dens.

### *Differential preservation of the mammoth remains in a fluvial environment*

Once in the river, three principal factors evidently determined the selective preservation of skeletal remains; the degree to which epiphyseal fusion has taken place, the shape and the weight of the element.

#### *The role of epiphyseal fusion in the preservation of skeletal elements*

One of the principal difficulties encountered when working with elephantid bones from Pleistocene sites, even where they are very numerous, is to obtain sufficient numbers of fully-grown post-cranial bones that can be measured. This is principally because of the delayed process of epiphyseal fusion unique to this group.

Most large mammals have a relatively short growing period during which time the deciduous dentition is replaced by the permanent dentition at or soon after sexual maturity. During this time the post-cranial bones are lengthening and thickening and their epiphyses fuse fairly synchronously as sexual maturity is reached. This is not the case with elephants. Dental replacement is a continuous process until the last molar (M3) comes into wear (at around 30 years of age in the African elephant). Epiphyseal fusion takes place sequentially well into middle age, the bones of females fusing at an earlier age than those of males. The sequence of fusion in the limb bones of the African elephant is shown in Table 4.8. With minor differences, Lister (1999) found a similar pattern of age and epiphyseal fusion among male Eurasian woolly mammoths.

The fact that the process of post-cranial fusion in elephantids extends well into maturity creates several difficulties from the point of view of the preservation and analysis of elephant/mammoth assemblages of palaeontological or archaeological interest. Unfused epiphyses detach relatively soon after death and the exposed end of the diaphysis, primarily composed of cancellous tissue, is vulnerable to weathering and other damage. This characteristic of delayed fusion in many bones significantly reduces the chances of unfused bones surviving in the fossil record and those diaphyses that do survive are frequently broken and difficult to measure. The younger the animal, the less robust is the bone. The degree of bone preservation at Stanton Harcourt appears to correspond to the sequence of epiphyseal fusion shown in Table 4.8. Each of the pair of pelvic bones is comprised of three parts – ilium, ischium and pubis. These three elements form the acetabulum and are the first bones of the postcranial skeleton to fuse which very likely contributes to the fact that the pelvis is the best represented skeletal element. Also well represented are the proximal scapulae, which do not have an epiphysis.

Limb bones with fused epiphyses are far less common than diaphyses without epiphyses and the latter are generally extensively damaged. The bones in the best state of preservation are those where epiphyses have completely fused to their shafts.

*The role of weight and shape in preservation of the mammoth remains*

The smaller the skeletal element, the greater its chance of being buried, destroyed or carried away from the death site as the carcass decomposes. Larger bones, particularly of the size of a mammoth, are more likely to remain at or fairly near to the death site. The fact that the bones of the Stanton Harcourt mammoths are in fluvial deposits suggests that they may have moved from the exact places where the animals died. However, the lack of evidence for major flood conditions suggests that these are the remains of animals that died reasonably close to the riverbank and became incorporated into the sediments by seasonal flooding or channel bank erosion. In a moderately flowing river (such as is suggested for Stanton Harcourt), smaller bones are liable to have been transported downstream. Consequently, the assemblage is dominated by the larger, heavier elements - skulls, lower jaws, shoulder blades and pelvic bones.

Weight is obviously a key factor in retarding movement downstream but so is shape. It is highly probable that the very high representation of tusks is due to their spiral shape, causing them to 'corkscrew' and become embedded in the sediment (Figure 4.12). Similarly, the relatively high number of pelvic bones is likely to be due to their large curved surfaces and protuberances. The pelvis (innominate) in proboscideans is sufficiently robust to ensure that it is generally well represented at fossil sites. Of the 57 pelvic bones at Stanton Harcourt, 21 are reasonably complete: 15 right and 6 left.

With the exception of the virtually complete right pelvis SH7/119, Figure 4.27 reveals a fairly consistent pattern of preservation at Stanton Harcourt. The acetabulum (the socket for the head of the femur) together with the shaft of the ilium is the best-preserved part of the pelvis. The pubic bone, being narrow and slender has generally broken away. The wing of the ilium, although fairly substantial, becomes thinner at its widest part and its relatively thin cortex encloses porous bone tissue. These characteristics combined result in significant damage to the ilium in almost all specimens. Furthermore, the epiphysis of the ilium is among the last bones to fuse (36+ years) and detaches from the wing of the ilium during the decay of the carcass. The loss of this strong, protective edge right across the wing of the ilium accelerates its disintegration. Only one specimen at Stanton Harcourt has the epiphysis of the ilium preserved intact (SH7/119). This pelvis is part of a complex of bones believed to represent a single mammoth at part of the site where relatively little post-depositional disturbance has resulted in exceptional preservation of bones with some still in articulation (Group E in Chapter 2).

The diaphysis of the femur is a very compact bone and, even though epiphyseal fusion is relatively late, the length, density and weight of this limb probably ensured that the Stanton Harcourt femora became submerged and protected by the river sediments even if the proximal and distal epiphyses had detached.

The scapula does not have a proximal (articular) epiphysis and, apart from the distal margin, is a large blade which, if embedded in sediment seems to preserve well. By comparison with the pelvic bones, the scapulae are few (five); these are reasonably complete except for damage to the distal margin (Figure 4.20). Agenbroad (2004) suggests that, although the blade of the scapula is a comparatively robust bone, its flat disc-like shape would enable it to be more easily carried away in water. Those that survived at Stanton Harcourt were found spine facing downwards. They possibly sank into the river soon after deposition and the spine acted like a keel, preventing further movement.

The humerus is a less robust bone than the femur. Although the distal epiphysis is one of the first bones to fuse with the diaphysis, which should give this element a better chance of survival, the proximal epiphysis is comprised of porous cancellous bone with a thin cortex and is particularly vulnerable. Apart from one damaged complete specimen with both fused epiphyses, humeri at Stanton Harcourt are represented almost exclusively by diaphyses.

The disproportionate representation of the lower limb bones is interpreted as a factor of relative weight and, to some extent, shape. The ulna is far better represented than the tibia, radius or fibula. Even so, only one of the 16 ulnae is complete and the majority of the rest are represented by the proximal end. The fact that the distal end is almost always missing is very likely due to the distal epiphysis being the last bone to fuse in both males and females. It seems that the greater weight and morphological contours of the proximal ulna enhanced its chances of not being carried downstream with the other lower limbs, vertebrae and bones of the feet. All but one of the six radii are incomplete, being either fragmentary or diaphyses lacking (unfused) epiphyses. Only one of the five tibiae is complete. The fibula is a relatively slender bone compared with the tibia and is represented by only two specimens, both lacking (unfused) epiphyses.

Ribs are commonly represented as might be expected because there are 40 ribs per skeleton but also perhaps because their curvature enabled them to become embedded in the river sediments. However, vertebrae and bones of the feet - carpals, tarsals, metapodials and phalanges - are rare. These are all relatively spongy bones, small and/or round, and probably easily carried away in the river currents.

## Population structure of the Stanton Harcourt mammoth assemblage

This section considers the ontogenetic ages of the mammoths and how the age range of the individuals represented might compare with the live herds from which they derived. Reference is made to studies of African elephant herd structure, mortality, and post-depositional factors affecting bones at death sites. The similarity between mammoths and extant elephants in terms of the structure and density of their bones, and the growth patterns of molars, tusks and post-cranial bones is well documented. It is widely assumed that their social structure and behaviour would have been similar too (Haynes 1991; Sukumar 2003; Lister and Bahn 2007). While comparisons with the modern studies are informative, they are not used without noting the caveat of Klein and Cruz-Uribe (1984) that the setting in which the studies are carried out might differ significantly from that in which the fossil assemblage accumulated and that they are usually relatively brief (perhaps weeks or months) compared with the unknown period represented by a fossil assemblage such as Stanton Harcourt.

### *Determining ontogenetic age in mammoths*

An estimate of the age at death of the mammoths might be made from their molars or their post-cranial bones. As discussed with reference to Table 4.8, age determination from the skeleton is based upon the fact that growth in elephants continues well into adulthood and epiphyses of long bones fuse with their respective diaphyses in sequence (Roth 1984; Haynes 1991; Lister 1999). Epiphyseal fusion in males completes at a more advanced age than in females and, since sex determination of postcrania is rarely possible in fossil assemblages (perhaps only when the bones are associated with the pelvis or cranium), age categories based on postcrania are necessarily very broad.

Far greater detail of age determination is possible from the molars. The growth and replacement of mammoth dentition takes place in the same way as in living elephants. There are 24 molariform teeth (six in each jaw) each of which is comprised of a number of parallel plates (lamellae) of dentine, covered with enamel and joined by cement. These molars develop in sequence with the first forming towards the anterior of the jaw and other developing progressively behind. Of these six teeth, three are already formed at birth, the first two of which erupt and come into wear in the first few weeks of life. Only one or two molars in each jaw are in use at any one time. As the anterior tooth moves forward, the occlusal surface is worn down and resorbed at the roots to be replaced by the successive tooth. Each tooth is larger than its predecessor and has a greater number of lamellae. The first three molariform teeth are referred to by various authors as milk molars (mm2-mm4), deciduous molars (dm2-dm4), deciduous premolars (dp2-dp4) or molars (M1-M3). In this study the first three molars are designated dP2-dP4 and the last three molars are M1-M3 (upper dentition) and dp2-m3 (lower dentition) in accordance with Lister and Sher (2015).

The process of replacement described above enabled Laws (1966) to divide a sample of almost 400 lower jaws of African elephants into 30 age classes according to the state of eruption and wear of the molars (Table 4.17). A potential lifespan of 60 years was based on data collected for African and Asian elephants (*Loxodonta africana* and *Elephas maximus*) and chronological ages assigned to the 30 groups. This scheme provided a useful basis on which many researchers have assigned approximate age at death to elephants from a variety of contexts e.g. culled, poached, archaeological and fossil. Although Laws referred to the mean age of the elephants in each category it has since become common practice to use the term 'AEY'. Although this term is widely deciphered as 'African elephant years', Haynes *et al.* (2018) point out that the initials AEY stand for 'African-equivalent years', an analytic tool introduced by vertebrate palaeontologist J. Saunders (1977), who intended the term to denote 'mean ages' assignable to *M. columbi* in cases where the stages of dental wear and progression were similar to those identified in *Loxodonta africana*. Here we use AEY as a convenience for assigning ontogenetic ages to the mammoths based on placement in the Laws system that correlates *L. africana* ontogenetic ages and dental wear.

Various authors have since corrected or refined Laws' scheme partly because long-term studies have determined that some elephants live beyond 70 years and because some variation is noted in the age at which teeth erupt, wear, and are replaced (Jachmann 1988; Lister 2009; Lee *et al.* 2012; Stansfield 2015; Haynes 2017). However, when Laws' scheme is compared, for example, with those of Lee *et al.* and Stansfield (*op. cit.*)., it is evident that the differences

| LAWS' AGE CLASS | STATE OF ERUPTION AND WEAR OF TEETH | Laws (1966) Mean age (AEY) | Lee *et al.* (2012) Age allocated (years) | Stansfield (2015) Age allocated (years) |
|---|---|---|---|---|
| I | dp2 protruding; no wear | 0 | < 3 months | 0-0.24 |
| II | dp2 and dp3 erupted; slight wear | 0.5 | 0.25 | 0.36-0.8 |
| III | dp2 well worn; dp3 moderate wear; dp4 slight wear | 1 | 1.25 | 1-1.6 |
| IV | dp2 lost; dp3 well worn; dp4 moderate wear | 2 | 2.3 | 1.8-2 |
| V | dp3 well worn; dp4 moderate wear; M1 in crypt | 3 | 4 | 2.5-3.5 |
| VI | dp3 almost disappeared; dp4 well worn; M1 unworn | 4 | 4-5 | 4-5 |
| VII | dp3 lost; dp4 well worn; M1 erupted but unworn | 6 | 5-7.8 | 4.5-6 |
| VIII | dp4 anterior edge eroding; M1 in wear | 8 | 5-9.7 | 6-7.5 |
| IX | dp4 almost completely eroded; M1 early wear | 10 | 7.8-9.7 | 7.5-11.5 |
| X | dp4 almost disappeared; almost all M1 in wear | 13 | 8.5-9.7 | 10-13 |
| XI | dp4 usually lost; all lamellae of M1 in wear | 15 | 9.7-16 | 13-15 |
| XII | M1 erosion of anterior edge; M2 just in wear | 18 | 14-16 | 15-17.5 |
| XIII | M1 progressive erosion; M2 in wear | 20 | 16-20 | 16-19 |
| XIV | M1 erosion from both ends | 22 | 16-24 | 19-22 |
| XV | M1 eroding further; M2 enamel loops complete | 24 | 21.5-31.5 | 20.5-25 |
| XVI | M1 almost disappeared; M2 all lamellae in wear | 26 | 20-27 | 22-26.5 |
| XVII | M1 socket; M2 moderate wear; M3 visible | 28 | 21.5-31.5 | 26.5-30 |
| XVIII | M2 slight erosion of anterior edge; M3 just in wear | 30 | 27-31.5 | 25-32 |
| XIX | M2 eroding; M3 early wear (2-3 lamellae) | 32 | 31.5-35 | 29-34 |
| XX | M2 eroding; M3 continuing to wear | 34 | 31-35 | 32-36 |
| XXI | M2 almost gone; M3 half worn | 36 | 30-37 | 34-40 |
| XXII | M2 socket only | 39 | 35-37 | 40-42 |
| XXIII | M3 erosion of anterior edge begins | 43 | 41-48 | 45 |
| XXIV | M2 vestige of socket; M3 all except last few lamellae in wear | 45 | 46.5-48.5 | 45-52 |
| XXV | M3 slight erosion of anterior edge | 47 | 41-47 | 45-56 |
| XXVI | M3 further erosion of anterior edge | 49 | 46.5-48.5 | 48-60 |
| XXVII | M3 anterior one-third eroded away | 53 | 46.5-67.5 | 52-66 |
| XXVIII | M3 few enamel loops remain | 55 | 47-68 | 60-68 |
| XXIX | M3 all but one or two enamel loops confluent | 57 | 62-68 | 68-70 |
| XXX | M3 less than 15cm remain rooted | 60 | 62-73 | 68-70+ |

Table 4.17 Tooth wear related to age in the African elephant *Loxodonta africana.* Table compiled from Laws (1966), Stansfield (2015) and Lee *et al.* (2012).

| UPPER DENTITION | | |
|---|---|---|
| | Isolated molars, complete or more or less complete | 75 |
| | Partial molars | 22 |
| | Skulls or maxillae with dentition | 7 |
| | | |
| LOWER DENTITION | | |
| | Isolated molars, complete or more or less complete | 27 |
| | Partial molars | 16 |
| | Mandibles with dentition, 12 of which had both left and right molars | 23 |

Table 4.18 The representation of upper and lower mammoth teeth at Stanton Harcourt

between Laws' age assignments and the revised versions pertain principally to the last two age categories (Table 4.17). The purpose here of using a scheme based on tooth eruption and wear is to present a broad picture of the age range of the individuals represented at Stanton Harcourt and, as discussed below, the variation recorded for older elephant categories does not affect the allocation to age categories or perceived population structure of the Stanton Harcourt assemblage. Furthermore, as Laws' data has been used previously to assign ages to the mammoths from La Cotte de St Brelade (Scott 1986 and *in press*), comparability is maintained.

Table 4.18 summarises the distribution of upper and lower mammoth dentition from the site. Details of these teeth are presented in Tables 4.19 and 4.20. Where possible the teeth are allocated to Laws' age classes and an estimated age (AEY) is given. These data were then used to construct an age profile for the assemblage.

### *Elephant and mammoth populations and age profiles*

Palaeobiologists and archaeologists have long recognised the interpretive possibilities of constructing fossil age profiles - graphs showing the frequencies of different age classes of a fossil sample. Most frequently, the age profiles are interpreted with reference to two theoretically expectable models: 'catastrophic' and 'attritional' (Klein and Cruz-Uribe 1984). A catastrophic profile is one in which successively older age classes contain progressively fewer individuals. This is typically the age profile of a living mammal population, one which is stable in size and structure. This kind of age profile, diagnostic of nonselective mortality, would become fixed in the fossil record if the entire population had been wiped out by a catastrophic event such as a flood or an epidemic. In an attritional profile, prime age adults are underrepresented compared to their presence in healthy populations and the very young and old are overrepresented – these sectors being more prone to accidents, disease, starvation, and other routine mortality factors.

Given their comparability in demography and growth characteristics, Haynes (1987, 2017) suggests that cautious generalisations about modern elephants may be applied to extinct taxa and uses modern elephant population data and recent death assemblages as a tool for interpreting fossil proboscidean age profiles. His results are discussed here and compared with those from Stanton Harcourt.

Death events or death processes are age-selective among modern African elephants. Stable proportions of different age classes are maintained by the dynamic interplay of the forces that allow population growth and the forces that keep the numbers down – principally resources and degree of predation. In stable or growing elephant populations, there will be a large percentage of young animals and fewer middle-range sub-adults. Old and declining populations will have mostly adults. During periods of environmental stress (drought, habitat compression, floods) sub-adult elephants suffer a higher mortality rate than any other age class because adults have built up a greater resilience to such conditions (Conybeare and Haynes 1984; Lindeque 1991). Furthermore, proportions of sex- and age-specific selectivity change through the course of die-offs caused by environmental factors. In the early stages, weaned but sub-adult elephants are most vulnerable (2-8 years old). If adverse conditions continue the increasingly vulnerable age classes are younger, unweaned animals followed by older female adults (over 40 years). As conditions continue to deteriorate, adult males begin to succumb.

Assuming the potential longevity of an elephant to be about 60 years, the Stanton Harcourt mammoth dentition is divided into five equal age categories considered meaningful in terms of an elephant's life history according to Haynes (1987 and pers. comm.).

1. Animals of 0-12 years are growing vigorously, are not sexually mature and, although weaned by about 2 years of age, remain in the mixed herd.
2. Animals between 12 and 24 years of age continue to grow but less rapidly. Epiphyses begin to fuse in some bones. Young males leave the mixed herd and form bachelor groups.
3. Animals aged 24-36 years are in their last stage of growth since all diaphyses have fused by 32 in females and 36 in males (Table 4.15).
4. After the age of 36, growth has effectively ceased: this fourth age group is made up of prime age individuals.
5. This group includes older animals from 48 years of age to 60+.

The Stanton Harcourt mammoth dentition is divided into these five categories (Figure 4.35) based on tooth wear related to age as in Tables 4.19 and 4.20). There is evidently no significant difference between the representations of upper *vs.* lower dentition and mature animals in both groups are better represented than the young. Discussion on the possible significance of these age profiles is based on the combined upper and lower molars.

As a model for the possible interpretation of fossil proboscideans, Haynes compares the age class frequencies in two samples of African elephants where the live population structure is recorded or the cause of death is known (Figure 4.36):

a. A simulated stable elephant population. Based on observations, this is a situation in which the herd is neither increasing nor decreasing in size and different age classes remain fairly consistent due to a balance between resources and predators. The young are best represented (die more frequently) due to factors discussed above; sub-adults and young mature individuals have a lower death rate; there is a decrease in deaths among older individuals. Figure 4.36 shows a simulated population based on a multi-year average of more than 8,000 elephants culled from a number of matriarchal

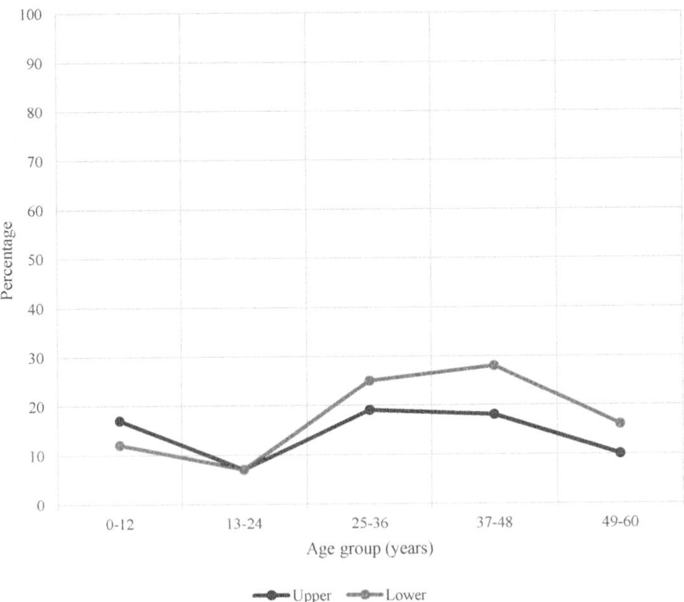


Figure 4.35 Stanton Harcourt mammoth dentition assigned to age groups based on tooth wear related to age as described in the text. For the upper molars n=71; for the lower molars n=88

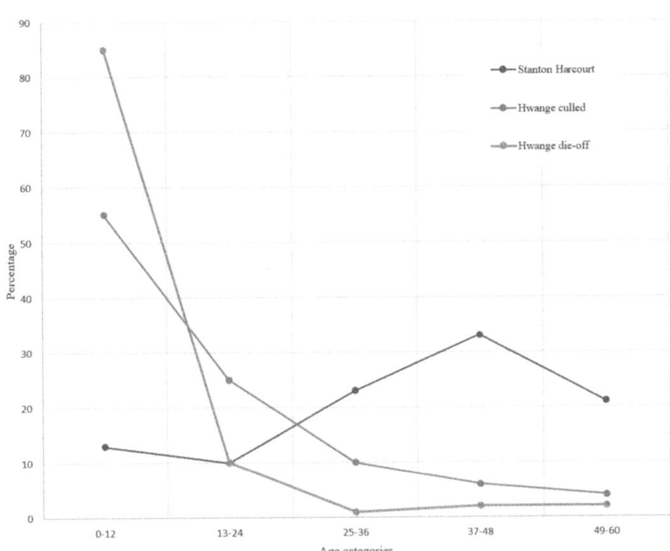


Figure 4.36 Mortality profile of the Stanton Harcourt mammoths compared with mortality profiles for African elephants. The simulated Hwange population is the combined data of the multi-year culled family groups to which is added the estimated number of males not culled. The die-off profile is constructed from data on deaths of herds affected by drought over multiple years (Haynes 1987 and unpublished data pers. comm.)

herds at Hwange, Zimbabwe. These were mixed herds (i.e. adult male groups were not included) and all age groups were indiscriminately killed.To create the observed elephant populations at the time of culling, the estimated number of males that were not culled is added to the data in Figure 4.36 (Haynes pers. comm.).

b. Drought affected herd. The deaths of 242 elephants from 4 remote locations in Hwange, Zimbabwe, were recorded following a short-lived but severe drought. In this situation, the young were least able to travel the distances between grazing and diminishing water resources and very high mortality resulted. To a lesser extent, this applied to the very old and their death rate can be seen to have increase slightly by comparison with populations not under duress.

When the Stanton Harcourt data are included, the most obvious feature is the low percentage of young animals (0-12 years) resulting in the assemblage being dominated by mature and old individuals. It is interesting then to compare the Stanton Harcourt mammoth age profile and the simulated 'stable' African elephant age profiles with profiles for two Pleistocene mammoth assemblages from North America and Poland (Figure 4.37).

The American mammoth profile is the amalgamation of material from Late Pleistocene silt ('muck') deposits from around Fairbanks, Alaska (Haynes 1987; Dixon 1999). These are believed to have accumulated in aeolian deposits over several millennia. Although artefacts are also recorded from some localities, the mammoth and other animal remains are regarded as of natural rather than archaeological origin. Of more than 40 individuals represented in Figure 4.37, 42% fall into the first age category, mirroring the high mortality expected in a stable elephant population. The assemblage from Kraków Spadzista Street B (KSB), on the other hand, is from an intensively occupied Upper Palaeolithic site with numerous stone tools (Lipecki and Wojtal 1995). Here, as with the Alaskan assemblage, most of the mammoths are young. KSB is of particular interest in that not only are 80% of a minimum of 118 mammoths in the youngest two age categories, but there are high frequencies of the smallest bones such as carpals, tarsals and sesamoids. While such a profile and body-part representation might correspond to a single catastrophic event, the archaeological context and the relatively small area ($150^2$ m) and the carbon dates led Lipecki and Wojtal (*op. cit.*) to conclude that the mammoths probably died of natural causes at or very near the excavated location and were accumulated over time by the hunters. More recent analysis casts a slightly different light on this mammoth accumulation (Haynes, Klimowitz and Wojtal 2018). The period of accumulation is suggested to be decades rather than centuries during which short term rapid climatic oscillations are evident at the site. Isotopic studies indicate that the mammoths (some of which can be shown to have come from the same subpopulation) died during a cold period. Haynes *et al.* speculate that the cold might have caused groups of mammoths periodically to shelter on the hillside near the archaeological site. It might also have depressed their metabolism, perhaps for hours, a form of hibernation known to be an effective coping mechanism among large mammal, and some may thus have died.

By comparison with the age representation common to evidently stable proboscidean populations (both modern and fossil), young mammoths are notably under-represented at Stanton Harcourt. There is every reason to suppose that mammoth populations resembled those of elephants and thus, by analogy with observations on African elephants, most deaths of mammoths probably occurred on land (rather than by drowning) and the young would

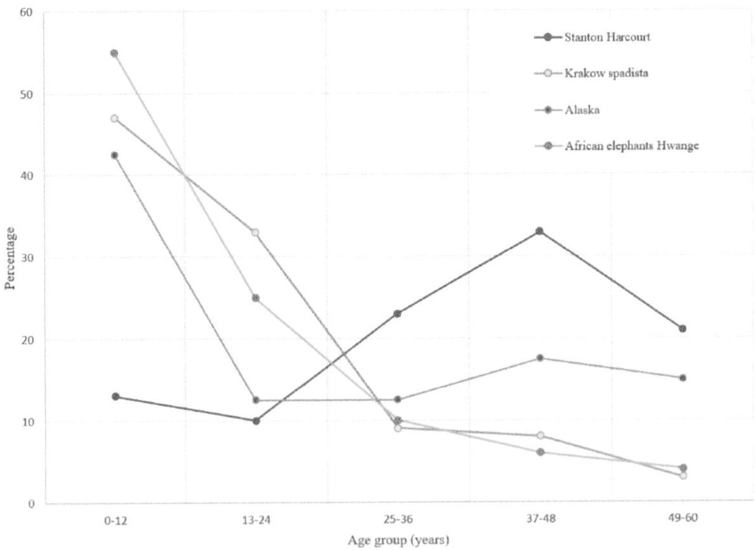


Figure 4.37 Stanton Harcourt mammoth age profile compared with data for a stable population of African elephants and Pleistocene mammoth assemblages from Poland and Alaska

have suffered higher rates of mortality than the old. One (or a combination) of factors might therefore account for the relatively low numbers of young individuals at Stanton Harcourt. Firstly, soon after death, the smaller, less robust remains of younger animals might have been more quickly destroyed by carnivores and weathering than those of older individuals and thus had a lesser chance to survive into the fossil record. Secondly, once the remains of an animal that died marginal to the river become incorporated into the fluvial deposits, smaller body parts were more likely to have been transported away from the site of death. Although the good preservation of many faunal and vegetational remains from Stanton Harcourt suggests minimal disturbance after deposition in some areas, fluvial activity almost certainly accounts for the relative scarcity of the smaller bones of all the mammalian species represented. This would include the molars of younger mammoths and thus weight the age profiles in favour of older animals.

There is an interesting third explanation that might partially account for the age profile of the Stanton Harcourt mammoths. Sukumar (2003) noted that among living elephants, a population declining from climatic and associated vegetational change could be expected to show decreased growth in body size, increased age of sexual maturity, and decreased fecundity. He suggests that this would translate into an age structure, preserved in the fossil record, with a relative predominance of older individuals. Although the biological data from Stanton Harcourt are indicative of interglacial conditions, there are several indicators for cooling climatic conditions in the later phases of deposition of the fossil-bearing sediments (see Chapter 6). Other MIS 7 sites with the small form of *M. trogontherii* also evidence change in vegetation and/or cooler climate indicators that herald the onset of the MIS 6 cold stage. At Latton, for example, where the small mammoth is also the most common vertebrate, there are no deciduous woodland species among the fauna or flora, and the temperate freshwater mollusc *Corbicula fluminalis* is absent (Scott and Buckingham 2001; Lewis *et al.* 2006). This might suggest reduced plant productivity on which the mammoths depended. The small size of the Stanton Harcourt mammoths relative to steppe mammoths of earlier stages has already been discussed but it is interesting to note that they were even smaller than those from other MIS 7 sites such as Ilford (Lister & Scott in prep; Scott & Lister in prep). The factors that brought about the progressive size reduction in the Late Middle Pleistocene *M. trogontherii* are probably complex, but dwarfing is often seen as a response to diminishing resources (e.g. Guthrie, 1984; Haynes, 1991; Roth, 2001). It is possible, therefore, that the extreme size reduction at Stanton Harcourt is a signal of population under stress, and therefore with reduced fecundity.

In sum, the destruction or removal of younger mammoths by means of one or more of the post-depositional processes discussed earlier in this section probably largely accounts for the profile shown in Figure 4.37. Perhaps, however, the age structure is an indication that the number of young individuals was actually relatively low because the species was not coping with the climatic and vegetational changes that signified the approaching cold stage.

| | SITE | REF | SQ | SQ | L/R | PART | TOOTH | LAWS' CATEGORIES | EST. AGE (AEY) |
|---|---|---|---|---|---|---|---|---|---|
| **Isolated upper dentition** | | | | | | | | | |
| | 1 | 11 | C4 | C5 | R | CO | M3 | XIX | 32 |
| | 1 | 12 | D6 | D6 | L | CO | dp3 | V | 3 |
| | 1 | 13 | D6 | D6 | L | CO | dp4 | V | 3 |
| | 1 | 55 | E6 | E6 | R | CO | M2 | XV | 24 |
| | 1 | 87 | F10 | F10 | R | CO | dp4 | VII | 6 |
| | 1 | 91 | G9 | G10 | L | CO | M1 | XII | 18 |
| | 1 | 108 | F5 | F5 | R | CO | M3 | XXVI | 49 |
| | 1 | 109 | F4 | F4 | L | CO | M3 | XXII | 39 |
| | 1 | 119 | H10 | H10 | R | CO | M3 | XIX.5 | 32 |
| | 1 | 151 | D12 | D12 | L | CO | M3 | XXVIII | 55 |
| | 1 | 158 | H12 | H12 | L | CO | M1/M2 | IX/XV | 10 or 24 |
| | 1 | 171 | C13 | D13 | R | CO | M3 | XIX | 57 |
| | 1 | 189 | E13 | E13 | L | CO | M3 | XX | 34 |
| | 1 | 196 | E14 | F14 | L | CO | M3 | XXVIII | 55 |
| | 1 | 222 | F13 | F13 | R | CO | M3 | XXVI | 49 |
| | 1 | 223 | C15 | C15 | L | CO | M3 | XX | 34 |
| | 1 | 229 | AA15 | BB15 | R | CO | M3 | XXII | 39 |
| | 1 | 230 | AA15 | AA15 | L | CO | M3 | XXII | 39 |
| | 1 | 292 | J14 | J14 | R | CO | M3 | XXI | 36 |
| | 1 | 310 | GG3 | GG3 | L | CO | M3 | XXVI | 49 |
| | 1 | 403 | - | - | L | CO | M1/M2 | XI/XVII | 15 or 28 |
| | 2 | 33 | F22 | F22 | R | CO | M2 | XVII | 28 |
| | 3 | 16 | K20 | K21 | R | CO | M3 | XXVI | 49 |
| | 4 | 4 | H24 | H25 | L | CO | M2 | XVII | 28 |
| | 4 | 26 | L35/L36 | M35/36 | R | CO | M3 | XXIV | 45 |

| | SITE | REF | SQ | SQ | L/R | PART | TOOTH | LAWS' CATEGORIES | EST. AGE (AEY) |
|---|---|---|---|---|---|---|---|---|---|
| | 4 | 67 | C43 | C43 | R | CO | M3 | check | 49 |
| | 4 | 72 | C42 | C42 | R | CO | M3 | XXIX | 57 |
| | 4 | 89 | P37 | P37 | L | CO | M3 | XXII | 39 |
| | 4 | 94 | I50 | I50 | L | CO | M3 | XVIII | 30 |
| | 4 | 106 | D43 | D44 | L | CO | M3 | XXX | 60 |
| | 5 | 13 | - | - | L | CO | M3 | XXIV | 45 |
| | 5 | 21 | A29 | A29 | L | CO | M2/M3 | | |
| | 5 | 58 | O20 | O20 | L | CO | ?M2 | | |
| | 5 | 59 | X33 | X33 | L | CO | M2 | XIX | 32 |
| | 5 | 62 | A49 | A49 | L | CO | M3 | XXVI | 49 |
| | 5 | 70 | D40 | D40 | R | CO | M2 | XVII | 28 |
| | 5 | 71 | B38 | B38 | R | CO | M3 | XXIV | 45 |
| | 6 | 21 | B27 | B27 | L | CO | M2 | XVII | 28 |
| | 6 | 32 | B26 | B26 | L | CO | dp4/M1 | V/IX | 3 or 10 |
| | 6 | 66 | R31 | R31 | L | CO | M3 | XVII | 28 |
| | 6 | 90 | D29 | D29 | L | CO | M3 | XXII | 39 |
| | 6 | 91 | D28 | D28 | R | CO | dp4/M1 | V/IX | 3 or 10 |
| | 6 | 100 | Q33 | R33 | L | CO | M3 | XXIX | 57 |
| | 6 | 126 | P36 | P36 | L | CO | M3 | XIX | 32 |
| | 6 | 131 | K32 | K32 | L | CO | M3 | XIX | 32 |
| | 6 | 138 | F31 | F31 | L | CO | M2/M3 | | |
| | 6 | 142 | U10 | U10 | R | CO | M1/M2 | XVIII/XIV | 30 or 22 |
| | 6 | 154 | G33 | G33 | L | CO | M3 | XXIII | 43 |
| | 6 | 155 | R25 | R25 | L | CO | M3 | XXI | 36 |
| | 6 | 211 | G31 | G31 | R | CO | M3 | XXIX | 57 |
| | 6 | 262 | E32 | E32 | L | CO | M2 | XVI | 26 |
| | 6 | 267 | G16 | G16 | L | CO | M3 | XXII | 39 |
| | 6 | 290 | A11 | A11 | R | CO | M3 | XXIII | 43 |
| | 6 | 302 | A10 | A10 | L | CO | M3 | XXIII | 43 |
| | 7 | 27 | J17 | J17 | L | CO | M3 | XIX | 32 |
| | 7 | 45 | F3 | F3 | L | CO | M3 | XXII | 39 |
| | 7 | 103 | P15 | P15 | L | CO | M3 | XXV | 47 |
| | 7 | 122 | I50 | !50 | R | CO | M3 | XXIII | 43 |
| | 7 | 177 | J17 | J18 | L | CO | M3 | XXIII | 43 |
| | 7 | 220 | C2 | C2 | L | CO | M3 | XXV | 47 |
| | 7 | 238 | V21 | V21 | R | CO | M3 | XXVII | 53 |
| | 7 | 240 | I13 | I13 | L | CO | M1/M2 | XII/XVIII | 18 or 30 |
| | 7 | 249 | Q9 | Q9 | L | CO | M3 | XXIII | 43 |
| | 7 | 253 | S17/S18 | R18/S18 | L | CO | M2 | XV | 24 |
| | 7 | 284 | S19 | S19 | R | CO | M3 | XXVII | 53 |
| | 7 | 286 | L21 | M22 | L | CO | M3 | XXVIII | 55 |
| | 8 | 22 | Y11 | Y11 | R | CO | M3 | XXIII | 43 |
| | 14 | 43 | D7 | D7 | R | CO | M3 | XXIII | 43 |
| | 14 | 44 | E7 | E7 | L | CO | M1 | IX | 10 |
| | 14 | 45 | E14 | E14 | R | CO | M1 | XII | 18 |
| | 14 | 47 | A3 | A3 | R | CO | M3 | XXII | 39 |

| | SITE | REF | SQ | SQ | L/R | PART | TOOTH | LAWS' CATEGORIES | EST. AGE (AEY) |
|---|---|---|---|---|---|---|---|---|---|
| | 14 | 51 | U6 | U6 | R | CO | dp4 | VI | 4 |
| | 15 | 2 | R2 | R2 | R | CO | M3 | XXI | 36 |
| | 15 | 13 | X10 | X10 | L | CO | M3 | XXIII | 43 |
| | 3 | 9 | J19 | J19 | L | PART | M2/M3 | | |
| | 4 | 19 | J27 | J27 | R | PART | M3 | XXII | 39 |
| | 4 | 27 | - | - | | PART | | | |
| | 6 | 194 | O29 | O29 | L | PART | M2/M3 | XII/XIX | 18 or 32 |
| | 6 | 205 | N26 | N27 | L | PART | M1/M2 | VII/XII | 28 or 18 |
| | 1 | 65 | A35 | A35 | | PART | | | |
| | 1 | 2 | C2 | C3 | R | PART | M3 | | |
| | 5 | 5 | C17 | C17 | | PART | ?M2 | | |
| | 6 | 119 | P27 | P27 | R | PART | M3 | XXIII | 43 |
| | 6 | 279 | T16 | T16 | L | PART | M1/M2 | medium wear | |
| | 7 | 30 | M14 | N14 | L | PART | M3 | XXVII/XVIII | |
| | 7 | 176 | U15 | U15 | L | PART | M3 | XXIII | 43 |
| | 1 | 335 | F19 | F19 | L | ANT | M1 | | |
| | 6 | 239 | X48 | X48 | L | ANT | M1/M2 | | |
| | 6 | 241 | W38 | W38 | R | ANT | M3 | XX | 34 |
| | 7 | 185 | B3 | B3 | R | ANT | M1/M2 | | |
| | 7 | 214 | B5 | B5 | | ANT | M1/M2 | very worn | |
| | 1 | 378 | - | - | L | POST | M3 | XXIII | 43 |
| | 4 | 7 | C22 | C22 | L | POST | dp4/M1 | | |
| | 4 | 99 | H49 | H49 | R | POST | M3 | XXIV | 45 |
| | 6 | 37 | spoil | spoil | | POST | | | |
| | 7 | 235 | Y22 | Y22 | R | POST | M3 | | |
| Partial skulls with dentition | | | | | | | | | |
| | 1 | 241 | AA15/ BB16 | BB16/ BB16 | L+R | CO | M3 | XXII | 39 |
| | 6 | 174 | O30 | P30 | L | PART | dp4 | IX/X | 10 or 13 |
| | 7 | 7 | L14 | M14 | L+R | CO | M2+M3 | XIII | 20 |
| | 7 | 37 | J17 | J18 | L+R | CO | dp4 | VIII | 8 |
| | 15 | 36 | X8 | X9 | L+R | CO | M2+M3 | XX | 34 |
| | 4 | 124 | B50 | B1 | R | CO | M3 | skull still in | fibreglass |
| | 1 | 242 | I12 | J12 | L | CO | M3 | XXVIII | 55 |

Table 4.19 Upper dentition of mammoths from Stanton Harcourt assigned to categories with estimated age at death based on African elephant data (Laws 1966). Where a tooth is damaged or has been lost, no age estimate is given

| | SITE | REF | SQ | SQ | L/R | PART | TOOTH | LAWS' CATEGORIES | EST. AGE (AEY) |
|---|---|---|---|---|---|---|---|---|---|
| Isolated lower dentition | | | | | | | | | |
| | 1 | 80 | BB11 | BB11 | R | CO | dp4 | VIII | 8 |
| | 1 | 232 | F15 | F15 | L | CO | m3 | XXIII | 43 |
| | 1 | 365 | HH18 | HH18 | L | CO | m3 | XXVI | 49 |
| | 1 | 397 | c.DD2 | c.DD2 | | CO | | | |
| | 2 | 34 | D29 | D29 | R | CO | m3 | XIX | 32 |
| | 2 | 50 | L34 | L34 | R | CO | m3 | XXV | 47 |
| | 4 | 18 | I27 | I27 | L | CO | m3 | XXV | 47 |
| | 4 | 140 | C39 | C39 | R | CO | m3 | XIX | 32 |
| | 5 | 75 | B31 | B31 | L | CO | m3 | XXI | 36 |
| | 5 | 2 | B18 | B18 | | CO | | | |
| | 6 | 4 | S37 | S37 | R | CO | m3 | XXV | 47 |
| | 6 | 101 | S33 | S33 | R | CO | m3 | XXVII | 53 |
| | 6 | 152 | Q35 | Q35 | R | CO | m2 | XV | 24 |
| | 6 | 153 | I36 | I36 | R | CO | m3 | XVIII | 30 |
| | 6 | 186 | F32 | G32 | R | CO | m3 | XX | 34 |
| | 6 | 238 | Y48 | Y48 | L | CO | m2 | XIII | 20 |
| | 6 | 242 | H30 | H30 | L | CO | m2 | XIX | 32 |
| | 7 | 10 | L4 | L4 | R | CO | m3 | XXVII | 53 |
| | 7 | 96 | G7 | G7 | R | CO | m3 | XXIII | 43 |
| | 7 | 128 | O16 | O16 | L | CO | m1/m2 | X | 13 |
| | 7 | 244 | V26 | W26 | R | CO | m3 | XXIII | 43 |
| | 12 | 1 | K32/K33 | L32 | R | CO | m3 | XXIII | 43 |
| | 14 | 29 | J5 | J5 | L | CO | m3 | XIX | 32 |
| | 15 | 6 | N2 | N2 | L | CO | m3 | XXVI/XXVII | 49/53 |
| | 15 | 26 | V7 | V7 | L | CO | m3 | XXVIII | 55 |
| | 15 | 37 | Y8 | Y9 | R | CO | m3 | XXVI | 49 |
| | 23 | 1 | | | L | CO | m3 | XXVI | 49 |
| | 1 | 268 | J13 | J13 | L | ANT | dp4/m1 | worn | |
| | 1 | 269 | J11 | J11 | L | ANT | m1 | unworn | 6 or under |
| | 4 | 52 | A46 | A46 | L | ANT | m1 | unworn | 6 or under |
| | 5 | 26 | A42 | A42 | R | ANT | m1/m2 | unworn | |
| | 7 | 23 | M16 | M16 | L | ANT | m1/m2 | unworn | |
| | 1 | 367 | FF3 | FF3 | | PART | dp4/m1 | | |
| | 4 | 81 | O27 | O27 | | PART | indet. | | |
| | 6 | 240 | Y45 | Y45 | R | PART | m3 | XXI | 36 |
| | 6 | 268 | H13 | H13 | R | PART | m2/m3 | | |
| | 7 | 303 | | | R | PART | m3 | | |
| | 14 | 30 | J12 | J12 | | PART | indet. | | |
| | 15 | 27 | K7 | K7 | L | PART | m2 | worn | |
| | 15 | 41 | X9 | X9 | L | PART | m2/m3 | unworn | |
| | 15 | 54 | V13 | V15 | R | PART | m3 | mw | |

| | SITE | REF | SQ | SQ | L/R | PART | TOOTH | LAWS' CATEGORIES | EST. AGE (AEY) |
|---|---|---|---|---|---|---|---|---|---|
| | 7 | 257 | T20 | T20 | R | POST | m3 | XXV | 47 |
| | 15 | 53 | | | R | POST | dp4/m1 | | |
| **Mandibles with dentition** | | | | | | | | | |
| | 1 | 22 | BB3 | AA3 | L | CO | m3 | XXVI | 49 |
| | 1 | 180 | G5 | H5 | L+R | CO | m3 | XXV | 47 |
| | 1 | 237 | AA15 | AA16 | L+R | CO | m3 | XXIII | 43 |
| | 1 | 239 | F14/F15 | G14/G15 | L+R | CO | m3 | XXVII | 53 |
| | 2 | 12 | I29 | I30 | L+R | CO | m1 | XI | 15 |
| | 3 | 20 | | | L | CO | m1 & m2 | XV | 24 |
| | 4 | 1 | I23 | I23 | L+R | CO | m3 | XXVI | |
| | 4 | 37 | A41 | A41 | L+R | CO | m3 | STOLEN | |
| | 4 | 95 | G49 | H49 | L+R | CO | m3 | XXVI.5 | |
| | 6 | 22 | B27 | B27 | L | CO | m3 | XXIV | 45 |
| | 6 | 123 | P37 | P37 | L+R | CO | m3 | XXIII | 43 |
| | 6 | 168 | N36 | O36 | L | CO | m3 | XXIII | 43 |
| | 6 | 233 | N36 | N36/O36 | L+R | CO | dp3 | III/IV | 1 or 2 |
| | 6 | 253 | J31 | J31 | L+R | CO | m2 & m3 | XXII | 39 |
| | 6 | 281 | S19 | S19 | L | PART | m3 | XXIII | 43 |
| | 7 | 1 | I11 | I11 | L+R | CO | m3 | XXII +HALF | 41 |
| | 7 | 54 | F14 | F14 | L | CO | m3 | XXVII | 53 |
| | 7 | 203 | B1 | B2 | L+R | CO | m3 | XXVI | 49 |
| | 7 | 289 | H8 | H9 | R | CO | m3 | XIX | 32 |
| | 7 | 256 | Q20 | R20 | R | PART | m3 | XXV | 47 |
| | 7 | 259 | T10 | T10 | L | CO | m3 | XXII | 39 |
| | 7 | 310 | S17/S18 | R18/S18 | | PART | indet. | | |
| | 8 | 45 | W1 | W1 | R | CO | dp4 | IV/V | 2 or 3 |

Table 4.20 Lower dentition of mammoths from Stanton Harcourt assigned to categories with estimated age at death based on Laws' African elephant data (1966). No age estimate is given in cases where a tooth is damaged or has been lost.

# Chapter 5

# Vertebrates other than Mammoths at Stanton Harcourt

The large vertebrates in the Stanton Harcourt assemblage are listed in Table 5.1. As the mammoth remains are so numerous, they are discussed in Chapter 4. This Chapter is devoted to the three species of carnivore (wolf, brown bear and lion and the remaining species of herbivore (straight-tusked elephant, horse, bison and red deer).

## The carnivores

| | |
|---|---|
| *Canis lupus* Linnaeus 1758 | wolf |
| *Ursus arctos* Linnaeus 1758 | brown bear |
| *Panthera spelaea* Goldfuss 1810 | cave lion |

The carnivore remains are listed in Table 5.2. The representation of wolf, brown bear and lion at other British MIS 7 sites is shown in Table 5.3. In her survey of British MIS 7 vertebrates, Schreve (1997) records jungle cat *Felis chaus*, wild cat *Felis sylvestris*, leopard *Panther pardus*, fox *Vulpes vulpes* and various mustelids (all in very small numbers) from several MIS 7 localities. Although spotted hyaena *Crocuta crocuta* is also listed by Schreve, it is argued (Scott 2007) that hyaena was not an element of the carnivore guild in Britain during MIS 7.

In Table 5.3 the numbers (NISP) of wolf, bear and lion are also shown as a percentage of all vertebrate species recorded at each locality. This highlights how consistently low their frequencies are by comparison with the herbivores which is generally the case with open site bone assemblages. Conversely, this is less true of cave sites: the relatively high frequencies of wolf at Hutton Cave, bear at Oreston Upper Cave and lion at Bleadon and Oreston very likely reflect deaths of carnivores while denning in these caves.

### ***Wolf*** - *Canis lupus*

Wolf is represented at Stanton Harcourt by a single metapodial, a left MC II (Figure 5.2). While the small size of this specimen might have suggested it was of *Canis mosbachensis*, by late MIS 7 this species has disappeared from the Pleistocene record and *Canis lupus* makes its earliest appearance at British sites (Flower 2014, 2016). Although little can be said about one specimen, it is interesting to consider the information available for the MIS 7 wolf from other British Pleistocene sites. Notably, it was relatively small, which has important implications.

Flower (*op. cit.*) estimates body mass for Late Pleistocene *C. lupus* to have been 35.81 ± 1.59kg (for Britain: 36.25 ± 1.59kg, for mainland Europe: 34.23 ± 1.64kg). Compared to the earlier Pleistocene canids, the Late Pleistocene *C. lupus* was up to a third larger, making it distinctly different in its prey choices and competitive interactions with other large predators. However, unlike the earlier Pleistocene canids, which appear relatively stable in body size through time,

| Species | NISP | % |
|---|---|---|
| *Canis lupus*, wolf | 1 | 0.09 |
| *Ursus arctos*, bear | 10 | 0.9 |
| *Felis spelaea*, lion | 3 | 0.3 |
| *Palaeoloxodon antiquus*, straight tusked elephant | 57 | 4.9 |
| *Mammuthus trogontherii*, steppe mammoth | 922 | 80 |
| *Equus ferus, horse* | 34 | 2.9 |
| *Cervus elaphus*, red deer | 4 | 0.3 |
| *Bison priscus*, bison | 125 | 10.8 |
| | **1156** | 100 |

Table 5.1 Large vertebrates from Stanton Harcourt: the number of identifiable bones of each species (NISP) and the percentage they represent

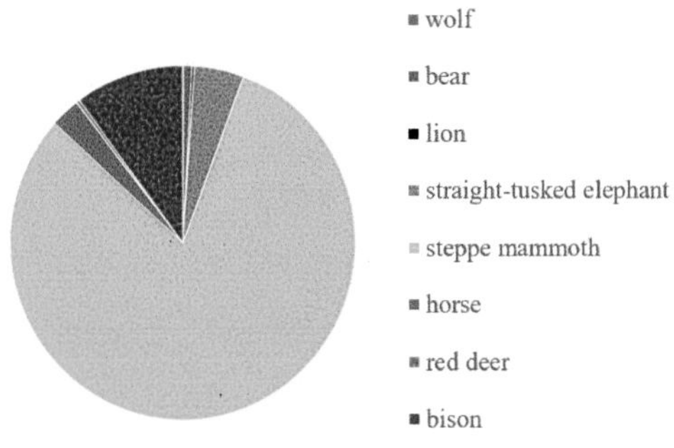


Figure 5.1 Proportional representation of large vertebrate species at Stanton Harcourt

| Species | Site | REF | L/R | Element |
|---|---|---|---|---|
| | | | | |
| *Canis lupus* | 7 | 24 | L | metacarpal (MCII) |
| | | | | |
| *Ursus arctos* | 1 | 115 | | lumbar vertebra |
| | 2 | 4 | | first phalanx |
| | 2 | 17 | | lumbar vertebra |
| | 2 | 20 | R | femur |
| | 5 | 8 | R | mandible with p2, p3, p4, m1 |
| | 6 | 105 | R | calcaneum |
| | 6 | 202 | R | part mandible without dentition |
| | 6 | 245 | L | part mandible without dentition |
| | 6 | 263 | L | mandible with canine |
| | 6 | 263 | R | mandible with canine, p3, p4, m1 |
| | 7 | 165 | | metapodial |
| | | | | |
| *Felis spelaea* | 2 | 27 | R | premaxilla with canine and P1 |
| | 4 | 22 | | metapodial |
| | 4 | 110 | L | mandible with p2, p3, p4, m1 |

Table 5.2 Representation of wolf, bear and lion at Stanton Harcourt

Flower found temporal variation in body size between MIS 7, 5a and 3 with the MIS 7 wolves being the smallest: MIS 7 at an estimated 34.03 ± 1.73kg, MIS 5a at 39.85 ± 1.64kg and MIS 3 at 35.40 ±1.63kg. These variations in body size are correlated with temporal variation in diet, which itself was related to differences in climate (openness of the terrain), prey diversity and competition.

Flower suggests that competition for resources, perhaps combined with palaeogeographical restrictions caused by island status, may have led to a rapid response in body size reduction

| | **Wolf** | ***Canis lupus*** | **Bear** | ***Ursus arctos*** | **Lion** | ***Panthera spelaea*** | SOURCE OF DATA |
|---|---|---|---|---|---|---|---|
| | NISP | % of total assemblage | NISP | % of total assemblage | NISP | % of total assemblage | |
| **OPEN SITES** | | | | | | | |
| Aveley | x | | x | | x | | * |
| Bielsbeck | 5 | 5.2 | 3 | 3.12 | 7 | 7.29 | * |
| Brundon | 1 | 0.36 | 2 | 0.72 | 3 | 1.08 | * |
| Crayford brickearths | 18 | 1.3 | 12 | 0.86 | 36 | 2.6 | * |
| Great Yeldham | 0 | | 3 | 12.5 | 0 | | * |
| Harkstead | 0 | | 0 | | 0 | | * |
| Ilford | 3 | 0.17 | 2 | 1.19 | 16 | 0.95 | * |
| Itteringham | 1 | 0.16 | 0 | | 0 | | * |
| Latton | 0 | | 0 | | 0 | | # |
| Lexden | 0 | | 0 | | 0 | | * |
| Lion Pit Tramway | 0 | | | 1.4 | 0 | | * |
| Marsworth | 44 | 5.27 | | 3.71 | 37 | 4.43 | * |
| Northfleet | x | | | | x | | * |
| Selsey | 0 | | | | 0 | | * |
| Sible Heddingham | 0 | | | | 0 | | * |
| Stanton Harcourt | 1 | 0.09 | | 0.9 | 3 | 0.3 | # |
| Stoke Goldington | 0 | | 0 | | 0 | | * |
| Stoke Tunnel | 17 | 3.66 | 11 | 2.37 | 19 | 4.09 | * |
| Stone | 0 | | 0 | | 0 | | * |
| Stutton | 0 | | 0 | | 5 | 3.81 | * |
| Strensham | 0 | | 0 | | 0 | | * |
| | | | | | | | |
| **CAVES** | | | | | | | |
| Bleadon | 35 | 1.22 | 22 | 0.77 | 380 | 13.34 | * |
| Hindlow | 0 | | 0 | | 0 | | * |
| Hutton | 188 | 22.62 | 0 | | 4 | 0.48 | * |
| Oreston Upper | 1 | 0.38 | 18 | 6.87 | 24 | 9.16 | * |
| Pontnewydd | x | | x | | x | | * |
| Tornewton | x | | x | | x | | * |

x = present, no numbers available  * = data from Schreve (1997)  # = K. Scott's own data

Table 5.3 Representation of wolf, lion and bear at British MIS 7 sites.

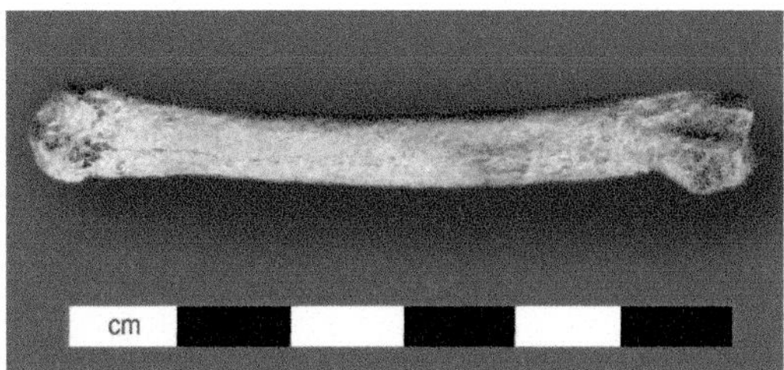


Figure 5.2 Wolf metacarpal (SH7-24: left MCII) from Stanton Harcourt

in British wolves. A rise in sea level in MIS 7a (Candy and Schreve 2007) may have influenced wolf body size by constraining species numbers and movements in the relatively smaller area of island Britain. In terms of the carnivores, although diversity is much reduced by late MIS 7 compared with the early Middle Pleistocene, in addition to wolf, there was lion (*Panthera spelaea*), and smaller carnivores included red fox (*Vulpes vulpes*) and wild cat (*Felis sylvestris*). Brown bear (*Ursus arctos*) was also present and, importantly, there was a new competitor in the form of *Homo neanderthalensis*, ubiquitous across southern and central Britain. However, in Flower's opinion, the presence of abundant lion dominating the carnivore community in the late Middle Pleistocene is a more likely reason for *C. lupus* to have decreased in size.

Wolves have shown themselves to be adaptable to a range of environments throughout the Pleistocene and Holocene of the northern hemisphere – forest, steppe, desert and tundra (Mecozzi and Lucenti 2018). They are gregarious and prey upon a variety of small and large mammals. Modern *C. lupus* hunts a wide range of prey (Flower 2014, 2016) with elk (*Alces alces*) the largest animal taken (400-800kg). Other large ungulates include wapiti (*Cervus canadensis*: 240-454kg), reindeer (*Rangifer tarandus*: 91-272kg) and red deer (*Cervus elaphus*:76-111kg). Medium and small prey taken include wild boar (*Sus scrofa*: 50- 200kg), white-tailed deer (*Odocoileus virginianus*: 18-136kg), roe deer (*Capreolus capreolus*: 17-23kg), Eurasian beaver (*Castor fiber*: 11-30kg) and hare (*Lepus sp.*: 1.2-5kg).

An interesting point to emerge from Flower's analysis of wolf cranio-dental measurements is that late MIS 7 *C. lupus* had comparatively weaker jaws than the later Pleistocene (MIS 5a) and modern wolves. It was therefore less adapted for fast flesh slicing and more adapted for non-flesh food crushing. Flower concludes that while MIS 7 *C. lupus* was certainly able to hunt prey much larger than itself, aided by co-operative hunting, the combination of its smaller body size and the presence of much larger predators, may have resulted in avoiding tackling the very large herbivores that modern wolves can bring down. In summary, the presence of larger predators including Neanderthals likely exerted competitive pressure on *C. lupus* during late MIS 7. However, the incorporation of non-flesh foods into the diet during this interglacial may have allowed *C. lupus* a degree of flexibility and enabled it to better resist high levels of competition.

***Bear*** *– Ursus arctos*

Brown bear remains are few at the site (Figure 5.3, Table 5.2).

Paleontological evidence regarding the appearance of brown bears in Europe is equivocal. Kurtén (1968) concluded that the species entered Europe during the mid-D-Holsteinian (approximately 230 kyBP) but, more recently, fossil remains of putative brown bears dating to approximately 500 kyBP have been identified (Davison et al. 2011). Irrespective of exactly when brown bears appeared in Europe, they evidently co-existed with the cave bear *Ursus spelaeus* until the Late Pleistocene when the latter became extinct (Pacher and Stuart 2009). In Britain, the cave bear *Ursus deningeri* is known from early Middle Pleistocene deposits and was replaced by *Ursus spelaeus* after the Anglian glaciation. The brown bear did not enter Britain until MIS 9, when it completely supplanted *U. spelaeus* (Schreve 2001; Schreve and Currant 2003).

Brown bear is relatively common in cave assemblages throughout the British Middle and Late Pleistocene during both warm and cold stages. In a morphometrical study of bear remains from British caves, Pappa (2014) compares the length-width ratios of the upper M2 and the lower m1 (carnassial) of the Pleistocene specimens with those of modern brown bears. The results for the lower m1 are of interest here as this tooth is consistently used to estimate body size of carnivores. Pappa concludes that the majority of bears from the British Pleistocene sites were larger than modern *Ursus arctos* although individuals from temperate stages were smaller and comparable to modern brown bear individuals from high latitudes (Figure 5.4 Table 5.4). Although only two bear mandibles with dentition were recovered at Stanton Harcourt, the carnassial measurements show them to concur with Pappa's results: modern brown bears and those from a temperate stage such as MIS 7 are smaller than their Late Pleistocene counterparts. As noted above, this is the same development as noted for the Pleistocene wolves.

Today, the brown bear occupies a wide variety of habitats from tundra to temperate forests. Its presence in Britain in association with herbivores of cold open landscapes (woolly mammoth, woolly rhinoceros and horse), as well as with those of temperate conditions, shows it to have been adaptable to a range of environments. As regards their diet, brown bears have evolved a generalist omnivore strategy. Although they possess all the morphological traits of carnivores, in many ecosystems their diet comprises primarily plant matter. Temperature and snow conditions are reported to be the most important factors determining the composition of brown bear diet (Bojarska and Selva 2012). Populations in locations with deeper snow cover, lower temperatures and lower productivity consume significantly more vertebrates, fewer invertebrates and less fruit from deciduous trees, such as acorns and other nuts. Brown bear populations from temperate forest biomes have the most diverse diet. The temperate conditions of MIS 7 would have enabled the bears to forage for plants, tubers and berries. However, as is true of their modern counterparts, they probably also preyed upon winter-weakened or old aged ungulates such as red deer (*Cervus elaphus*) and their calves. They also scavenge carrion of larger ungulates and prey upon small mammals such as squirrels, voles and lemmings (Flower 2014).

Based on the ungulate prey and carrion utilisation of modern brown bears, competition was likely between *U. arctos* and the other carnivores (lion and wolf) based on targeting similarly medium and large sized prey. Regarding competition between these three predators, Flower (*op. cit.*) makes an interesting suggestion: that the Late Pleistocene *U. arctos* was larger and more carnivorous than its modern counterpart or the MIS 7 brown bears, raising the possibility that, as in the present interglacial, the smaller MIS 7 *U. arctos* may also have had a more varied diet, thereby relieving competitive pressure from lion and wolf during climatically favourable periods.

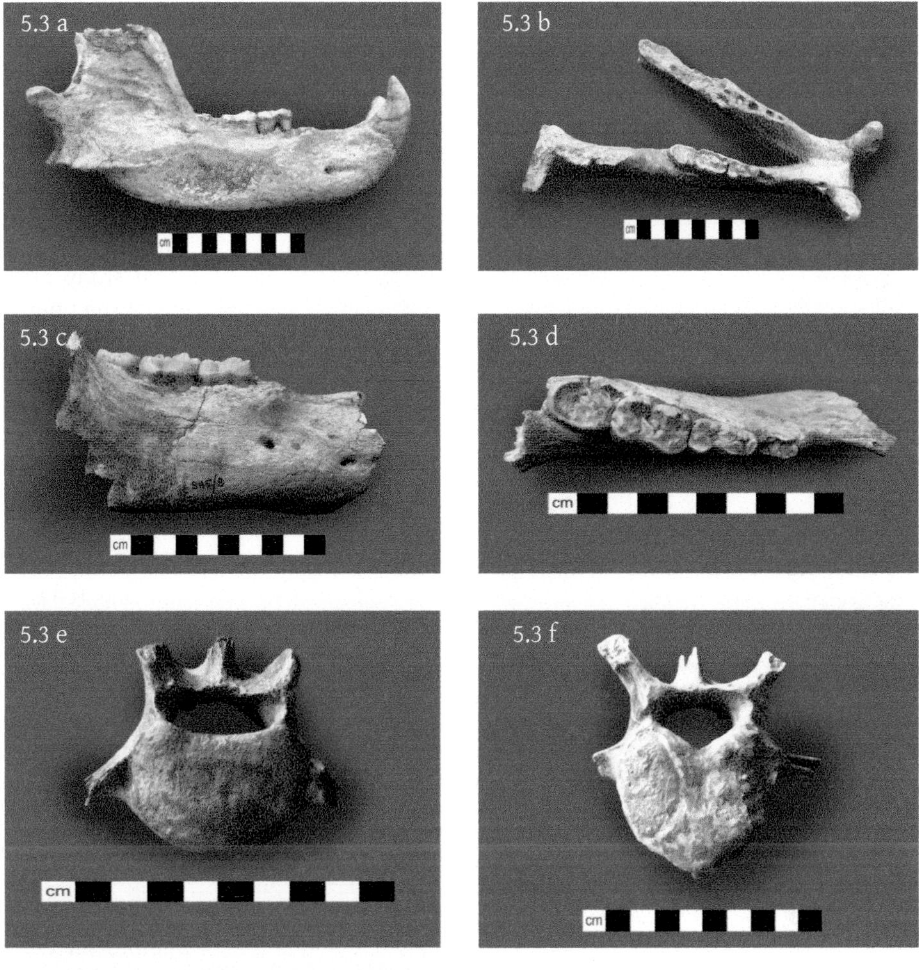


Figure 5.3 Bear remains from Stanton Harcourt. Top row: mandible SH6-263; upper middle row: mandible SH5-8; lower middle row lumbar vertebrae SH2-17 and SH1-115; bottom row: first phalanx SH2-4

***European cave lion*** - *Panthera spelaea*

All Pleistocene and modern lion specimens are assigned to the genus *Panthera* but there is little consensus as to the number of distinct species or the extent of overlap of their distributions. Phylogenetic analysis of ancient and modern DNA sequences show that European fossil lions form a clade that is most closely related to the extant lions from Africa and Asia. However, these data also show that the European cave lion sequences represent lineages that were

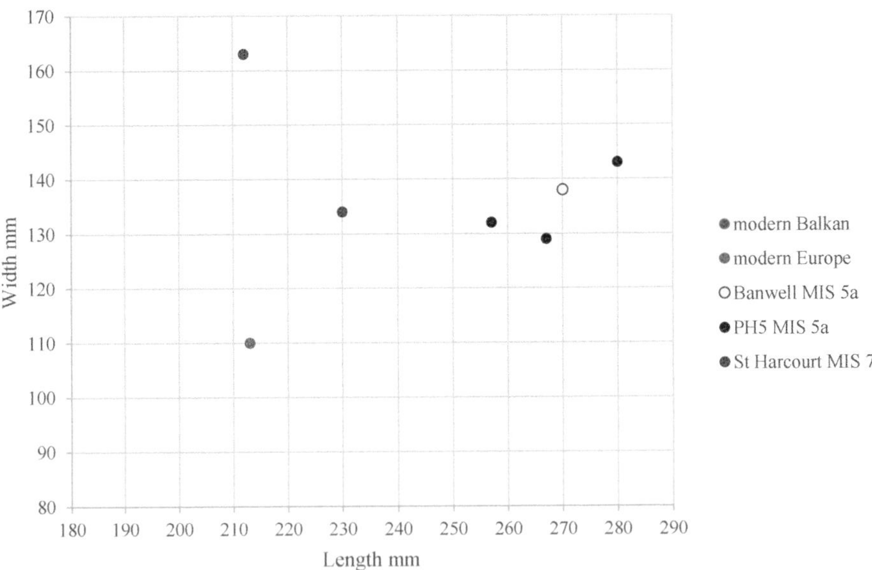


Figure 5.4 Comparison of lower m1 measurements for *Ursus arctos*. Data for Banwell Bone Cave, modern Europe and the Balkans are from Pappa (2014) and shown as the mean for m1 in each sample. Points for specimens from Picken's Hole (PH5) and Stanton Harcourt are K.Scott's measurements

| Modern | Balkan * | Modern | Europe * | Banwell | MIS 5a * | Picken's Hole | MIS 5a # | Stanton Harcourt | MIS 7 # | Bleadon | MIS 7 * |
|---|---|---|---|---|---|---|---|---|---|---|---|
| | | | | | | | | | | | |
| length | breadth | length | breadth | length | breadth | length | breadth | length | breadth | length | breadth |
| | | | | | | | | | | | |
| 223 | 140 | 213 | 110 | 270 | 138 | 280 | 143 | 230 | 134 | 270 | 140 |
| | | | | | | 257 | 132 | 212 | 163 | | |
| | | | | | | 267 | 129 | | | | |

Assemblages marked with * show the mean of the m1 length and breadth measurements for bears published by Pappa (2014).
Assemblages marked with # indicate data of K. Scott

Table 5.4 Lower m1 measurements for bear *Ursus arctos*

isolated from lions in Africa and Asia since their dispersal across Europe at about 600 ky B.P., as they are not found among their sample of extant populations. Although the European cave lion has commonly been regarded as a subspecies (Kurtén 1968; Turner 1984) differences in cranial and dental anatomy between the cave lion and the living *P. leo* are now regarded as sufficient to justify specific status to the cave lion (references in Stuart and Lister 2011).

The more or less continuous presence of the cave lion in the European fossil record indicates that it was a successful predator throughout the Middle and Late Pleistocene, becoming extinct almost synchronously across northern Eurasia within a few hundred years of the onset of the first part of the Lateglacial Interstadial (Greenland Interstadial GI-1e, or Bølling) which occurred at ca. 14.7 cal ka BP. The youngest dated lions are a canine from Zigeunerfels, Sigmaringen, Germany dated at 12,375 ±50 $^{14}$C BP, 14378 cal BP and the skeleton of a cave lion from Le Closeau, northern France, with a date of 12,248 ± 66 $^{14}$C BP, 14,141 cal BP on a metacarpal (Stuart and Lister 2011).

Lion is present in the British fossil record in both warm and cold stages from the Hoxnian until the Late Devensian, indicating that open habitats and an abundance of large prey were probably its main ecological requirement (Stuart 1982). It is interesting to note that it is historically and commonly referred to in the literature as the 'cave' lion although lion remains from open context are far more numerous than those from caves. Exceptions are, for example, approximately 20 individuals discovered in Wierzchowska Cave in southern Poland (Kurtén 1968) and a minimum of 14 in Bleadon Cave, Somerset, England (Schreve 1997) but such concentrations of lion are rare. Table 5.3 shows the representation of wolf, brown bear and lion at the 27 sites ascribed to MIS 7 in Britain of which only 6 are caves. Evidently, access to a cave for denning purposes was not essential to the lions.

At all these sites, carnivore numbers are low by comparison with the herbivores. Accordingly, the remains of lion from Stanton Harcourt are few, consisting of a premaxilla, and a metapodial and a well-preserved mandible (Table 5.2 and Figure 5.5).

There has long been a general perception that lions from British MIS 7 sites are very large – larger than those from Late Pleistocene sites and larger than modern comparative specimens (Schreve 1997). However, this is not evidently a simple chronological development. During the Late Middle Pleistocene (MIS 7/6), smaller forms co-existed with large forms and during MIS 5, at the beginning of the Late Pleistocene, both large and smaller individuals are recorded (Marciszak *et al.* 2104). Among extant African lions, considerable variation in size is determined by two factors: geography and sexual dimorphism. On average, lions from southern Africa are larger than those from eastern regions and males are larger than females. In eastern Africa males average about 170kg and females about 120kg. In southern Africa males average 190 – 225kg (Turner and Antón 1997).

Given the above considerations, it would require large samples in order to generalise about the size of the Pleistocene lion in Britain. Comparison of the lower carnassial (m1) would provide a basis from which to compare MIS 7 lions with those from other contexts but there is a scarcity of measurable specimens from reliable stratigraphic context in Britain (Table 5.3). There is only one complete mandible from Stanton Harcourt (SH4-110) and another from Marsworth (Figures 5.5 and 5.6). There are also three partial mandibles from Crayford. The measurements of the dentition of these specimens are given in Table 5.5 from which the carnassial (m1) is selected to compare with measurements taken African lions where sex (and sometimes height and girth) are known. Based on the length of the lower carnassial Figure 5.7 illustrates that African lions are generally larger than lionesses. The Stanton Harcourt and Marsworth individuals and the three specimens from Crayford are as large or a little larger than the African males. While acknowledging that the British Pleistocene sample is very small, based on the data for modern African lions where carnassial length and shoulder height is recorded, a shoulder height of approximately 1.1m is suggested for SH4-110 (Figure 5.8).

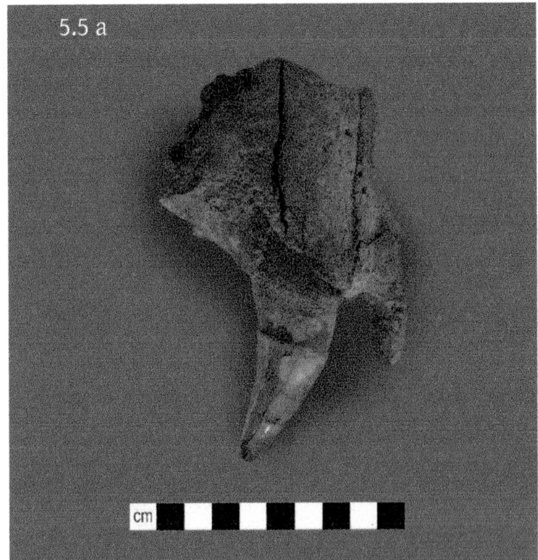


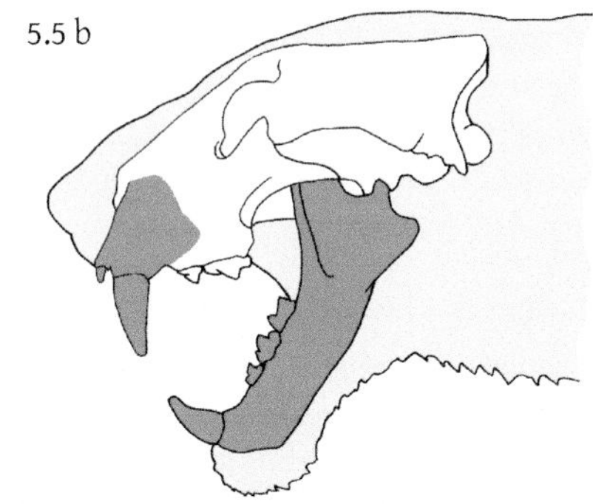


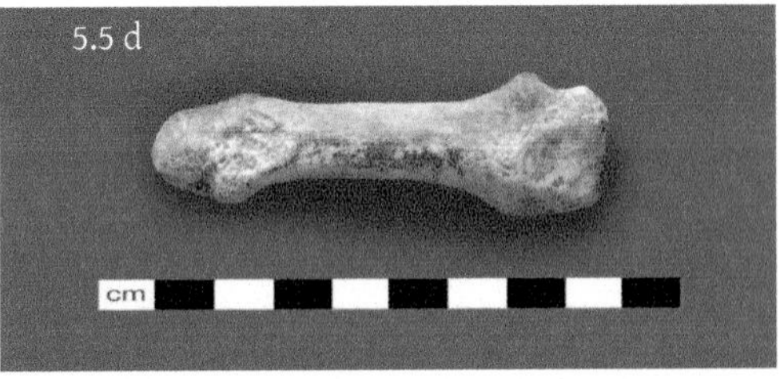

Figure 5.5. Remains of lion *Felis spelaea* from Stanton Harcourt. Diagram (top right) shows relative positions of premaxilla SH2-27 (5.5a) and left mandible SH4-10 (5.5c) Figure 5.5d shows metapodial SH4-22

Figure 5.6 Right mandible of lion from Marsworth (MIS 7) Specimen AYBCM : 1980.544.L7/2. Photograph and measurements of this specimen were kindly supplied by M. Palmer, Buckinghamshire County Museum

| Catalogue Number | Locality | Country | Latitude | Sex | Side | Wear stage | M1 length | M1 width | M1 crown height | length tooth row | Shoulder height (cm) | Girth (cm) |
|---|---|---|---|---|---|---|---|---|---|---|---|---|
| **Pleistocene *P. spelaea*** | | | | | | | | | | | | |
| SH4-110 | Stanton Harcourt | England | | | L | MW | 30.5 | 15.5 | 18.5 | 82.2 | | |
| | | | | | | | | | | | | |
| | Marsworth | England | | | R | MW | 33.4 | | | | | |
| | | | | | | | | | | | | |
| NHM M5031 | Crayford Spurnell coll. | England | | | L | EW | 32.7 | 16.6 | | 80.4 | | |
| NHM M5032 | Crayford Spurnell coll. | England | | | R | MW | 30.8 | 16.3 | | 76.8 | | |
| NHM M5030 | Crayford Spurnell coll. | England | | | R | MW | 29.1 | 16 | | | | |
| **African *P. leo*** | | | | | | | | | | | | |
| FMNH 35742 | Mababe Flats | Botswana | 18.5 | F | R | EW | 26.9 | | 16.5 | | | |
| FMNH 35743 | N'Kate | Botswana | 20 | F | R | MW | 24.3 | | 15 | | | |
| BMNH 31215 | Mababe Flats | Botswana | 18.5 | F | R | EW | 23.8 | 11.8 | 14.9 | | | |
| BMNH 31216 | Kabulubula | Botswana | 17.5 | F | R | EW | 24.5 | | 14.7 | | | |
| FMNH 35739 | Gomodimo Pan | Botswana | 20 | M | R | EW | 25.3 | | 17 | | | |
| FMNH 35741 | Mababe Flats | Botswana | 18.5 | M | R | MW | 27 | | 15.7 | | | |
| BMNH 31214 | Mababe Flats | Botswana | 18.5 | M | R | MW | 29.8 | | 18.4 | | | |
| SAM 3983 | Lake Ngami | Botswana | 20.37 | M | R | EW | 29.5 | | 19.2 | | | |
| PCM MM473/ NN184 | | Chad | 9.3 | F | R | EW | 27.3 | | 14.8 | | | |
| PCM NN185/ MM474 | | Chad | 9.3 | F | R | MW | 23.3 | | 16.5 | | | |
| PCM NN270MM475 | | Chad | 7.48 | F | R | MW | 23.4 | 11.7 | 14.1 | | | |
| PCM A146/ MN462 | | Ethiopia | 12.3 | F | R | EW | 24 | | 14.3 | | | |
| FMNH 20756 | Athi Plains | Kenya | 2.59 | F | R | MW | 25 | | 14.1 | | | |
| FMNH 20758 | Athi Plains | Kenya | 2.59 | F | R | EW | 24.2 | | 14.9 | | | |
| BMNH 3113 | Mbuyuni | Kenya | 3.25 | F | R | EW | 24.6 | | 15.9 | | | |
| BMNH 31132 | Nairobi | Kenya | 1.17 | F | R | EW | 25.1 | | 15.6 | | | |
| BMNH 143211 | Guaso Nyiro R. | Kenya | 1.09 | F | R | MW | 23.9 | | 14.4 | | | |
| FMNH 20757 | Athi Plains | Kenya | 2.59 | M | R | EW | 27.5 | | 16.8 | | | |
| FMNH 20762 | Athi Plains | Kenya | 2.59 | M | R | EW | 27.6 | | 17.5 | | | |
| BMNH 31125 | Hallagedud | Kenya | | M | R | EW | 25.9 | | 15.5 | | | |
| BMNH 31131 | Nairobi | Kenya | 1.17 | M | R | MW | 25.8 | | 16.5 | | | |
| BMNH 3212291 | Bua R. | Malawi | 12.9 | M | R | MW | 27.5 | | 15.8 | | | |

| Catalogue Number | Locality | Country | Latitude | Sex | Side | Wear stage | M1 length | M1 width | M1 crown height | length tooth row | Shoulder height (cm) | Girth (cm) |
|---|---|---|---|---|---|---|---|---|---|---|---|---|
| BMNH 121212 | Port Herald | Malawi | 16.47 | M | R | EW | 27.2 | | 16.2 | | | |
| TM 978 | Mwanza R/ Chikvawe | Malawi | 15.37 | M | R | | 27.6 | | 18 | | | |
| TM 765 | Pafuri R. | Mozambiq | 22.27 | F | R | | 26.3 | | 18.5 | | | |
| TM 927 | Boror | Mozambiq | | M | R | | 27.8 | | 18.3 | | | |
| TM 964 | Pafuri R. | Mozambiq | 22.27 | M | R | | 29.6 | | 17.4 | | | |
| SAM 39302 | Otavi District | Namibia | 19.31 | F | R | MW | 26.1 | | 15.9 | | | |
| TM 1023 | Odangna | Namibia | 19 | M | R | | 27.2 | | 17.5 | | | |
| TM 1024 | Odangna | Namibia | 19 | M | R | | 30.6 | | 18 | | | |
| TM 5604 | Mkusi R/ Mbombo | Natal | 3.35 | M | R | | 28 | | 17.7 | | | |
| PCM S2/ MN461 | | Somalia | | M | R | EW | 26 | | 14.2 | | | |
| BMNH 327686 | Shebeyli R/ Ogaden | Somalia | 0.12 | M | R | EW | 26.2 | | 14.8 | | | |
| TM 3187 | via zoo | Sudan | | F | R | | 28.7 | | 17 | | | |
| BMNH 1937784 | Khartoum | Sudan | 15.33 | M | R | MW | 25.5 | | 16.5 | | | |
| TM 3186 | | Sudan | | M | R | | 24.8 | | 15.9 | | | |
| FMNH 33480 | Serengeti | Tanzania | 2.5 | F | R | EW | 23.9 | | 14.9 | | | |
| FMNH 35132 | Serengeti | Tanzania | 2.5 | F | R | EW | 23.9 | | 14.9 | | | |
| BMNH 353143 | Serengeti | Tanzania | 2.5 | F | R | EW | 23.3 | | 14.6 | | | |
| BMNH 321141 | Tabora | Tanzania | 6.49 | F | R | EW | 24.8 | | 15.5 | | | |
| BMNH 32451 | Kwa-Ku-Chinja | Tanzania | 3.41 | F | R | EW | 24.9 | | 14.9 | | | |
| FMNH 33479 | Serengeti | Tanzania | 2.5 | M | R | EW | 27.4 | | 15.6 | | | |
| FMNH 35131 | Serengeti | Tanzania | 2.5 | M | R | MW | 27.8 | | 16.6 | | | |
| BMNH 353142 | Serengeti | Tanzania | 2.5 | M | R | EW | 27.7 | | 15.6 | | | |
| TM 1026 | Kruger NP | South Africa | 24 | F | R | | 27.8 | | 18 | | | |
| TM 385 | Pietersburg | South Africa | 23.54 | F | R | | 24.6 | | 15.8 | | | |
| TM 4402 | Satara/Sabi Res | South Africa | 24.29 | F | R | | 26.8 | | 15.7 | | | |
| TM 868 | N | South Africa | 24.36 | F | R | | 25.5 | | 16 | | | |
| TM 4403 | Satara/Sabi Res | South Africa | 24.29 | M | R | | 25 | | 17 | | | |
| TM 869 | N | South Africa | 24.36 | M | R | | 30.8 | | 19.1 | | | |
| PCM U29/ MN466 | | Uganda | 1.15 | F | R | EW | 24.5 | | 16 | | | |
| BMNH 344115 | Mt. Elgon | Uganda | 1.07 | F | R | EW | 23.5 | | 15.8 | | | |
| BMNH 352143 | Gomba/ Entebbe | Uganda | 0.3 | F | R | EW | 27.9 | | 17.3 | | | |
| BMNH 311729 | Mulema | Uganda | 0.34 | F | R | EW | 24.9 | | 15.9 | | | |
| BMNH 311730 | Mulema | Uganda | 0.34 | F | R | MW | 25.1 | | 15.6 | | | |
| BMNH 311731 | Mulema | Uganda | 0.34 | F | R | EW | 23.9 | | 13.7 | | | |
| BMNH 311732 | Mulema | Uganda | 0.34 | F | R | EW | 24.1 | 11.9 | 13.9 | | | |

| Catalogue Number | Locality | Country | Latitude | Sex | Side | Wear stage | M1 length | M1 width | M1 crown height | length tooth row | Shoulder height (cm) | Girth (cm) |
|---|---|---|---|---|---|---|---|---|---|---|---|---|
| PCM U161/ MN469 | | Uganda | 0.2 | F | R | EW | 24.8 | | 14.3 | | H=79 | Girth =84 |
| PCM U22/ MN463 | | Uganda | 1 | F | R | MW | 25.9 | | 16 | | H=109 | Girth =86 |
| PCM U107/ MN468 | | Uganda | 0.2 | M | R | MW | 28.6 | | 19.2 | | H =102 | Girth =119 |
| PCM U385/ MN470 | | Uganda | 4 | M | R | EW | 28.2 | | 15.9 | | H= 95 | Girth =94 |
| PCM U386/ MN471 | | Uganda | 4 | M | R | EW | 26.9 | | 17.2 | | H= 95 | Girth =94 |
| PCM U27/ MN464 | | Uganda | 1.15 | M | R | MW | 28.2 | | 17.3 | | | |
| PCM U28/ MN465 | | Uganda | 1.15 | M | R | EW | 30 | | 18.4 | | | |
| BMNH 344113 | Mt. Elgon | Uganda | 1.07 | M | R | EW | 26.1 | | 15.9 | | | |
| BMNH 344114 | Mt. Elgon | Uganda | 1.07 | M | R | EW | 25.1 | | 16 | | | |
| BMNH 352144 | Gomba/ Entebbe | Uganda | 0.3 | M | R | MW | 29 | | 16.5 | | | |
| BMNH 311728 | Mulema | Uganda | 0.34 | M | R | MW | 26.9 | | 16.2 | | | |
| BMNH 31418 | Masindi/ Bunyoro | Uganda | 1.41 | M | R | MW | 23.2 | | 14.3 | | | |
| BMNH 31419 | Bunyoro | Uganda | 1.4 | M | R | EW | 24.6 | | 15.6 | | | |
| BMNH 19372121 | Gomba/ Entebbe | Uganda | 0.3 | M | R | EW | 27.6 | | 17 | | | |
| PCM C401/ MN472 | | Zaire | 0.4 | F | R | EW | 23.9 | | 16.6 | | | |
| RG 2122 | Rutshuru | Zaire | 1.11 | F | R | EW | 26 | | 18 | | | |
| RG 2264 | Kategela | Zaire | 5.01 | F | R | EW | 23 | 12 | 14.2 | | | |
| RG 19232 | Kilwa/L. Mocro | Zaire | 9.15 | M | R | MW | 29.2 | | 17.5 | | | |
| RG 19233 | Luombwa/ N'gai R. | Zaire | 12.13 | M | R | EW | 28.4 | | 18 | | | |
| RG 2121 | Rutshuru | Zaire | 1.11 | M | R | EW | 29.2 | | 18.1 | | | |
| BMNH 18935201 | Umfuli R/ Mashona | Zimbabwe | 17.31 | M | R | MW | 27.2 | | 16.6 | | | |

Table 5.5 Carnassial (m1) measurements for Pleistocene and African lions. Data for African lions kindly supplied by Prof. R.G. Klein (Stanford University)

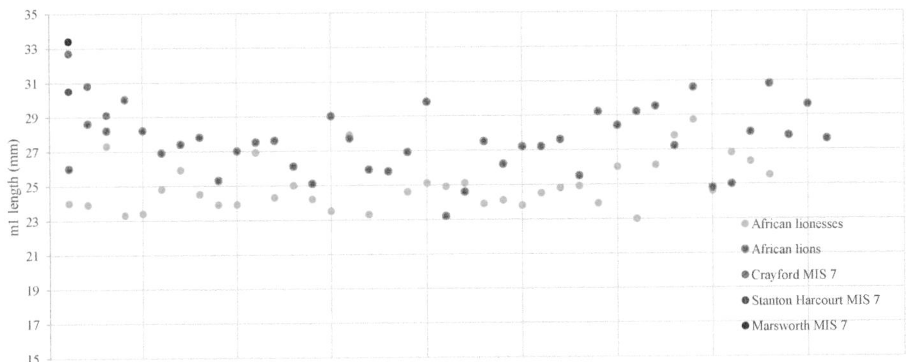


Figure 5.7 Lion carnassial (m1) length for modern African lions compared with British Pleistocene specimens (see Table 5.5 for measurements)

Figure 5.8 Estimating the relative heights of Neanderthal man and the lion *Felis spelaea* from Stanton Harcourt (SH4-22). With thanks to Jersey Heritage and sculptor Elizabeth Daynes for permission to reproduce the photograph of the statue of a Neanderthal man.

Lions appear frequently in European cave art in the late glacial. An interesting feature noted by Turner and Antón (*op. cit.*) and other authors is the absence of a mane in all these depictions although Diedrich (2012) cites at least one depiction that he suggests shows the mane of a male. Whether the European male cave lion indeed had no mane, or the artists chose to illustrate females remains unknown.

## The herbivores

| | |
|---|---|
| *Palaeoloxodon antiquus* Falconer and Cautley 1845 | straight-tusked elephant |
| *Equus ferus* Boddaert 1785 | horse |
| *Cervus elaphus* Linnaeus 1758 | red deer |
| *Bison priscus* Bojanus 1827 | steppe bison |

As illustrated in Figure 5.1 and Table 5.1, mammoths comprise 80% of the vertebrate fauna. In much lower frequencies are bison (11%), straight-tusked elephant (5%), and horse (3%). Red deer is represented by only 3 specimens (0.3%). This assemblage confirms all other biological data from the site indicating a temperate climate and a primarily open grassland habitat with some woodland in the vicinity.

### ***Straight-tusked elephant*** - *Palaeoloxodon antiquus*

The fragmentary post-cranial remains of straight-tusked (forest) elephant at Stanton Harcourt are summarised in Table 5.6 and Figure 5.9. Molars and parts of mandibles are somewhat better represented (Table 5.7 and Figure 5.10). There are three almost complete tusks and three partial tusks.

Straight-tusked elephant is a consistent faunal element of all post-Anglian interglacials, where it is usually found in association with evidence for temperate forest (Stuart 1982, 2005). In Western and Central Europe the straight-tusked elephant appeared during the early middle Pleistocene, the Galerian (Azzaroli *et al.*, 1988) and probably MIS 17 in Britain (Stuart and Lister, 2001; Parfitt *et al.*, 2005). It is common in European fossil assemblages during the Holsteinian (MIS 11) and Eemian (MIS 5e) interglacials and became extinct during the late Eemian in Central Europe and in the early Weichselian (MIS 4) in the Mediterranean (Italy) (Stuart 1991, 2005). The possibility that *P. antiquus* also survived in North-western Europe (Netherlands) well into the Last Cold Stage (MIS 3) is reported by Mol *et al.* (2007).

The European continental evidence shows that the straight-tusked or forest elephant was adapted to a temperate, humid climate. Its range covered most of Europe during successive interglacials and it was associated with a variety of vegetation types. In general, however, the vegetation consisted of expanses of mixed temperate forest with occasional open grassland (Davies 2002). This indicates an adaptability not unlike that of extant elephants. *Loxodonta africana* (African elephant) and *Elephas maximus* (Asian elephant) feed on a large quantity of forage, about 100–300 kg per day. Their diets comprise a mixture of browse and grass. African elephants may feed on more than 80 different plant species (Eltringham 1992). They prefer zones which provide a rich variety of food. The Asian species generally inhabits grassland or forests, including open grassy glades and transition areas rich in several kinds of plants (reference?). In some African regions, grazing can be very important. Accordingly, extant elephants, both African and Asian, can be regarded as mixed feeders with complex diets. The percentage of browse varies widely, depending on the availability of different kinds of plants and on climatic conditions. The amount of browse eaten by African elephants increases during

| | SKELETAL ELEMENT | NISP | L/R |
|---|---|---|---|
| **Cranial** | | | |
| | occipital condyle | 1 | R |
| | premaxilla, part | 2 | |
| | fragment | 1 | |
| | upper molars, complete | 2 | L |
| | mandible, with dentition | 1 | |
| | without dentition | 4 | |
| | lower molars, complete | 5 | 1/4 |
| | all molars, partial + frags. | 13 | |
| | tusks, complete | 3 | |
| | section | 3 | |
| **Post-cranial** | | | |
| | Vertebrae | | |
| | atlas | 1 | |
| | axis, complete | 1 | |
| | partial | 2 | |
| | cervical, complete | | |
| | partial | | |
| | thoracic, complete | | |
| | partial | 1 | |
| | lumbar | | |
| | sacral | | |
| | caudal | | |
| | Scapula | | |
| | complete | | |
| | proximal | | |
| | fragments | 1 | R |
| | Humerus | | |
| | complete | | |
| | proximal | | |
| | diaphysis | 1 | R |
| | distal | | |
| | Radius | | |
| | complete | | |
| | proximal | 1 | |
| | diaphysis | | |
| | distal | | |
| | Ulna | | |
| | complete | | |
| | prox. + diaphysis | 1 | R |
| | distal + diaphysis | 1 | L |
| | Pelvis | | |
| | complete | 1 | R |
| | ilium | 1 | R |
| | acetabulum/ilium | 1 | R |
| | acetab/ischium/pubis | 1 | L |
| | pubis | | |
| | fragments | 2 | |
| | Femur | | |
| | complete | | |
| | proximal | | |
| | diaphyseal frags. | 3 | R |
| | distal condyle | 1 | R |
| | Patella | | |
| | Tibia | | |
| | complete | | |
| | proximal | | |
| | diaphysis | | |
| | distal | | |
| | Fibula | | |
| | distal + diaphysis | 1 | |
| | Tarsals | | |
| | calcaneum | | |
| | astragalus | | |
| | naviculo-cuboid | | |
| | Other carpal/tarsal | | |
| | Metacarpal | | |
| | Metatarsal | | |
| | Phalanges | | |
| | Ribs and fragments | | |

Table 5.6 Skeletal remains of straight-tusked elephant *Palaeoloxodon antiquus* at Stanton Harcourt

| SITE | REF | TOOTH | U/L | L/R | NO. of PLATES | NO. of PLATES in WEAR | WIDTH (mm) | LENGTH COMPLETE TOOTH | EXISTING LENGTH | LAWS CATEGORY | AEY (mean age) | COMMENTS |
|---|---|---|---|---|---|---|---|---|---|---|---|---|
| 1 | 96 | M3 | U | L | 16 | 6 | 7.55 (P2) | 23.5 | | XXII | 18 | complete |
| 1 | 338 | m3 | L | R | 10+ | 3 | 5.81 | | 21 | XXII | 18 | almost complete; missing ?3 plates |
| 1 | 144 | m3 | L | R | 4++ | 4 | 7.8 (P3) | | 7.85 | | | anterior |
| 1 | 224 | M3 | U | L | 17 | 9 | 7.15 (P5) | 24.5 | | XXI | 36 | complete |
| 1 | 276 | m3 | L | R | 17 | 17 | | | | | | no record |
| 1 | 418 | M3 | U | R | 16 | 3 | 7.3 (3) | 20.5 | | | | missing posterior talon |
| 2 | 56 | m1 | L | L | 10++ | 10 | 6.29 (P4) | | 13 | IX | 10 | same tooth as SH14-42 |
| 4 | 24 | m2 | L | R | 11 | 11 | 6.15 (P7) | 20 | | XVIII | 30 | complete |
| 6 | 244 | m2 | L | R | ++6 | | 6.6 | | 10.2 | | | sampled |
| 4 | 45 | m2 | L | R | 4 + 5 | 4 + 5 | 7.5 | | | | | sampled |
| 5 | 49 | - | | | | | | | | | | many fragments |
| 5 | 74 | ?m2 | L | - | ++3++ | 3 | 7.4 | | 6 | | | 3 central plates |
| 5 | 65 | | | L | ++2++ | 2 | | | | | | 2 central plates |
| 6 | 198 | m2 | L | L | 10 | 10 | 6.55 (P6) | 20.5 | | XII | 18 | complete; possibly 11 plates |
| 6 | 303 | ?m3 | L | - | ++4++ | 4 | 6.5 | | 3.9 | | | central |
| 7 | 47 | M2 | U | L | 12+ | 3 | 6.4 (P1) | | 16 | XIII | 20 | almost complete; missing ?1 plate |
| 7 | 198 | ?dp4 | L | L | +4++ | 4 | 5.68 | | 5.58 | ?IX | | anterior of tooth; similar to SH14-42 but possibly dp4 |
| 14 | 42 | m1 | L | L | 13 | 11 | 5.5 (P3) | 18 | | X | 13 | complete; 1st plate eroding |
| 14 | 38 | ?m2 | L | - | ++2++ | 2 | 6.6 | | 3.3 | | | 2 central plates |

Table 5.7 Dentition of straight-tusked elephant *Palaeoloxodon antiquus* from Stanton Harcourt

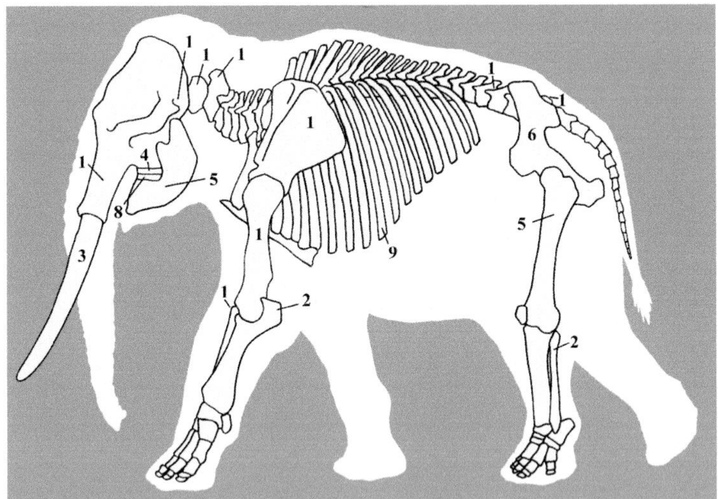


Figure 5.9 Remains (NISP) of straight-tusked elephant at Stanton Harcourt

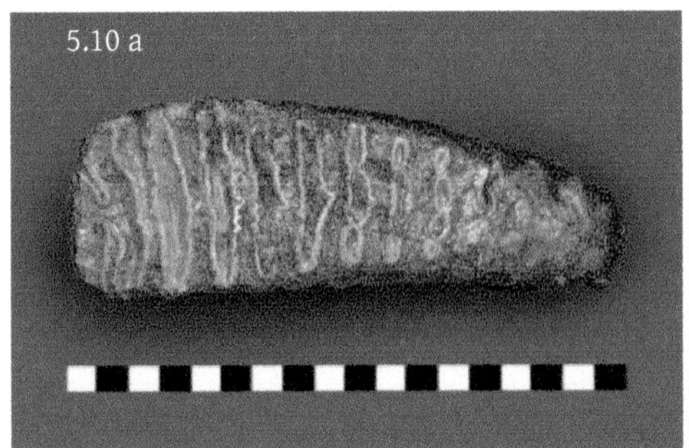

Figure 5.10 Dentition of straight-tusked elephant *Palaeoloxodon antiquus* from Stanton Harcourt. 10a and 10b: occlusal and buccal views of lower left m1 (SH14-42). 10c and 10d: buccal and occlusal views of upper left M3 (SH1-96)

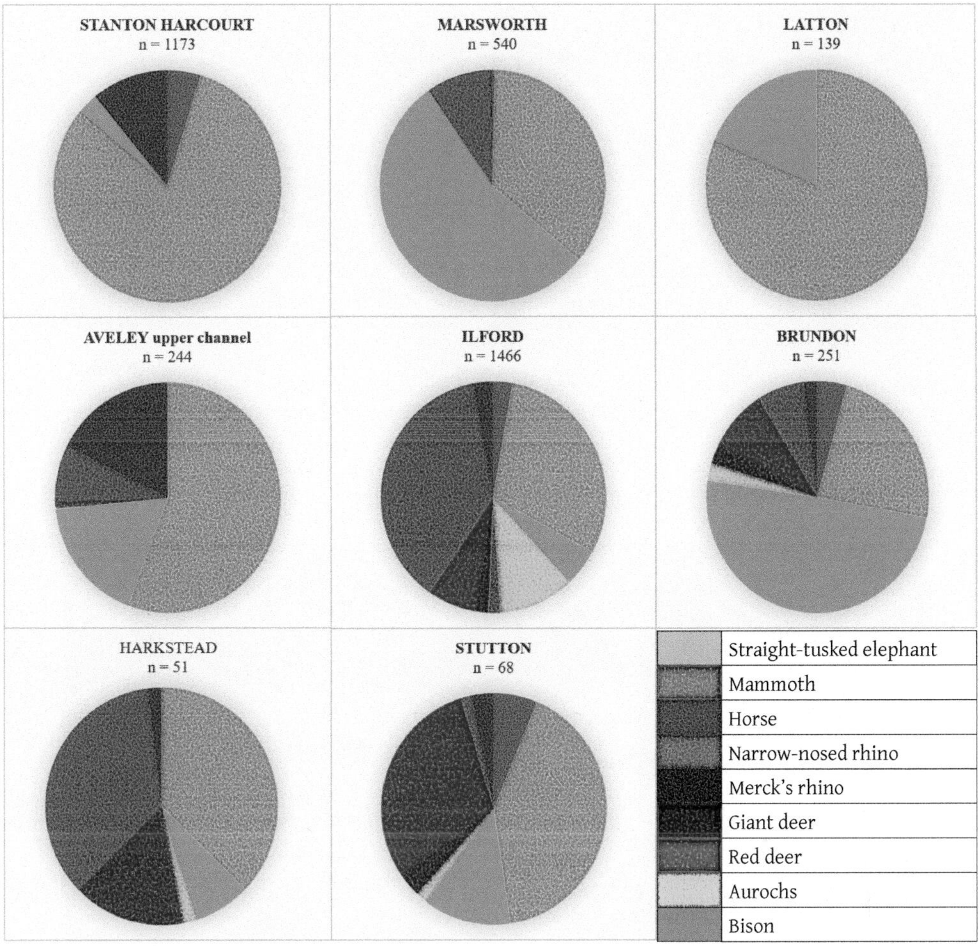

STANTON HARCOURT
n = 1173

MARSWORTH
n = 540

LATTON
n = 139

AVELEY upper channel
n = 244

ILFORD
n = 1466

BRUNDON
n = 251

HARKSTEAD
n = 51

STUTTON
n = 68

| | |
|---|---|
| | Straight-tusked elephant |
| | Mammoth |
| | Horse |
| | Narrow-nosed rhino |
| | Merck's rhino |
| | Giant deer |
| | Red deer |
| | Aurochs |
| | Bison |

Figure 5.11 Representation of herbivores at MIS 7 sites. Data for Stanton Harcourt, Marsworth and Latton is K.Scott's; data for other sites from Schreve (1997).

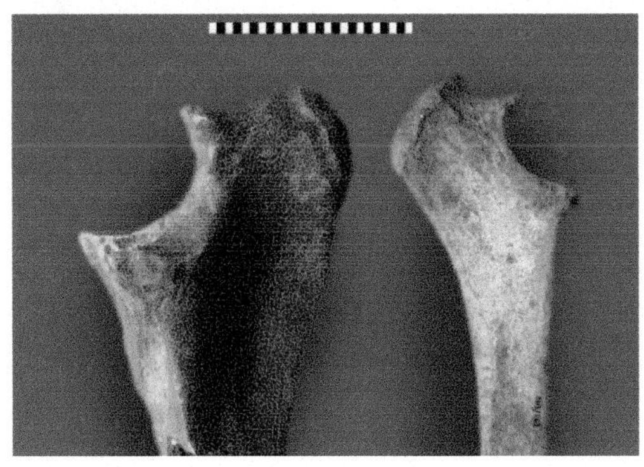

Figure 5.12 Proximal ulnae of elephant (SH1-282) and mammoth (SH6-162)

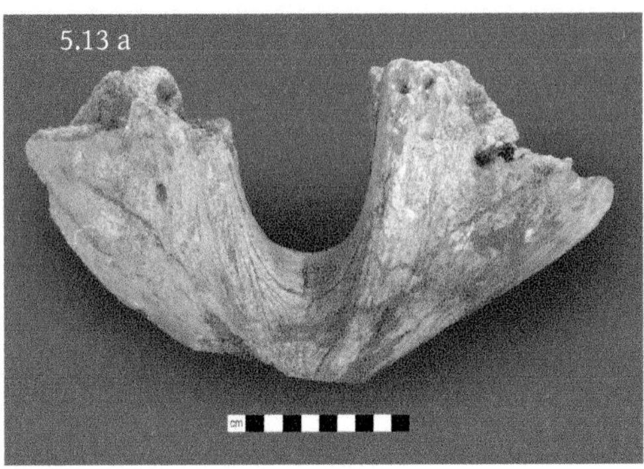


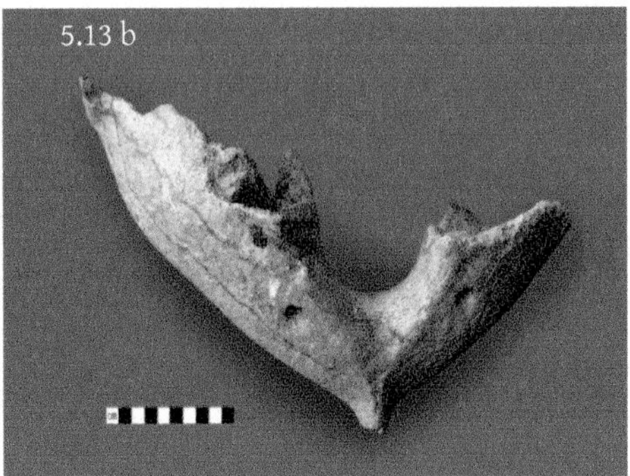


Figure 5.13 Partial mandible of straight-tusked elephant (SH7-290) compared with mammoth mandible (SH1-103) showing the much deeper, thicker definition of the mandibular symphysis in the elephant.

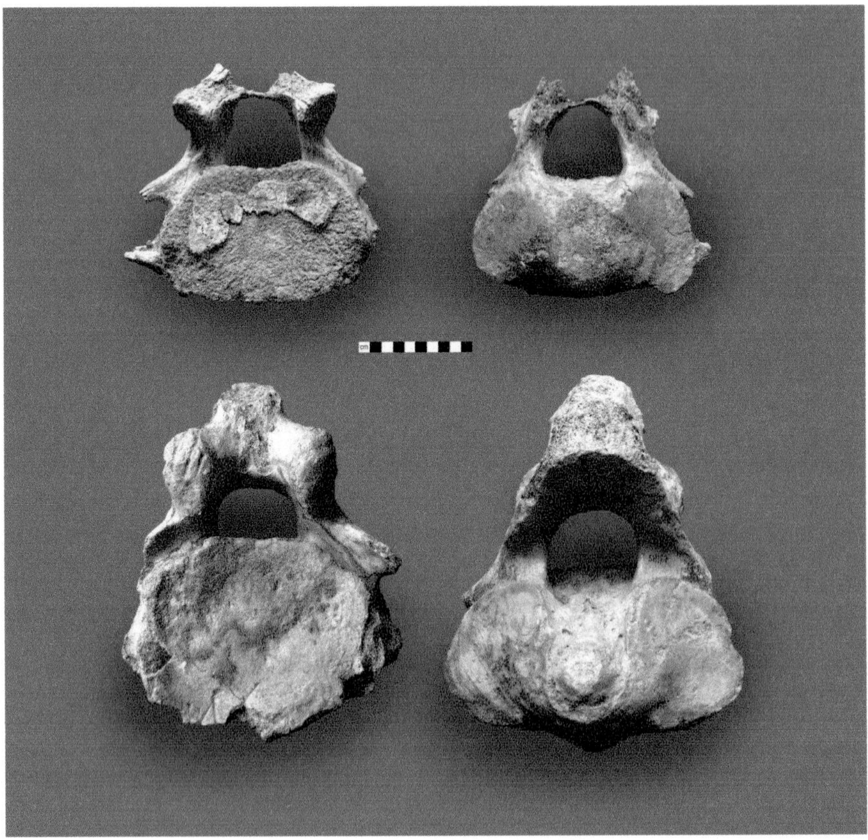

Figure 5.14 Posterior and anterior views of axis vertebrae of mammoth (SH4-126 above) and of straight-tusked elephant (SH1-7)

the dry season, when grass tends to wither, while during the wet season, grass consumption notably increases. During the dry season Asian elephants also eat more browse, while in the wet season, browse and grass consumption are almost equal (Eltringham, *op. cit*).

The British record for MIS 7 is interesting in that *P. antiquus* occurs in small numbers at many sites contemporaneously with the far more numerous steppe mammoth *M. trogontherii.* Of 25 assemblages of MIS 7 age with mammoth, 13 also have straight-tusked elephant (Scott in prep.). This is the only interglacial in which two members of the proboscideans are recorded from the same stratigraphic context. The environmental evidence for MIS 7 indicates a mixture of open grassland and deciduous woodland. Schreve (2001a, 2001b) suggests that the early part of the interglacial (MIS 7e) was more forested than the mid and later phases of the interglacial (MIS 7c and MIS 7a). The absence of absolute dates and good stratigraphic records makes it speculative as to when in MIS 7 most of the vertebrate assemblages accumulated but the degree of open *vs.* woodland habitat is likely to have varied either regionally or temporally. Examples of this variation are shown in Figure 5.11. The mammoth is always a significant component (as it is in almost all the MIS 7 assemblages) regardless of whether the environmental data indicate more wooded or more open habitat.

The late Middle Pleistocene *P. antiquus* was considerably larger than the contemporary MIS 7 mammoth. Shoulder height of the MIS 7 *M. trogontherii* is estimated to have ranged between c. 2.4 and 2.8 m (Scott in prep.). A mass of c. 2.3-3.5 tonnes is suggested by comparison with extant *Elephas maximus* in a comparable range of shoulder height. The average fully grown male *P. antiquus* in optimal conditions is calculated to have been around 4 m at the shoulder with a mass of 13 tonnes (Larramendi 2014). Post-cranial remains of both species at Stanton Harcourt are illustrated in Figures 5.12 - 14.

***Horse*** - *Equus ferus*

The complicated taxonomy of the Pleistocene equids is reviewed by Schreve (1997). Although '*Equus caballus*' is used by some authors to describe Pleistocene caballine horses this term is generally considered appropriate for the domesticated horse. Thus, *Equus ferus* is applied to all British later Middle Pleistocene and Late Pleistocene caballine equids discussed here.

Horse is represented at Stanton Harcourt by only 36 identifiable elements, almost half of which (14) are isolated teeth (See Table 5.8 and Figure 5.15). Apart from one tooth and one unfused bone, all the remains represent adult horses. The remains are distributed fairly evenly across the site but are less than a third as numerous as bison (Table 5.9). This more probably reflects their relative abundance in the area rather than the effect of differential preservation as the post-cranial bones of both species, especially those of mature individuals, are particularly robust and generally survive well in the fossil record. Herd size in wild horses is influenced by habitat structure and food availability. The basic wild horse herd consists of an alpha male with five or six unrelated females and their offspring, groups of bachelors, or solitary males (Bennett and Hoffmann 1999). The low frequencies of horse remains at Stanton Harcourt perhaps suggest similarly small herds in the vicinity.

*Equus ferus* in the European Middle and Late Pleistocene was evidently adaptable to a variety of tundra, grassland, and open forest conditions in both cold and temperate climates. Although

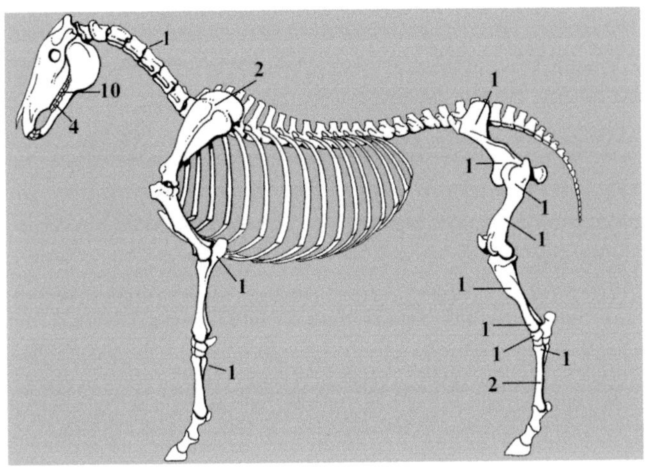


Figure 5.15 Representation of skeletal remains of horse at Stanton Harcourt (ribs were present but not counted)

| | | HORSE | BISON |
|---|---|---|---|
| **Skeletal element** | | NISP | NISP |
| Horn core | | | 26 |
| Cranial fragments | | | 5 |
| Occiptal condyle | | | 1 |
| Maxillary dentition | isolated teeth | 10 | 2 |
| Mandibular dentition | in situ | | 2 |
| | isolated teeth | 4 | |
| Mandibular fragments | | | 2 |
| Dental fragments | | | 9 |
| Vertebrae | atlas | | 4 |
| | axis | | |
| | cervical | 1 | 5 |
| | thoracic | | 7 |
| | lumbar | | 5 |
| | sacrum | | 3 |
| | caudal | | 3 |
| | indet frags | | 9 |
| Scapula | | 2 | |
| Humerus | complete | | 1 |
| | proximal | | 1 |
| | diaphysis | | |
| | distal | | 1 |
| Radius | complete | | 2 |
| | proximal | | 2 |
| | diaphysis | | 1 |
| | distal | | 1 |
| Ulna | | 1 | 3 |
| Carpals | | | 1 |
| Pelvis | complete | 1 | 2 |
| | acetabulum | 1 | 2 |
| Femur | complete | 1 | 1 |
| | proximal | 1 | |
| | diaphysis | 1 | |
| | distal | | |
| Patella | | | 1 |
| Tibia | complete | | 2 |
| | proximal | | |
| | diaphysis | 1 | 1 |
| | distal | 1 | |
| Tarsals | calcaneum | | 4 |
| | astragalus | 1 | 1 |
| | naviculo-cuboid | | 2 |
| Metacarpal | complete | | 1 |
| | proximal | | 2 |
| | distal | | 2 |
| Metatarsal | complete | 2 | 4 |
| | proximal | 1 | |
| | distal | | |
| Metapodial fragments | | | |
| Metapodial 2nd/4th | splint | 1 | |
| Phalanges | 1st | | 1 |
| | 2nd | | 3 |
| | 3rd | | |
| Indet. Fragments | | x | x |
| Ribs and fragments | | x | x |

NISP = number of identified specimens; x indicates present but not counted

Table 5.8 Representation of horse *Equus ferus* and bison *Bison priscus* at Stanton Harcourt

| | | HORSE | | BISON |
|---|---|---|---|---|
| SITE | | NISP | | NISP |
| | | | | |
| 1 | | 14 | | 53 |
| 2 | | 4 | | 5 |
| 3 | | 1 | | 1 |
| 4 | | 3 | | 20 |
| 5 | | 2 | | 4 |
| 6 | | 10 | | 23 |
| 7 | | 1 | | 23 |
| 8 | | 1 | | 4 |
| 14 | | | | 1 |
| 15 | | | | 1 |
| | | | | |
| **TOTAL** | | **36** | | **135** |

Table 5.9 Distribution of horse *Equus ferus* and bison *Bison priscus* across the site

| SITE | REF | L/R | 1. medio-lateral width (mm) | 2. anterio-posterior width (mm) | 3. distal epiphysis width (mm) | 4. length (mm) | EST. HEIGHT AT WITHERS |
|---|---|---|---|---|---|---|---|
| | | | | | | | |
| S. HARCOURT | 1-259 | L | 62 | 58 | 59 | 28 | 149.44 |
| | 2-31 | L | 56 | 46 | | | |
| | 1-176 | L | 58 | 50 | 59 | 29 | 154.57 |
| | | | | | | | |
| ILFORD | 45322 | R | 65 | 52 | 62 | 30 | 159.9 |
| | 45321 | L | 61 | 55 | 57 | 29 | 154.57 |
| | 45319 | R | 58 | 52 | 57 | 28 | 149.44 |
| | | | | | | | |
| SUTTON | SC1 | R | 57 | 45 | 53 | 26 | 138.58 |
| COURTENAY | SC125 | R | 47 | 46 | | 26 | 138.58 |
| | SC181 | R | 47 | 41 | | | |
| | SC128 | R | 52 | 43 | | | |
| | SC276 | R | 46 | 48 | 48 | 26 | 138.58 |
| | SC266 | R | 51 | 45 | 53 | 27 | 143.91 |
| | SC279 | L | 50 | 43 | 48 | 27 | 143.91 |
| | SC267 | R | 47 | 42 | 43 | 27 | 143.91 |
| | SC264 | R | 48 | 43 | | | |
| | SC268 | R | 52 | 45 | | | |
| | SC273 | L | 53 | 42 | | | |
| | SC221 | L | 55 | 46 | 52 | 26 | 138.58 |
| | SC1171 | L | 51 | 43 | 47 | 26 | 138.58 |
| | SC1125 | L | | | 47 | 27 | 143.91 |

Table 5.10 Metatarsal measurements for horse from Stanton Harcourt and Ilford (MIS 7) and Sutton Courtenay (MIS 3). The estimated height at the withers of the individuals represented is calculated (as discussed in the text) by multiplying the maximum length of the metatarsal by 5.33. See Figure 5.20.

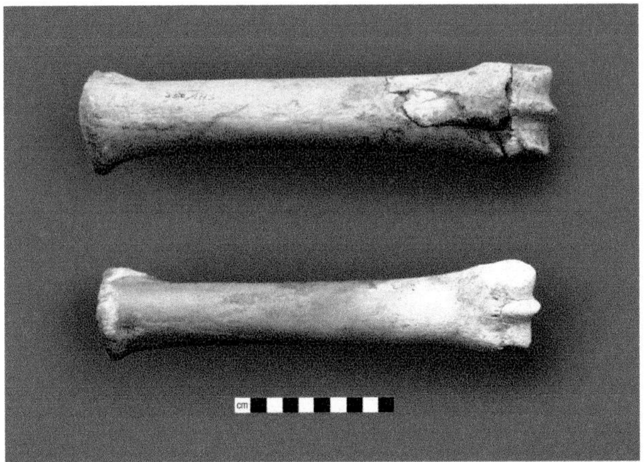


Figure 5.16 Horse metatarsal from Stanton Harcourt (above) compared with the largest from Sutton Courtenay (MIS 3)

Figure 5.17 Horse distal tibiae from Stanton Harcourt (left) and Sutton Courtenay

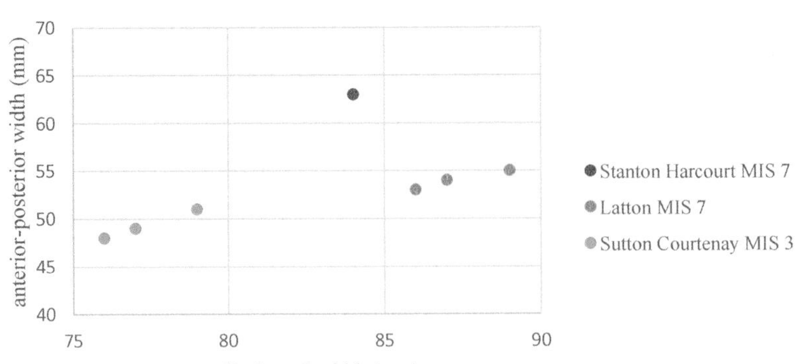


Figure 5.18 Comparison of horse distal tibia measurements from Stanton Harcourt and Latton (MIS 7 sites) with those from Sutton Courtenay (MIS 3)

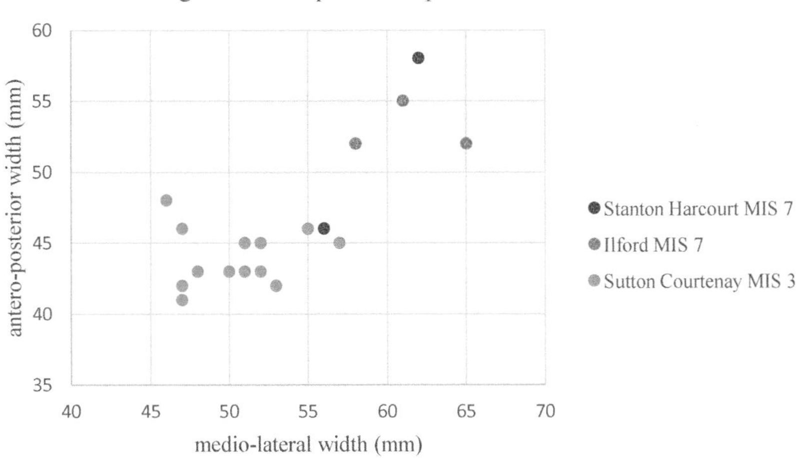


Figure 5.19 Comparison of horse proximal metatarsal measurements from Stanton Harcourt and Ilford (MIS 7 sites) with those from Sutton Courtenay (MIS 3)

| SITE | REF | L/R | 1. Medio-lateral width | 2. Antero-posterior width |
|---|---|---|---|---|
| STANTON HARCOURT | SH7-3 | L | 84 | 63 |
| | | | | |
| LATTON | LQ-25 | R | 86 | 53 |
| | LQL-15 | R | 87 | 54 |
| | LQH | L | 89 | 55 |
| | | | | |
| SUTTON COURTENAY | SC-52 | L | 76 | 48 |
| | SC-94 | R | 77 | 49 |
| | SC-196 | R | 79 | 51 |

Table 5.11 *Equus ferus* distal tibia measurements (mm). See Figure 5.19.

size decrease is usually believed to occur in response to climatic warming in accordance with Bergmann's Rule, Eurasian horses during the Middle Pleistocene (MIS 7) tended to be larger than in the colder Late Pleistocene. Equid post-cranial elements from Stanton Harcourt and other MIS 7 sites (Ilford and Latton) are compared with those from deposits of Last Glacial (MIS 3) age at Sutton Courtenay to demonstrate the comparatively large size of the MIS 7 horses (Tables 5.10, 5.11 and Figures 5.16, 5.17, 5.18, 5.19). Various authors discuss size changes during the Middle and Late Pleistocene evolution of the horse. As they are sometimes described in the literature as 'relatively large' or 'relatively small' it is interesting to consider what this might mean in terms of the actual size (shoulder height) of the Pleistocene horses.

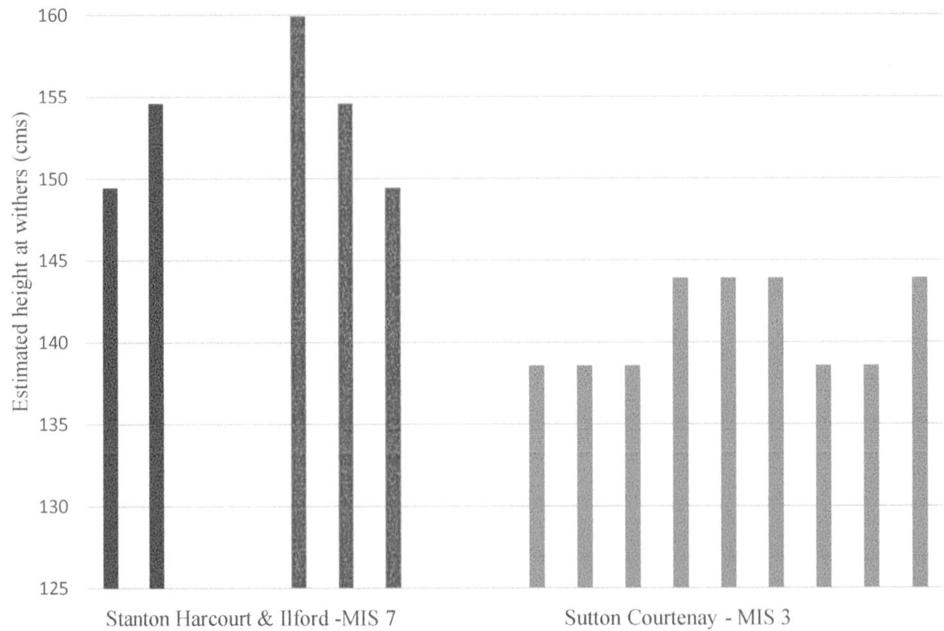


Figure 5.20 Comparison of estimated height at withers for *Equus ferus* based on metatarsals from MIS 7 and MIS 3

Chrószcz *et al.* (2011) review various methods of estimating the height of horses at the withers and find a strong correlation between the length of metapodial bones and withers height. The formulae are i) for the 3rd metacarpal: maximum length multiplied by 6.41 and ii) for the 3rd metatarsal: maximum length multiplied by 5.33. The Pleistocene samples to which these are applied are small and the ratio of males to females in the fossil samples is not known. Horses, however, display little sexual dimorphism or adult variation in the post-cranial skeleton (van Asperen 2011). Metatarsal measurements are given in Table 5.10 from which the height at the withers of the MIS 7 horses is estimated to range between 149 cm and 159 cm (Figure 5.20). This equates to between 15 and 16 hands – the height of a medium sized domestic horse. The MIS 3 metapodia represent slightly smaller horses with a mean height at the withers of 14 hands, which in equestrian terms would be categorised as a large pony.

As regards the likely appearance of the Pleistocene horses Pruvosta *et al.* (2014) discuss whether the representations of animals in European cave art have the potential to provide first-hand insights into the physical environment that humans encountered thousands of years ago and the phenotypic appearance of the animals depicted. Exact numbers of Upper Palaeolithic sites with animal depictions are uncertain because of ongoing debates regarding the taxonomic identification of some images and the dating of some. Where animal species can be confidently identified, horses are depicted at most sites with more than 1,250 documented depictions (about 30% of more than 30 mammal species illustrated). These range from the Early Aurignacian of Chauvet (c. 31 ky BP) to the Late Magdalenian sites (16–11 ky BP) in France and Spain and from the Iberian Peninsula to the Ural Mountains. Depictions are commonly in a caricature form that slightly exaggerates the most typical 'horsey' features.

Although taken as a whole, images of horses are often quite rudimentary in their execution, some detailed representations from both Western Europe and the Ural Mountains are realistic enough potentially to represent the actual appearance of the animals when alive. In these cases, attributes of coat colour may also have been depicted with deliberate naturalism, emphasizing colours or patterns that characterised contemporary horses. For example, the brown and black horses dominant at Lascaux and Chauvet, France, phenotypically match the extant coat colours bay and black.

Pruvosta *et al.* (op. cit.) cite their earlier ancient DNA study (Ludwig, A. *et al.* 2009) of coat colouration in pre-domestic horses which produced evidence that the only phenotypes present in ancestral, pre-domestic horse populations were bay and black. Today, bay-dun is still found in the Przewalski horse *Equus ferus przewalskii*, which is listed as the last remaining wild horse by the International Union for Conservation of Nature (IUCN) and often discussed as a close relative of domestic horses. Its taxonomic status is controversial however and recent studies cited by Provosta *et al.* found that the Przewalski horse displays DNA haplotypes not present in modern or ancient domestic horses, suggesting that Przewalski horses are not directly ancestral to modern domestic horses. Nevertheless, independently of its taxonomic status, several lines of evidence suggest that the bay phenotype of the Przewalski horse represents an ancestral character. Firstly, several wild ass species, which undoubtedly represent wild equids, also show a bay-dun phenotype; and secondly, horses of this phenotype are depicted in remarkable detail in Palaeolithic cave paintings, such as at Lascaux and le Chauvet.

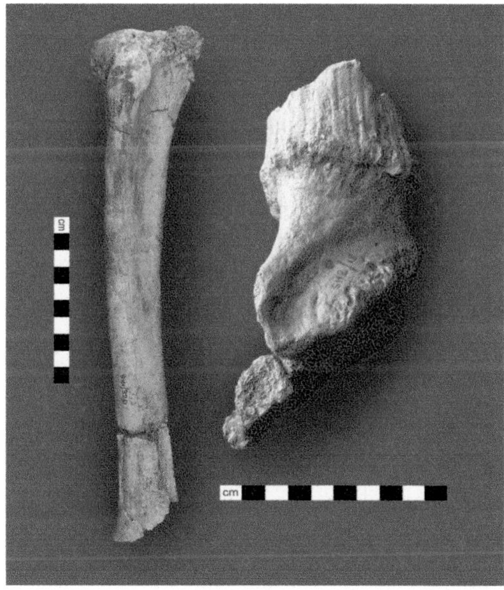


Figure 5.21 Tibia (SH1-320) and part of frontal with antler (SH2-18&19) of red deer *Cervus elaphus* from Stanton Harcourt

### ***Red deer*** - *Cervus elaphus*

Red deer is represented by only three specimens: the base of an antler attached to a frontal fragment, part of an antler base (from a separate individual) and a tibia Figure 5.21.

Although red deer are primarily grazers and are well adapted to open landscapes, they seek woodland for food and shelter, especially in the winter (Crane 2000). Although red deer is ubiquitous throughout the British Middle and Upper Pleistocene interglacials in association with temperate and coniferous forests (Lister 1990; Schreve 1997), their presence in assemblages of MIS7 age is variable (Figure 5.11). The fact that they are relatively well represented at some sites and absent or virtually absent at others may reflect the proximity or density of woodland in the immediate vicinity of the Pleistocene sites. This seems to be substantiated by the presence of aurochs *Bos primigenius* (a species associated with forested habitat) in assemblages where red deer are fairly numerous, and where a wide range of other species are often present. By contrast, Stanton Harcourt, Marsworth, Aveley (Upper Channel) and Latton have relatively few species - primarily indicative of open grassland conditions - and red deer is rare or absent.

***Steppe Bison*** - *Bison priscus*

Three species of large bovid are known from British Middle Pleistocene interglacial deposits: aurochs (*Bos primigenius* Bojanus 1827), woodland bison (*Bison schoetensacki* Freudenberg 1910), and steppe bison (*Bison priscus* Bojanus 1827).

*Bos primigenius* first appeared in Britain during the Hoxnian interglacial and was present during each successive interglacial (Stuart 1982; Schreve 1997) becoming extinct in the 17thC (Stuart *op.cit.*). Although probably primarily a grazer, the remains of aurochs are frequently associated with forest-adapted species suggesting a preference for a woodland habitat. Aurochs is recorded at several British sites of MIS 7 age generally in association with other species of woodland preference e.g. red deer *Cervus elaphus* and Merck's rhinoceros *Stephanorhinus kirchbergensis* (Schreve 1997).

The earliest members of the genus *Bison* appeared at the beginning of the Pleistocene in India and China and then spread from Asia to Europe and America. Fossil evidence shows that from the Middle Pleistocene onwards two species of bison existed in Europe: the steppe bison and the woodland bison (Marsolier-Kergoat *et al.* 2015). The woodland bison *Bison schoetensacki*, more scarcely represented by fossil remains than the steppe bison, had a body size and horns smaller than those of the steppe bison. On the Continental mainland *B. schoetensacki* is present until the end of the early Middle Pleistocene when it is suggested to have been replaced by *B. priscus* (Sala 1986). Sher (1997) on the other hand, records the contemporaneous existence of *B. schoetensacki* and *B. priscus* at German sites of this age. In Britain, *Bison schoetensacki* is recorded in association with regional pine-elm-birch woodland or mixed oak forest only during the Cromerian and does not otherwise appear in the literature for the British Pleistocene. After the Cromerian the only large bovids recorded from British fossil vertebrate assemblages are either *Bos primigenius* or *Bison priscus* (Schreve 1997).

The steppe bison *Bison priscus* commonly occurs in both interglacial and cold stage Pleistocene deposits. It had a wide geographic distribution (known as the great Pleistocene bison belt) extending from England to North America, as far south as Mexico, and from the archipelago of Novaya Zembla in the Arctic Ocean in the north of Russia to Spain, and the Caucasus (Marsolier-Kergoat *et al.* 2015). It appeared in Europe at the end of the early Middle Pleistocene but became extinct at the end of the last Ice Age, about 10,000 years ago (Kurtén 1968; Breda *et al.* 2010). Its descendant, the European bison today survives only as protected herds in Poland and the Caucasus, where it lives in mixed woodlands with some open areas (Krasińska and Krasiński (2002).

Distinguishing between these three species based on their dentition and post-cranial remains is reportedly problematic, principally because there is a wide range of morphological variability among the species and they are characterised by pronounced sexual dimorphism, the males being significantly larger. Despite these difficulties, various authors (e.g. Gee 1993; Olsen

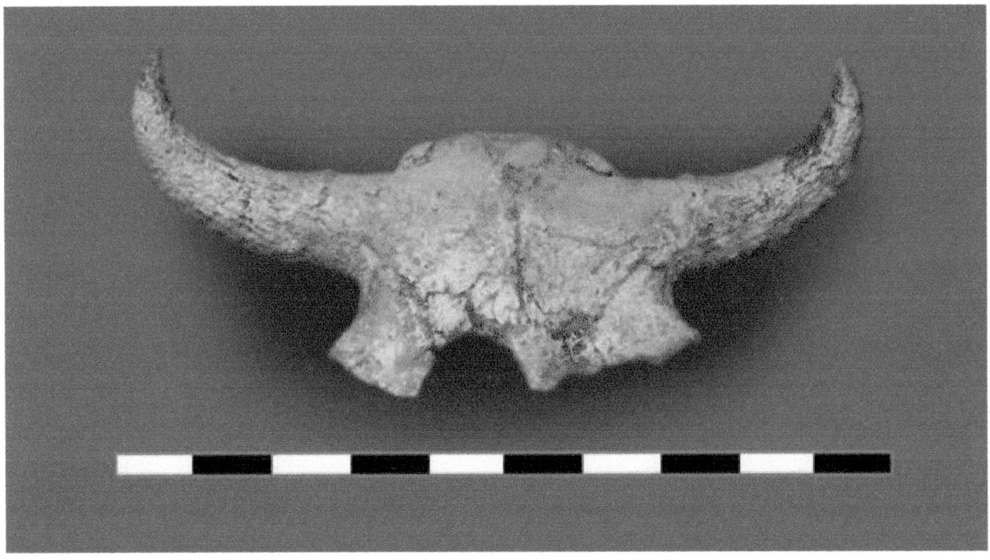

Figure 5.23 Partial skull of Stanton Harcourt bison *Bison priscus* (SH4-63)

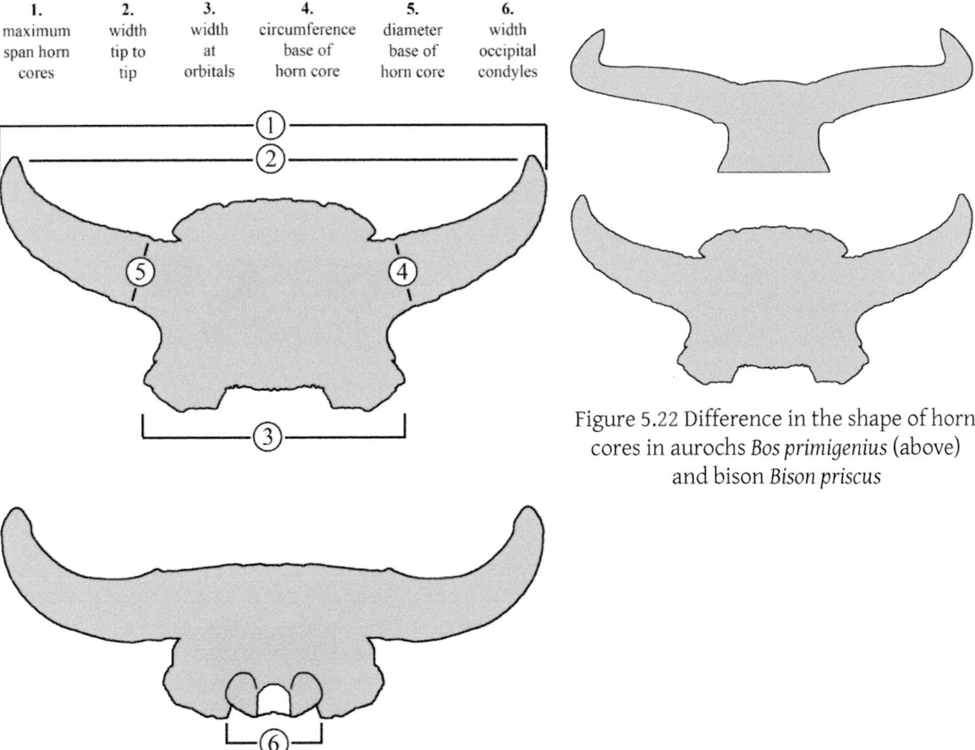


Figure 5.22 Difference in the shape of horn cores in aurochs *Bos primigenius* (above) and bison *Bison priscus*

Figure 5.24 Measurements taken on skulls of bison

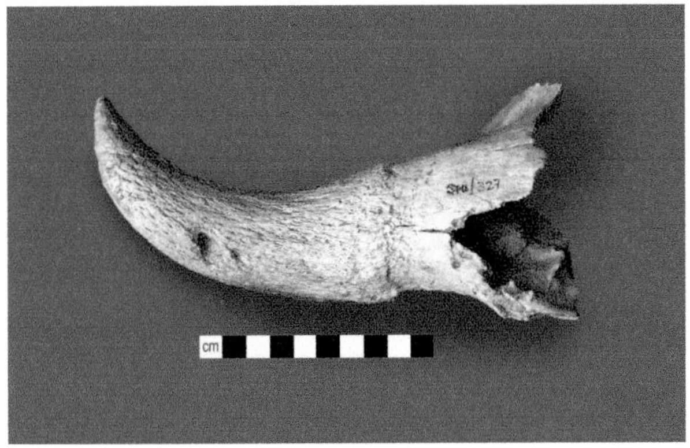


Figure 5.25 Right horn core of young bison (SH1-327)

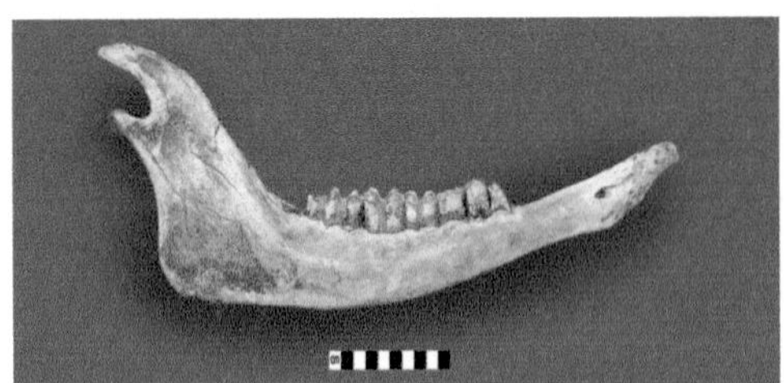

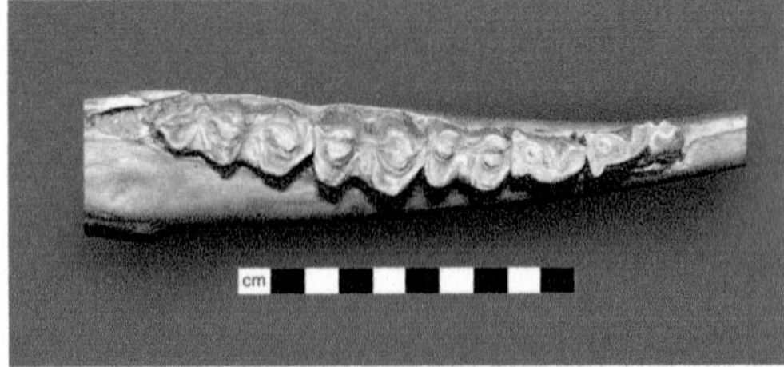


Figure 5.26 Lateral view (left) of bison right mandible (SH1-131) and close up of dentition (right)

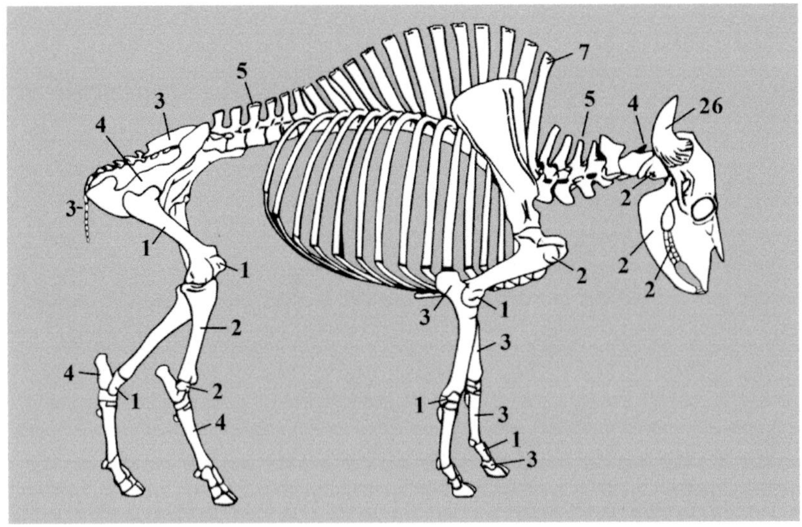


Figure 5.27 Representation of skeletal remains of bison at Stanton Harcourt (number of identified elements)

| SITE | REF | L/R | 1. maximum span horn cores | 2. width tip to tip | 3. width at orbitals | 4. circum-ference base of horn core | 5. diameter base of horn core | 6. width occipital condyles | COMMENT |
|---|---|---|---|---|---|---|---|---|---|
| 1 | 327 | R | | | | 19.5 | 6.27 | | juvenile horn core |
| 1 | 81 | L | | | | | 6.6 | | frontal with eye socket and h/c |
| 1 | 197 | R | | | | | 9.5 | | horn core |
| 1 | 206 | L+R | e72.0 | e67.0 | | e25.5 | 9.35 | | frontal with horn cores |
| 1 | 234 | L+R | 68 | 62 | | 29 | 8.6 | 13.5 | frontal with horn cores |
| 1 | 54 | L+R | | | | | | 15.1 | occiptal |
| 4 | 63 | L+R | 69 | 63 | | | | 15 | frontal with horn cores |
| 5 | 48 | L+R | 91.5 | 86 | e38 | in plaster | R=11.35 | in plaster | frontal with horn cores |
| 6 | 29 | R | | | | | 10.35 | | horn core |
| 6 | 146 | L+R | 75.3 | 68.2 | 36 | R=28.8 | R=9.28 | 13.6 | frontal with horn cores |
| 6 | 96 | R | | | | 31.5 | 9.26 | | horn core |
| 6 | 165 | L+R | 97 | 89.5 | | 35 | 10.28 | in f/ glass | frontal with horn cores |
| 7 | 140 | L | | | | 29.5 | 9.58 | | part frontal with horn core |
| 7 | 160 | L | | | | 29 | 9.45 | | part skull with occiptal and horn core |
| 7 | 116 | L+R | | | | 20.5 | 6.74 | | juvenile frontal and horn cores |
| 7 | 69 | R | | | | | e9.61 | | horn core, frontal and orbit |
| 8 | 2 | L+R | 73 | 69.5 | | R=30.5 | R=9.45 | | frontal with horn cores |
| 14 | 36 | R | | | | 19.5 | 6.63 | | horn core |

Table 5.12 Cranial measurements (mm) for bison *Bison priscus* from Stanton Harcourt

1960; Sher 1997; Wood 2004) have identified criteria which enable taxonomic distinctions to be made between the postcrania of *Bos* and *Bison*.

The skulls and horncores, however, of aurochs *Bos primigenius* and steppe bison *Bison priscus* are highly distinctive. In *Bos primigenius*, the horns are twisted in two planes, both forwards and upwards whereas in *Bison priscus*, the horns are angled upwards only (Figure 5.22). The horn cores from Stanton Harcourt are readily identifiable as those of *B. priscus* (Figure 5.23). In the complete absence of any cranial material referable to *Bos primigenius*, the post-cranial bovid remains are presumed also to be those of bison.

*Bison cranial and dentition*

There are no complete skulls in the assemblage, but horn cores attached to frontal bones are relatively well preserved. A minimum of 16 individuals is represented by 8 frontals with both horn cores, 3 left horn cores and 6 right. The largest 7 skulls have a horn span ranging from 68 cm to almost a metre across (Table 5.12 and Figure 5.24). Although the majority are of mature animals (as indicated by fused frontal bones), there are several younger individuals and at least one immature calf (Figures 5.25).

| SITE | REF | L/R | U/L | TOOTH | Occlusal length | Max. width of tooth | Width anterior cusp | Crown height anterior cusp | Length M1-M3 | Length P2-M3 | Max. length mandible | COMMENT |
|---|---|---|---|---|---|---|---|---|---|---|---|---|
| 1 | 131 | R | L | M3 | 459 | | 174 | | 1044 | 1605 | 4700 | Complete mandible, no incisors; P2, P3, P4, M1, M2, M3 |
| 1 | 131 | R | L | P4 | 235 | 146 | | | | | | |
| | | | | | | | | | | | | |
| 1 | 340 | R | L | M3 | 433 | | 181 | | 991 | | | No incisors; S, P3, S, M1-M3<br><br>Ice wedge cast split this mandible |
| | | | | | | | | | | | | |
| 4 | 132 | L | L | M3 | e425 | | 178 | | | | | |
| | | | | | | | | | | | | |
| 4 | 151 | R | U | M1/M2 | 325 | | 220 | 489 | | | | |
| | | | | | | | | | | | | |
| | | | | | | | | | | | | |
| | | | | | | | | | | | | |
| 7 | 221 | R | L | M3 | 437 | | 192 | | | | | |
| | | | | | | | | | | | | |
| 7 | 126 | | U | M1/M2 | | | | | | | | fragments upper M1/M2 |

Table 5.13 Dentition measurements (mm) for bison *Bison priscus* from Stanton Harcourt

Dentition is poorly represented. There are only 2 lower and 2 upper molars, two mandibular fragments without dentition and two complete right mandibles with dentition (Figure 5.26 and Table 5.13).

*Bison post-cranial*

By comparison with the cranial remains which represent a minimum of 16 individuals, the bison postcrania are relatively few (Table 5.14, Figure 5.27). In Table 5.14, in addition to the number of identifiable items per body part (NISP), the customary calculation of the minimum number of individuals they represent (MNI) is also given. Clearly, MNI from the post-cranial remains is not comparable to the number of individuals known from the crania. In a fluvial deposit such as this, with remains distributed over a wide area, the under-representation of post-cranial bison remains in relation to skulls is likely to be a preservation issue. Skulls are heavier than other body parts which would make them less transportable in the current of the river. Furthermore, the majority were discovered having come to rest 'face down' on the frontal bones, which is likely to have caused the horns/horncores to become embedded in the riverbed rendering the skull immoveable.

| Skeletal element | | NISP | MNI |
|---|---|---|---|
| | | | |
| Horn core | | 26 | 16 |
| Cranial fragments | | 5 | |
| Occiptal condyle | | 1 | 1 |
| Maxillary dentition | | 2 | 1 |
| Mandibular dentition | | 2 | 2 |
| Mandibular fragments | | 2 | 2 |
| Dental fragments | | 9 | |
| Vertebrae | atlas | 4 | 4 |
| | axis | | |
| | cervical | 5 | |
| | thoracic | 7 | |
| | lumbar | 5 | |
| | sacrum | 3 | |
| | caudal | 3 | |
| | indet frags | 9 | |
| Scapula | | | |
| Humerus | complete | 1 | 2 |
| | proximal | 1 | |
| | diaphysis | | |
| | distal | 1 | |
| Radius | complete | 2 | 3 |
| | proximal | 2 | |
| | diaphysis | 1 | |
| | distal | 1 | |
| Ulna | | 3 | 3 |
| Carpals | | 1 | |
| Pelvis | complete | 2 | 1 |
| | acetabulum | 2 | 2 |
| Femur | complete | 1 | 1 |
| | proximal | | |
| | diaphysis | | |
| | distal | | |
| Patella | | 1 | 1 |
| Tibia | complete | 2 | 1 |
| | proximal | | |
| | diaphysis | 1 | |
| | distal | | |
| Tarsals | calcaneum | 4 | 3 |
| | astragalus | 1 | 1 |
| | naviculo-cuboid | 2 | 2 |
| Metacarpal | complete | 1 | 3 |
| | proximal | 2 | |
| | distal | 2 | |
| Metatarsal | complete | 4 | 3 |
| | proximal | | |
| | distal | | |
| Metapodial fragments | | | |
| Phalanges | 1st | 1 | 1 |
| | 2nd | 3 | 1 |
| | 3rd | | |
| Indet. Fragments | x | | |
| Ribs and fragments | xx | | |

Table 5.14 Skeletal remains of bison *Bison priscus* from Stanton Harcourt

| SITE | REF | NUMBER | L/R | 1. Medio-lateral width | 2. Antero-posterior width | 3. Width distal epiphysis | 4. Length |
|---|---|---|---|---|---|---|---|
| **STANTON** | SH6 | 68 | R | | 55 | | |
| **HARCOURT** | SH1 | 116 | R | 87 | 54 | 86 | 236 |
| | SH6 | 47 | R | | | | |
| | SH7 | 246 | L | | | | |
| | SH1 | 31 | R | 92 | 49 | | |
| **THRUPP, UK** | TH | 104 | L | 82 | 49 | 79 | 212 |
| | TH | 237 | R | 65 | 37 | 65 | 200 |
| | TH | 328 | R | 77 | 44 | 76 | 210 |
| | TH | 142 | R | 66 | 37 | 69 | 217 |
| **YARNTON, UK** | YP | 36 | L | 67 | 31 | 70 | 190 |
| | YP | 157 | R | 70 | 38 | 61 | 211 |
| | YP | 131 | R | 64 | 38 | 57 | 191 |
| | YP | 339 | R | 63 | 36 | 66 | 200 |
| | YP | 134 | R | 76 | 39 | | 210 |
| | YP | 133 | R | 66 | 35 | | |
| | YP | 250 | L | 64 | 38 | | |
| | YP | 135 | L | 55 | 40 | 61 | 201 |
| | YP | 136 | L | 69 | 33 | 72 | 201 |
| **SUTTON** | SC | 1144 | L | 74 | 43 | | 195 |
| **COURTENAY** | SC | 7 | L | 62 | 35 | | |
| | SC | 310 | L | 68 | 38 | 69 | 210 |
| | SC | 1224 | L | 64 | 38 | | |
| | SC | 317 | L | 81 | 45 | 77 | 215 |
| | SC | 1139 | L | 64 | 39 | 64 | 202 |
| | SC | 321 | L | 60 | 35 | 61 | 205 |
| | SC | 1168 | L | 65 | 37 | | |
| | SC | 312 | L | 73 | 40 | 74 | 190 |
| | SC | 153 | L | 64 | 37 | 63 | 208 |
| | SC | 1127 | L | 73 | 45 | 77 | 205 |
| | SC | 327 | L | 85 | 48 | 87 | 230 |
| | SC | 315 | L | 81 | 45 | 77 | 237 |
| | SC | 143 | L | 86 | 48 | 87 | 223 |
| | SC | 313 | L | 73 | 44 | 76 | 212 |
| | SC | 1154 | L | 70 | 40 | | |
| | SC | 152 | L | 65 | 40 | 65 | 203 |
| | SC | 154 | R | 72 | 42 | 73 | 215 |
| | SC | 323 | R | 65 | 37 | | 205 |
| | SC | 1275 | R | 66 | 39 | | |
| | SC | 329 | R | 68 | 36 | | 215 |
| | SC | 195 | R | 74 | 45 | 73 | 237 |
| | SC | 325 | R | 88 | 49 | 90 | 241 |
| | SC | 154 | R | 73 | 42 | 73 | 218 |
| | SC | 40 | R | 78 | 47 | 78 | 230 |
| | SC | 326 | R | | | 77 | |
| | SC | 309 | R | | | 70 | |
| | SC | 307 | R | | | 78 | |
| | SC | 319 | R | | 38 | | |
| | SC | 335 | R | | | 819 | |

| SITE | REF | NUMBER | L/R | 1. Medio-lateral width | 2. Antero-posterior width | 3. Width distal epiphysis | 4. Length |
|---|---|---|---|---|---|---|---|
| | SC | 122 | R | | | 828 | |
| | SC | 50 | L | | 43 | | |
| | SC | 328 | L | | 66 | | |
| | SC | 46 | L | e 61 | 36 | 65 | 200 |
| | SC | 36 | R | | 32 | | |
| **SIBERIA** | MP PGPI 1222–34 | male | | 91 | 53 | 84 | 230 |
| various locations | | mean of 42 ♂ | | 90 | 52 | 91 | 243 |
| | | mean of 99 ♂ | | 90 | 51 | 93 | 241 |
| | **Total 142 ♂** | **mean** | | **90** | **52** | **89** | **238** |
| | | | | | | | |
| | | mean of 28 ♀ | | 79 | 46 | 80 | 237 |
| | | mean of 45 ♀ | | 78 | 44 | 79 | 234 |
| | **Total 73 ♀** | **mean** | | **79** | **45** | **80** | **235** |

Table 5.15 Bison metacarpal measurements (mm) from Stanton Harcourt (MIS 7), Thrupp and Yarnton (MIS 5a), Sutton Courtenay (MIS 3) and Siberia (MIS 5-3). Data from British sites: K. Scott; Siberian data from Shpansky *et al.* (2016)

| SITE | REF | NO | L/R | 1. Medio-lateral width | 2. Antero-posterior width | 3. Max. distal width | 4. Length |
|---|---|---|---|---|---|---|---|
| **STANTON** | SH1 | 193 | L | 73 | 70 | 85 | 285 |
| **HARCOURT** | SH7 | 76 | L | 73 | 73 | 81 | 310 |
| | SH8 | 73 | R | | 70 | 87 | 290 |
| | SH4 | 15 | L | | | 82 | 270 |
| | | | | | | | |
| **THRUPP** | TH | 115 | R | 65 | 62 | 71 | 261 |
| | TH | 270 | R | 60 | 58 | 68 | 255 |
| | TH | 108 | R | 54 | 44 | 65 | 248 |
| | TH | 193 | R | | | 70 | |
| | TH | 107 | R | | | 68 | 255 |
| | TH | 48 | R | 50 | 49 | | |
| | | | | | | | |
| **SUTTON** | SC | 102 | L | 60 | 58 | 71 | |
| **COURTENAY** | SC | 303 | L | 55 | 54 | 63 | 270 |
| | SC | 158 | L | 51 | 48 | | |
| | SC | 1281 | L | 49 | 51 | | |
| | SC | 1126 | L | 54 | 53 | 63 | 245 |
| | SC | 1274 | L | | | 60 | |
| | SC | 1291 | R | 60 | 57 | | |
| | SC | 6 | R | 51 | 55 | | |
| | SC | 301 | R | 51 | 52 | | |
| | SC | 204 | R | 60 | 55 | | |
| | SC | 291 | R | 66 | 65 | | |

Table 5.16 Bison metatarsal measurements (mm) from Stanton Harcourt (MIS 7), Thrupp (MIS 5a), Sutton Courtenay (MIS 3)

| SITE | REF | NUMBER | L/R | 1. Distal medio-lateral width | 2. Distal antero-posterior width | 3. Maximum length | 4. Proximal medio-lateral width | 5. Proximal antero-posterior width |
|---|---|---|---|---|---|---|---|---|
| | | | | | | | | |
| **STANTON** | SH7 | 60 | L | 104 | 76 | 525 | 128 | 135 |
| **HARCOURT** | SH4 | 75 | R | 92 | 63 | 480 | 139 | 120 |
| | | | | | | | | |
| **MARSWORTH** | 544 | J9/7 | R | 89 | 65 | | | |
| | 544 | K16/8 | R | 89 | 70 | | | |
| | 544 | I7/15 | L | 94 | 68 | | | |
| | 544 | J15/1 | L | 91 | 67 | | | |
| | | | | | | | | |
| **THRUPP** | TH | 109 | L | 73 | 56 | 420 | e115 | 121 |
| | TH | 23 | L | | | | 110 | 102 |
| | TH | 20 | L | 72 | 50 | | | |
| | TH | 39 | R | 80 | 58 | | | |
| | TH | 117 | R | 80 | 61 | 445 | e105 | |
| | TH | 105 | R | 77 | 52 | | | |
| | TH | 396 | L | | | | 133 | |
| | TH | 106 | R | | | 445 | 126 | 125 |
| | TH | 38 | R | | | | 126 | |
| | | | | | | | | |
| **SUTTON** | SC | 294 | L | 68 | 56 | | | |
| **COURTENAY** | SC | 172 | R | 71 | 55 | | | |
| | SC | 1228 | R | 74 | 51 | | | |
| | SC | 1229 | L | 80 | 62 | | | |
| | SC | 1217 | L | 74 | 55 | | | |
| | SC | 1152 | L | 67 | 49 | | | |
| | SC | 1267 | L | 77 | 57 | | | |
| | SC | 1166 | L | 77 | 52 | | | |
| | SC | 130 | L | 62 | 50 | | | |
| | SC | 207 | R | 72 | | | | |
| | SC | 224 | R | | | | | |
| | | | | | | | | |
| **SIBERIA** | | | | | | | | |
| Grigorievka | MP PGPI 1222-41 | male | | 86 | 64 | 440 | 140 | 132 |
| Krasniy Yar | | mean of 20 ♂ | | 91 | 67 | 488 | 140 | 123 |
| | **Total 21 ♂** | **mean** | | **89** | **66** | | | |
| Krasniy Yar | PM TSU 5/1443 | female | | 82 | 59 | 425 | 111 | 108 |
| Krasniy Yar | | mean of 15 ♀ | | 80 | 60 | 452 | 126 | 107 |
| | **Total 16 ♀** | **mean** | | **81** | **60** | | | |

Table 5.17 Bison tibia measurements (mm) from Stanton Harcourt and Marsworth (MIS 7), Thrupp (MIS 5a), Sutton Courtenay (MIS 3) and Siberia (MIS 5-3). Data from British sites: K. Scott; Siberian data from Shpansky *et al.* (2016)

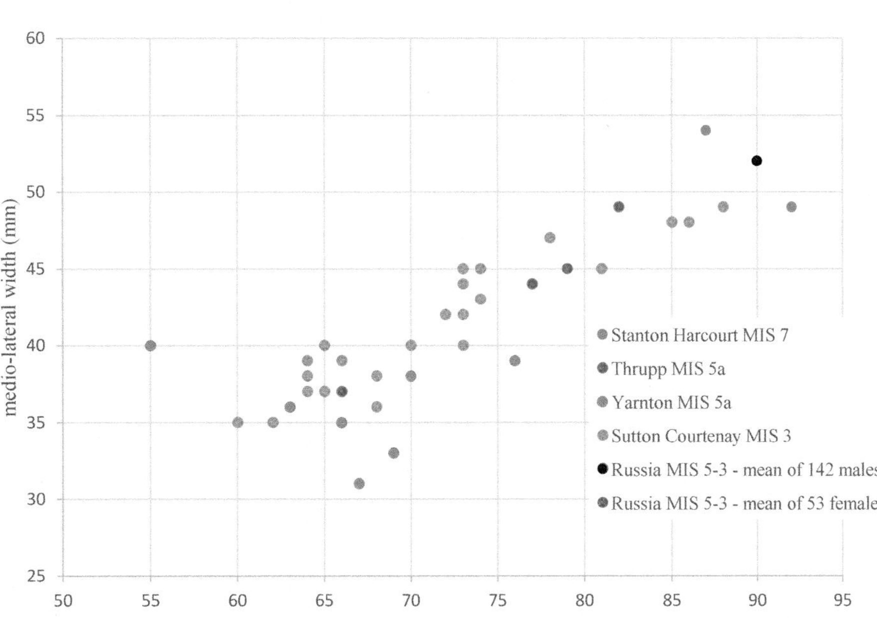


Figure 5.28 Bison proximal metacarpal measurements from Stanton Harcourt compared with those from various Late Pleistocene sites

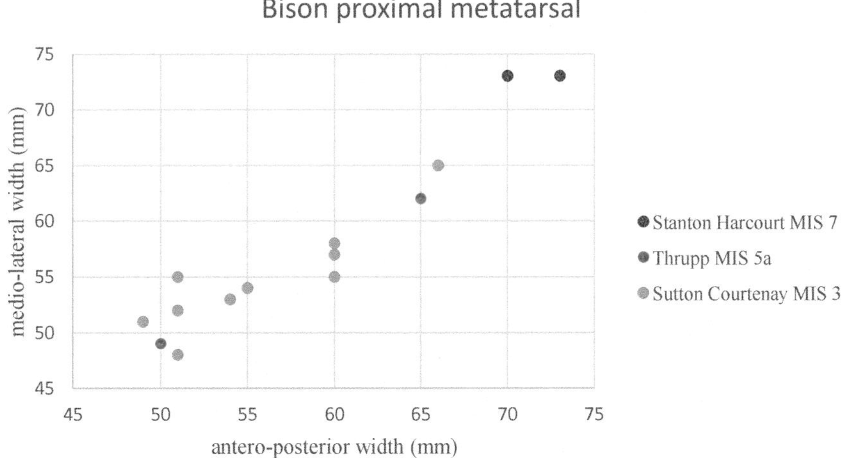


Figure 5.29 Bison proximal metatarsal measurements from Stanton Harcourt compared with those from Late Pleistocene British localities

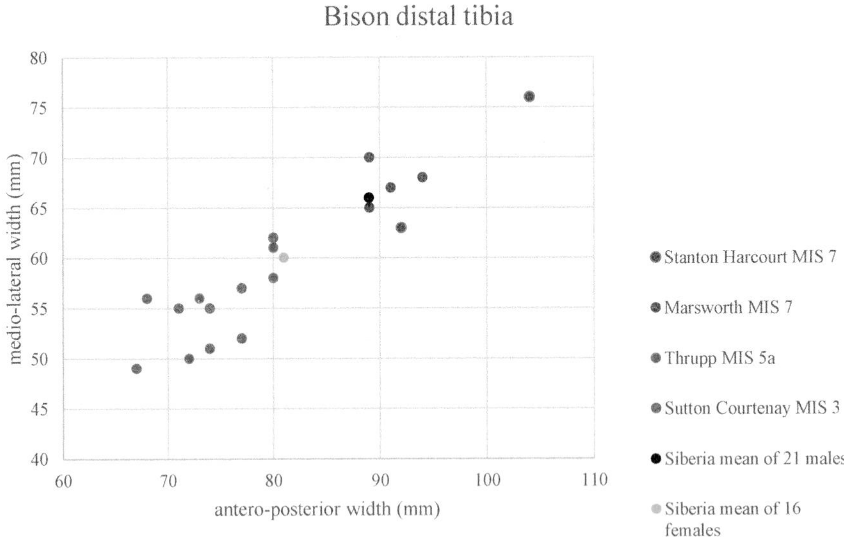


Figure 5.30 Bison distal tibia measurements from MIS 7 sites (Stanton Harcourt and Marsworth) compared with those from various Late Pleistocene sites

The MIS 7 steppe bison is likely to have been a formidable animal. The widespread horncores of the largest skulls from Stanton Harcourt span almost a metre. The addition of the horn sheaths would have increased the span by 20-30 cm. The robusticity of the metapodials and the tibiae at Stanton Harcourt enables size comparisons to be made with bison remains from other Pleistocene sites. Measurements of these elements are given in Tables 5.12, 5.15-5.17. In addition to Stanton Harcourt, measurements are given for *Bison priscus* from Marsworth (also MIS 7), from Late Pleistocene deposits nearby at Thrupp and Yarnton (both MIS 5a), Sutton Courtenay (MIS 3) and for some very large samples of Late Pleistocene age (MIS 5-3) from western Siberia (Shpansky *et al.* 2016). Although geographically remote, the Siberian data are interesting as the presence of partial skeletons provides a framework for estimating the size (height at the withers) of the Stanton Harcourt bison.

The metacarpals, metatarsals and tibiae from the above-mentioned localities are compared in Figures 5.28-5.30. At the British sites, it is evident that the MIS 7 *Bison priscus* at Stanton Harcourt and Marsworth was a substantially larger animal than the bison from the Late Pleistocene localities. Although bison are strongly sexually dimorphic, the small samples from Stanton Harcourt preclude any discussion on the ratio of males to females. However, it is noteworthy that the Stanton Harcourt specimens are comparable in size to the bulls in the very large sample of both sexes from Siberia. From the combined measurements of other skeletal elements, the height at the withers for these bulls is calculated to range from 1.8-1.9 m (Shpansky *et al. op. cit.*). These measurements equate to those published for free ranging European bison *Bison bonasus* where height at the withers averages 1.85 m for adult bulls and 1.67 m for cows. Body length reaches 3 m in bulls and 2.7 m in cows and weight in adults ranges from around 800 – 1,000 kg (Krasiñska and Krasiñski (2002).

## Small vertebrates

Despite numerous attempts to retrieve small vertebrates by sieving large bulk sediment samples, the results were negligible as described below.

### *Herpetofauna - reptiles and amphibians*

The following report on the herpetofauna is given by C. Gleed-Owen (1998):

Sixteen bulk samples of organic silt, sand and gravel, each weighing c.4-6kg, were sieved at 250 or 500μm. The residues were initially studied in order to extract molluscan remains in 1993, but all small vertebrate remains were also removed. The icthyofaunal remains identified by Irving were extracted from these samples. Each sample has produced a few fragmentary

| SAMPLE NO. | |
|---|---|
| 1041 | cf. *Rana* sp. |
| 1046 | cf. *Rana* sp. |
| 2002A | *Rana* sp. |
| 2002B | *T. vulgaris/helveticus, R. temporaria, Rana* sp. |
| 2003A | *Rana* sp. |
| 2003B | cf. *Rana* sp. |
| 2007 | Anura indet. |
| 2008A | Anura indet. |
| 2009 | Anura indet. |
| 2010 | cf. *Rana* sp. |
| 2011 | *R. cf. arvalis/dalmatina* |
| 2017 | Anura indet. |
| 4003 | Anura indet. |
| 5001 | *R.* cf. *arvalis/dalmatina* |
| 5004 | *Rana* sp. |
| 5005 | *R.* cf. *temporaria, Rana* sp. cf. green frog, *Rana* sp. |

Table 5.18 Herpetofauna from Stanton Harcourt (Gleed-Owen 1998)

| | |
|---|---|
| *Gasterosteus aculeatus* | three-spined stickleback |
| *Esox lucius* | pike |
| *Perca fluviatilis* | perch |
| *Anguilla anguilla* | eel |
| *Abramis brama* | bream |
| *Gobio gobio* | gudgeon |
| *Leuciscus leuciscus* | dace |
| *Squalius cephalus* | chub |
| *Rutilus rutilus* | roach |
| *Barbatula barbatula* | stone loach |
| *Salmo salar* | salmon |

Table 5.19 Icthyofaunal species identified in the Stanton Harcourt channel (B. Irving 1995)

anuran (frog and toad) remains, but they were disappointingly sparse. Most of the remains are not easily identifiable and few can be identified to species or even to genus level. There appears to be a taphonomic bias against terrestrial small vertebrates in the samples. Only a few poorly preserved microtine teeth were recovered, and the remains of amphibious species such as anurans were scant. Conversely, freshwater Mollusca were abundant and eleven fish species were identified. The herpetofaunal list identified is as follows: *T. vulgaris/helveticus, R. temporaria, R.* cf. *temporaria, R.* cf. *arvalis/dalmatina, Rana* sp. (cf. green frog), *Rana* sp., cf. *Rana* sp., Anura indet. The taxa recovered from each sample are listed in Table 5.18.

### *Icthyofauna - fish*

Eleven species of fish have been identified by B. Irving (Institute of Archaeology, University College London): three-spined stickleback, pike, perch, eel, bream, gudgeon, dace, chub, roach, stone loach, salmon. Eel fry travel from the Sargasso Sea to the British Isles on the Gulf Stream, indicating that its position was in interglacial mode (B. Irving, pers. comm.). The overall assemblage is typical of the upper reaches of the Thames today.

### *Avifauna - birds*

Of the 5 bird bones recovered from the excavations, only one is identifiable to species. This is SH1-273, the diaphysis of a L. humerus identified as that of a black-headed gull *Chroicocephalus ridibundus* (J. Cooper, Natural History Museum, London, pers. comm.). According to the field notes, this bone was 'wedged in mammoth skull SH1-242' in square I12. This skull is part of the virtually undisturbed bone group A in Bed 37 and the bird bone is at the interface of Beds 37 and 39 (see Chapter 2 and Figures 2.9 ad 2.10).

The black-headed gull is a small gull commonly found almost anywhere inland in the Palearctic including Europe and eastern Canada. In some regions these gulls migrate south in the winter but some birds reside year-round in the milder westernmost areas of Europe.

The remaining four bones are unidentifiable to species. SH6-81 and SH8-54 are mid-shaft limb fragments, SH8-28 is a partial coracoid and SH15-51 a carpometacarpus.

# Chapter 6

# The Climatic and Environmental Evidence

The climate and environment of southern Britain varied both temporally and regionally during the interglacial equated with MIS 7. Rapid sea-level rise early in the interglacial (MIS 7e) was synonymous on land with temperate climatic conditions and the establishment of deciduous forest. Following a period of lowered sea level and cooler climate (MIS 7d), the faunal and floral data for the rest of the interglacial indicate the continuation of a temperate climate and deciduous forest but with more open grassland than in MIS 7e, interrupted by a second interval of lower sea-level and cooler climate (Figure 6.1).

There has been much debate about the timing of any land bridges linking Britain to the continent during MIS 7 (Preece 1995; Candy and Schreve 2007; Pettitt and White 2012). It is agreed that sea level was high during MIS 7e and access to Britain from the Continental mainland would have been impossible, but opinion varies regarding the extent to which Britain remained an island for the rest of MIS 7. Sea-level fell during the two cool intervals (MIS 7b and 7d) of the interglacial but it is widely held that only during MIS 7d did sea-levels become sufficiently depressed for Britain to have become a peninsula and thus accessible again from the Continental mainland.

The presence of eel among the fish remains from the site (Table 6.5) suggests that during the part of the interglacial represented at Stanton Harcourt, the Gulf Stream and the Polar Front were in a similar position to today and that the English Channel was open at least to the Thames Estuary (Irving 1995). Small changes in solar output have a direct influence on the distribution of ocean currents in the North Atlantic (Bond *et al.* 2001) and the sea surface

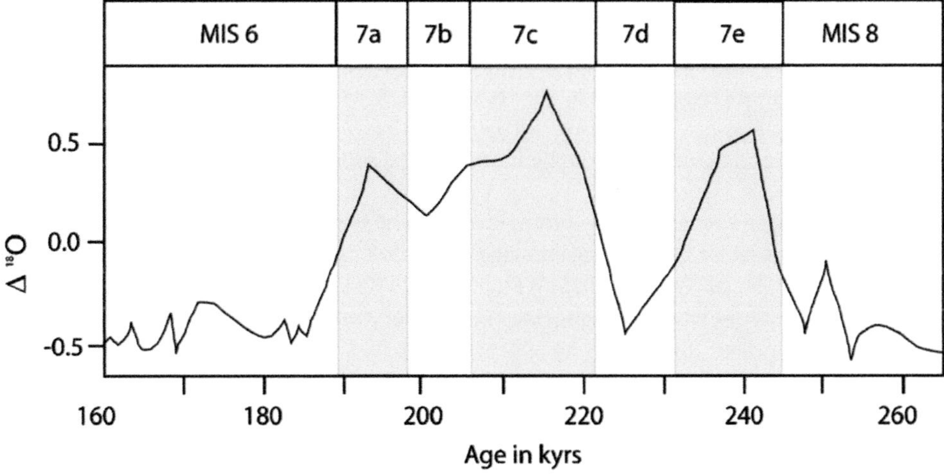


Figure 6.1 $\delta^{18}O$ curve for Marine Isotope Stage 7 showing warm substages MIS 7e, 7c and 7a (shaded) and cold substages MIS 7d and 7b (from Martinson et al. 1987, Candy and Schreve 2007)

temperature (SST) appears to be a critical factor in influencing changes in the mean climate, by influencing the distribution of pressure belts and precipitation (Barron and Pollard 2002). Warm sea temperatures in the North Atlantic would have given central and southern Britain a maritime climate, not dissimilar to present day Britain.

### Wood and other vegetation as climatic indicators

As described in Chapter 3, well preserved wood was abundant across the site and logs and branches were readily identifiable to species such as oak, willow and dogwood. As the excavations progressed, it became evident that the overall site represented a variety of local riverine environments. During the course of the excavation, a number of specialists took samples for analysis of what was available at the time. Therefore, not all sites were sampled by every specialist and the nature of the sample to some extent biased the result. Bulk sediment samples gave a bias towards grasses and trees with abundant pollen (such as hazel) while others (such as oak) were poorly represented in the bulk samples but very evident in the same deposit as logs and branches.

The extraordinary variety of trees and other plants is illustrated in Figures 6.2, 6.3 and 6.4. A list of all species identified is given in Table 6.1 with detailed analyses presented in Chapter 3: Table 6. Many hazelnuts were also excavated but were not counted and are thus not recorded in this table. The floral evidence from the site is broadly indicative of a temperate climate with species that are predominantly common in present-day Britain. The climatic interpretations of the specialists who analysed the samples are presented below.

#### *Wood*

Hather (1993) identified 254 pieces of wood across the site. These were predominantly twigs ranging from 3 to 7 years old, a few pieces of mature wood more than 10 years old, and mature oak 80-90 years old. He reported that most of the trees are deciduous which confirms interglacial conditions at the time of their deposition at Stanton Harcourt. All species still occur in northern, temperate Europe most typically in southern Britain or mid-France. Two species in particular emphasize the temperate nature of this part of the interglacial. Oak at Stanton Harcourt is well represented by the trunks and roots of mature trees as well as acorns which Hather described as a continental northern/central/southern European species. Hazel is represented by wood as well as nuts, the latter being a southern European species.

Although pioneer tree species such as birch, known to be one of the first species to appear at the start of an interglacial, the tree assemblage as a whole is dominated by species typical of stage III of an interglacial (Godwin 1975; West 1981, 1984)). Giving equal weight to fragments whether twigs or mature wood and ignoring the two very minor taxa *Betula* birch and *Juniper* juniper, the environment appears to have been dominated by *Fagus* beech (43.7%), *Carpinus* hornbeam (31.8%) and *Quercus* oak (24.5%). Percentages based on twig wood only match these figures closely. However, percentages based upon the much smaller number of fragments of mature wood reveal *Fagus* to be equal to *Quercus* (each 41.6%) with *Carpinus* a less dominant component (16%). One of the large pieces of oak (sample 14001) was estimated to be 80-90 years old.

Figure 6.2 Trees and shrubs represented at Stanton Harcourt

Figure 6.3 Flowers and grasses represented at Stanton Harcourt

Figure 6.4 Aquatics represented at Stanton Harcourt

| TREES |
|---|
| *Abies sp.* (Fir) |
| *Acer sp.* (Maple) |
| *Alnus* sp. (Alder) |
| *Betula* sp. (Birch) |
| *Carpinus sp.* (Hornbeam) |
| *Cornus sanguinea* L. (Dogwood) |
| *Corylus sp.* (Hazel) |
| *Corylus avellana L.* (Hazel / cobnut) |
| *Crataegus sp.* Hawthorn |
| *Fagus silvatica* (Beech) |
| *Fraxinus excelsior* (Ash) |
| *Juniperus* sp.(Juniper) |
| *Picea* (Spruce) |
| *Pinus* sp. (Pine) |
| *Prunus* sp. |
| *Prunus spinosa* L. (Blackthorn, sloe) |
| *Quercus sp.* (Oak) |
| *Rubus sp.* (Blackberry) |
| *Sambucus nigra* L. (Elder) |
| *Taxacaeae* (Yew) |
| *Taxacaeae* ? (conifer probably yew) |
| *Salix sp.* (Willow) |
| *Atriplex* sp. (Orache) |
| *Brassica* sp. |
| *Calluna* (Heather) |
| *Chenopodium* sp. (Goosefoot) |
| *Cirsium palustre* L. Scop (Marsh thistle) |
| *Compositae/Liguliflorae* |
| *Filipendula ulmaria L.* |
| *Gramineae*( grasses) |
| *Hippophae* (Sea buckthorn) |
| *Leontodon autumnalis* L. (Autumnal hawkbit) |
| *Lilliaceaea* |
| *Lychnis flos-cuculi* L. (Ragged robin) |
| *Polygonum* sp. (Knotweed or Knotgrass) |
| *Polygonum cf. lapathifolium* (Knotweed) |
| *Polygonum aviculare* (Knotgrass) |
| *Plantago media/major* (hoary or greater Plantain) |
| *Ranunculus acris* L. (Meadow buttercup), *repens* L. (Creeping buttercup) and *bulbosus L.* (Bulbous buttercup) |
| *Ranunculus batrachium*<br>*Ranunculus ranunculus sp. buttercup* |
| *Scabiosa columbaria* L. (Small scabious) |
| *Scleranthus annuus* L. (Annual knawel) |
| *Silene* sp. |
| *Solanum dulcamara* L. (Bittersweet, Woody nightshade) |
| *Stellaria cf. nemoram* Stitchwort |
| *Taraxacum* sp. (Dandelion) |
| *Thalictrum flavum* L. (Common meadow rue) |
| *Typha* (Bullrush) |
| *Valerianella dentata* (L.) Pollich (Narrow-fruited cornsalad) |

| HELIOPHYTES |
|---|
| *Bidens cernua* L. (Nodding bur-marigold) |
| *Cladium mariscus*(L.) Pohl. (Sedge) |
| *Lycopus europaeus* L. (Gipsy-wort) |
| *Myosotis* sp. (Forget-me-knot) possibly *(M.scorpioides* L.) and (*M. sylvatica* Hoffm.) |
| *Schoenoplectus lacustris* (L.) Palla. (Bullrush) |
| |
| **AQUATICS** |
| *Alisma plantago-aquatica* L. (Water plantain) |
| *Chara* sp. (Stonewort) |
| *Cyperaceaea* (sedges) |
| *Hippuris vulgaris* (Mare's-tail) |
| *Myriphyllum spicatum* (spiked water-milfoil) |
| *Potomogeton* sp. (Pondweed) possibly includes *P. compressus* L. - the grass wrack pondweed; broadleaf pondweed *P. natans* L. or Loddon pondweed - *P.nodosus* Poiret |
| *Ranunculus* (Batrachium) sp. (common water crowfoot/water buttercup) |
| *Sagittaria sagittifolia* L. (Arrowhead) |
| *Sparganium* erectum L. (Branched bur-reed) |
| *Zannichellia palustris* L. (Horned pondweed) |
| |
| **UNCLASSIFIED** |
| *Carex* sp. (sedge) |
| *Cyperaceae indet.(sedge)* |
| Fungal zoospores |
| Filicales |
| *Hieracium* sp. |
| *Hipericum* sp. |
| *Mentha* sp. |
| Moss spores |
| Moss fragments |
| *Pediastrum* |
| *Peridinium* |
| Pteridium *(algal cyst)* |
| Rosaceae |
| *Sigmopollis* |
| *Sphagnum* |
| *Spirogyra* |
| *Viola* sp. |

Table 6.1 Summary of vegetation from Stanton Harcourt (for details see Chapter 3)

Hather references Ellenberg (1988:158) on the association of oak, hornbeam and beech. Oak and hornbeam prefer wetter areas and beech drier conditions. Oak, hornbeam and beech can all tolerate a significant lowering of temperature in the winter but not in the spring. They can open buds at a temperature just below freezing. In central Europe today, the oak hornbeam association is rarely seen above 500m above sea level although beech woods can occur up to 1500 m. Oak/hornbeam woods are most common between 45 and 55° north requiring an annual rainfall of 500-600mm, a July temp of 17-19°C and a mean annual temperature of 9°C.

Ian Gourlay reported (2000) that most wood samples he analysed revealed narrow rings indicative of slow growth for most of their observable life. Narrow rings are typical of trees under stress caused by hot dry conditions or cold weather or short daylight hours and short growing season, such as in Scotland. As all the identified species still occur in northern temperate Europe, the latter situation is more likely. Such species in a prolonged hot dry climate would be likely to die after a year or two of such alien conditions, whilst they would be more likely to survive a long cold spell with reduced daylight. However, the samples Gourlay examined included mature oak trees (*Quercus robus* or *Quercus petraea*) up to 120 years old, indicating relatively stable, interglacial conditions.

***Pollen, seeds and spores***

Campbell (1993) had bulk sediment samples from Sites 1 and 2.

*Site 1 - pollen and spores*

Campbell took 5 samples, only 3 of which yielded pollen. These are from a light grey silty sand, the pollen is not abundant, and the preservation is poor. A large percentage of the seeds and spores were not identifiable, and many were likely Jurassic.

*Site 2 - seeds and spores*

Campbell analysed 6 bulk samples. Two contained wood fragments which he did not include in his analysis but he noted that these contained the most seeds. These were silty, sandy sediments and the general condition of the seeds ranged from very good to poor, with many crushed and fragmented seeds present. Below and above these gravel horizons contained fewer seeds, but the assemblages are essentially consistent throughout the deposit. This therefore probably reflects changes in hydrological conditions during deposition rather than any change in climate.

Campbell concluded that the assemblages from both sites represent a fully interglacial climate and environment. The local environment is discussed later in this chapter.

Robinson (2012) analysed seeds from a 500-gram bulk sample (1069) from Site 1. This sample was from a *Corbicula* bed and had been sieved through a .5-micron sieve for molluscs. This is a larger mesh diameter than is normally used for seed analysis and therefore there are likely, originally, to have been many more seeds in the sample.

Robinson's results concur with those of Campbell: that the seeds of wildflowers and other plants identified are common in the British Isles today, with the majority restricted to southern Britain.

## Climatic interpretation of the insect assemblages

Details of the insects identified by Coope (2006) are given in Chapter 3: Table 4. The main sample he analysed was from the vicinity of *in situ* roots and wood at Site 2. He was also able to identify the location of his original sample presented in the BAR report (Briggs *et al.* 1985). This was also at Site 2 but about 18m to the NE of the wood described in this volume. Coope observed no significant differences across the site and regarded the insect remains as one assemblage. He provided the following interpretation of their climatic significance:

> 'Almost all the species in this assemblage are still living in the British Isles, albeit some of them are very rare today and may possibly now be extinct. Thus, *Bembidion octomaculatum* and *Polistichus connexus* are very local and rare in the extreme south of England (Lindroth 1974). The British occurrences of *Rhysemus germanus* are known only from old specimens and records from Bristol to Swansea (Jessop 1986). The four non-British species all have predominantly southern distributions. *Oxytelus gibbulus* has a curiously disparate geographical range but its nearest present day occurrence is in the Caucasus Mountains (Hammond *et al.* 1979). The nominate form of *Aphodius carpetanus* is restricted to Spain but a variety is also known from Sicily (Balthasar 1964). The natural range of *Stomodes gyrosicollis*, a weevil that feeds on tripholium and other polygonaceae, reaches only as far north as eastern Austria and Czechoslovakia where it is confined to the plains. However, there is an isolated population in central France where it was introduced from Bosnia/Herzgovina, probably in lucerne brought in as fodder for horses during the Franco-Prussian war (Hoffman 1950).
>
> Since all these species are so varied in their ecological requirements - terrestrial and water beetles, predators and phytophages, specialists of many sorts - it is difficult to explain their presence by some peculiarity in availability of local habitats. Rather, their only unifying demand would appear to be a mutually acceptable climate.
>
> In addition to these climatically significant beetles, the caddisfly *Hydropsyche bulbifer* does not occur in Britain but is a central and southern European species that reaches only as far north as northern Germany (Wilkinson 1980).
>
> The fact that the insect fauna, viewed as a whole, requires a climate at least as warm as that of the present day, indicates that the Stanton Harcourt channel deposit accumulated under temperate conditions. This interpretation is supported by the presence of an oak dependent beetle species and also by the presence, within the deposit, of large oak trunks indicating that the climate had been temperate for long enough for the colonisation and maturation of a deciduous forest. The evidence shows conclusively that the Stanton Harcourt channel was filled under interglacial conditions.
>
> It is possible to make quantified estimates of the thermal climate at the time when the sediment accumulated by applying the Mutual Climatic Range (MCR) method to

> the coleopteran assemblage (Atkinson *et al.* 1987). In order to make estimates that are independent of those derived from the botanical data, only the predatory or general scavenging beetle species have been used here i.e. species whose food chains do not depend on the presence of particular macrophytes. There are 44 species in the Stanton Harcourt assemblage that are also on the MCR database. The overlap of the climatic envelopes of all these species was 100%. MCR estimates based on them gave the following figures where Tmax is the mean temperature of the warmest month (July) and Tmin is the mean temperature of the coldest months (January and February).
>
> Tmax between 17°C and 18°C
>
> Tmin between -2°C and 2°C
>
> Assuming a simple sinusoidal curve for the variation in mean monthly temperatures, it is possible from the above figures to give estimates for the mean temperatures for the other months in the year. It is likely that the actual temperatures may have been rather higher than these figures indicate since the most southerly species are not on the MCR database.
>
> It is not possible to quantify the mean monthly precipitation levels on the basis of the insect fauna. However, adequate rainfall must have been available to maintain a continuous flow of water in the river throughout the year and to sustain the marsh during the summer months when all the Coleoptera were active'.

In addition to the above, Coope (*pers. comm.)* observed that there were no obligate northern species in this assemblage in contrast to many other mammoth sites elsewhere in Britain.

## Climatic interpretation of the molluscs

The climatic interpretation of the molluscs from Stanton Harcourt is based on the analysis of the populations and species present (Keen 1992; Gleed-Owen 1998) and their isotopic composition (Buckingham 2004). Molluscs were abundant across the excavation site. Those identified in 13 samples from Sites 1,2,4 and 6 are shown in Chapter 3: Table 3. As noted in Chapter 3, the nomenclature of some species has changed since Keen and Gleed-Owen prepared their species lists. Both the current and former names are given in the Table but the updated nomenclature is used forthwith.

### *Populations and species*

Keen (1990) stresses the importance of Mollusc populations from fluvial sediments for regional climatic reconstruction, particularly if there are both freshwater and terrestrial species present. The presence of particularly thermophilous molluscs such as *Corbicula fluminalis, Potomida littoralis* and *Odhneripisidium moitessierianum*, which currently thrive in warmer climates than Britain, suggests that temperatures could have been higher, at least for part of the time that the sediments were accumulating. Articulated specimens of these molluscs were present across the excavation site wherever the interglacial sediments were encountered, which helped to delineate the extent of the Stanton Harcourt Channel.

The Site 1 and Site 4 samples were taken from within the point bar sequence described in Chapter 2. Sample 4003 came from the Corbicula bed at the top of Bed 58. One of the Corbicula from this layer (4003.2) was used in the isotope study discussed in the following section. Sample 1021a came from a very thin layer of slightly sandy, clayey gravel just below the quarry floor, from Bed 73, which is high in the bar sequence at Site 1, above a mammoth pelvis SH1/8 in Bed 39, a handaxe A4 in Bed 68 and a handaxe A1 in Bed 72. This sample is also stratigraphically above all the virtually undisturbed bone groups A, B, C, D and E and associated artefacts discussed in Chapter 2, indicating interglacial conditions during burial. Keen comments on sample 1021a that species such as *C. tridentatum*, *E. montana*, and *V. pygmaea* have a distinctly restricted northern range at present suggesting a climate little cooler than now while *P. amnicum*, *E. henslowana* and *O. moitessierianum* are usually temperate indicators.

The extinct species *C. crayfordensis*, seen at all the sampled sites, is only known in temperate environments. *C. fluminalis*, also excavated from many layers in the bar sequence and from all sampled sites, is mainly limited to sub-tropical areas of N. Africa and India at present and is generally regarded as a temperate indicator when it occurs in Britain. However, mass mortality is known to occur in *Corbicula fluminalis* populations when temperatures fall below 0°C for one week (Meijer and Preece 2000). Thus, in some instances, where large numbers of small articulated specimens were recorded on site, this might be indicative of cold winter temperatures at times (see section below on isotope studies).

The Site 2 samples were mainly from very thin layers or lenses of mainly silty sediment, in an area where there were *in-situ* tree roots and branches of wood. Clusters of articulated *Corbicula fluminalis* were observed during excavation near the tree roots. A large bulk sample (2002A) was taken from a shallow scour cut into the Oxford Clay, which was also picked for seeds and vertebrate bones. Sample 2007 was just above 2002A in the sequence at this Site and both these samples were below the main wood horizon. 2009 and 2010 were from the wood layer followed by 2008a which was a more extensive iron-stained sandy gravel with many *Gryphaea*. Sample 2011 was just above 2008A. The last sample in the sequence at this Site (2006) was taken from a very thin silty sand at the top of a sand filled scour which cut into all the earlier layers. This was just below remnants of MIS 6 sediment.

In an earlier exposure of the Stanton Harcourt Channel deposits, Cartledge (in Briggs *et al.* 1985) identified 23 molluscan taxa to species level and 10 more to genus. All these taxa were found in Keen's analysis except *Pisidium clessini* Neumayr, *Pisidium milium* Held and *Cepaea* sp. and the total number of taxa recorded rose to 48. Keen regarded this as wholly typical of the number to be expected of a fluvially derived assemblage formed under interglacial conditions. Many hand excavated molluscs, including some land snails, are being collated and have not been identified at this time.

### *Isotope studies*

Mollusc shells from the Stanton Harcourt Channel were used in an Oxygen and Carbon isotope study which formed part of a doctoral thesis (Buckingham 2004). Only an example is presented here with some new observations on the stratigraphic context. A summary is presented of the other data which will be published elsewhere.

This research is based on the observation that shell carbonate stores carbon and oxygen which are dissolved in the river water. This in turn reflects temperature and water conditions. Oxygen and carbon are made up of different isotopes. For example. rainwater contains a higher proportion of the lighter $O^{16}$ isotope than the heavier $O^{18}$ isotope. Similarly, the isotopes $C^{12}$ and $C^{13}$ can be extracted from shell carbonate. By measuring the proportions of these isotopes, it is possible to get a glimpse of the climate at the time that the mollusc was alive. River water in Britain at the present day has negative $O^{18}$ values.

The shells used for the isotope study were all in good condition and articulated where possible. Most molluscs that inhabit warm water have shells made up of aragonite which is a more soluble form of Calcium Carbonate. If they have survived in the fossil record then burial has often been soon after death or there has been some mineral replacement. Several specimens from Stanton Harcourt that were used in this study were tested to confirm that they had retained their original aragonite composition.

*Corbicula fluminalis* is a small non-marine mollusc that resembles a cockle, which is sometimes referred to as the Asiatic clam. It has a tough cross-lamellar shell (Kennedy *et al.* 1969) so has the potential for survival in the sediment, but disarticulation of this bivalve would occur soon after death. As discussed in Chapter 2, whole beds of articulated specimens are likely to be close to where they lived. *Potomida littoralis* from Stanton Harcourt were also often articulated when excavated. This mollusc more closely resembles a mussel. Its shell has a different structure and was more difficult to sample than the Corbicula as it was crumblier.

In total, 19 *C. fluminalis*, three *P. littoralis* and one *Odhneripisidium moitessierianum* were sampled from the Stanton Harcourt Channel for the isotope study from Sites 1,2,4,5,6,7,9 and 14. A few molluscs from other Pleistocene sites in Britain were also used for comparison and data from some modern specimens were collected. This study was in collaboration with laboratory technicians at Oxford University and some of the other specimens were sampled by Martin Brasier and Alison Jones from the Earth Sciences Department, Oxford University. Much of this data will be published elsewhere.

One example of the full data from a *Corbicula fluminalis* specimen from a known context in the Stanton Harcourt Channel and its climatic and environmental interpretation is discussed here.

Sample 4003.2 is an articulated specimen of *Corbicula fluminalis* (*Cf*) which was excavated with many other articulated *Cf* of a similar size, articulated *Potomida littoralis* (*Pl*) and some other small molluscs. The *P. littoralis* were clearly *in-situ* with their shell edges uppermost (Figure 6.5). These shells were in a very thin layer of dark grey silty, sandy gravel at the top of Bed 58 in the sedimentary sequence at Site 4. Below the shells, Bed 58 is composite and consists of many very thin alternating layers or lenses of sand or silty gravel which infill a scour immediately W of the 'plank' (see Chapter 2 Figure 3). The top of Bed 58 drapes this large piece of wood although there are fewer shells as the layer becomes thinner and dies out eastwards. The scour is located between a 'turtle' stone at position 1R of the right bank and the wood at the later bank position 2R and was probably the result of bank erosion and eddying at this locality. This scour gradually filled with sediment in low to moderate energy conditions. Shells were found at several levels. It is likely that these molluscs were living in a slightly sheltered area

Figure 6.5 Bed of *Potomida littoralis*

Figure 6.6 The eroded right bank of the River Evenlode

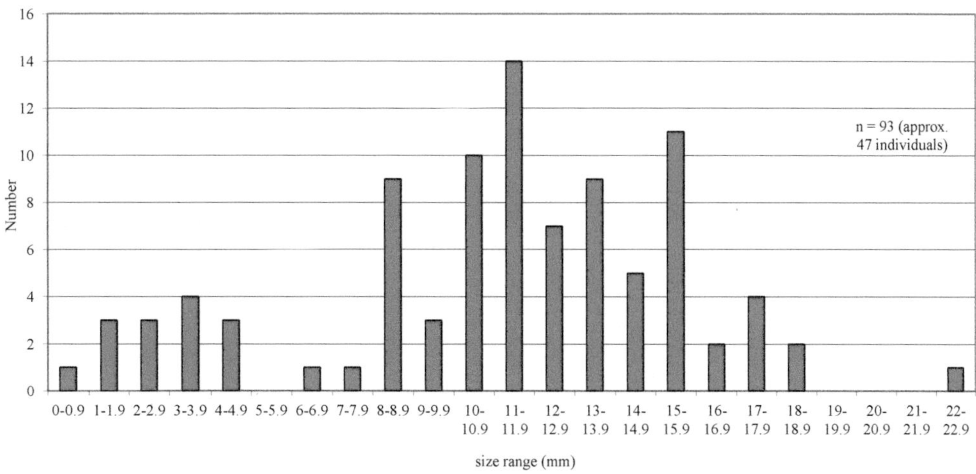


Figure 6.7 Size range of *Corbicula fluminalis* from sample 1069 from the top of Bed 58 at Site 1

but near to the *thalweg* for nutrients for their filter feeding. Similar scours, with sand or silty gravel being deposited in quiet water near tree roots, are located near the right bank of the nearby tributary of the Thames, the R. Evenlode, where the faster flowing water in the *thalweg* swings away from the bank (Figure 6.6).

When the section drawings were correlated and the stratigraphy at Site 1 was completed, it became apparent that although the Corbicula samples 4003.2, 1069.2, 1069.3 and 1065a were from slightly different locations, they were all from near the top of Bed 58 near the Site 1/Site 4 boundary. *Corbicula fluminalis* specimens from a bulk sediment sample 1069 from Bed 58 were measured and assessed for weathering and damage. All ages from juveniles to young adults were represented of at least 49 individuals and more than 50% of the shells were unweathered on the inside (Figure 6.7). There were also some seeds and nuts recovered from this sample (identified by M. Robinson of the Natural History Museum, Oxford: Chapter 3 Table 6).

It is interesting to note that the very thin sand/silty gravel layers of Bed 58 at Sites 1 and 4 were being deposited at approximately the same time as laminated sandy silts at Site 8 near the right bank at position R4 (Chapter 2 Figure 2.22). The *in-situ* partial mammoth carcass in Bone Group E discussed in Chapter 2 is partially in or on lenses of this bed, thus the interpretation of the shell isotopic data from Bed 58 is likely to indicate the climate when this carcass first came to rest at the edge of the channel. This was slightly later in the sedimentary sequence than the other hardly disturbed bone groups A-D in the Stanton Harcourt Channel, discussed in Chapter 2.

The right valve of *C. fluminalis* 4003.2 was tested to confirm that it was composed of aragonite and that no alteration to the shell had occurred (Allman and Lawrence 1972). After cleaning remnants of periostracum (the outer organic layer) from the left valve of 4003.2, samples of aragonite were taken from the outermost layer of the shell at closely spaced intervals along the length of the shell (Table 6.2). These samples represent growth increments. After preparation to remove any remaining organic residue, the oxygen and carbon isotopes were measured

| STANTON HARCOURT CHANNEL, Dix Pit, Stanton Harcourt | | | Collected 14.9.96 | | | Results 1.12.99 |
|---|---|---|---|---|---|---|
| *Corbicula fluminalis* | Site 4, B45, layer 11 on section drawing 96/21, layer 11 on section drawing 96/7 (top of Bed 58) | | | | | |
| Articulated | Thin, dark grey, silty, sandy gravel (same bed as in situ *Potomida littoralis* ) | | | | | |
| **Sample Number (CB)** | **Sample Number (Lab)** | **Distance from Umbo(cm)** | **$\delta^{13}C$** | **$\delta^{18}O$** | **Growth lines** | **No. of Results** |
| 4003.2 left valve | A99/6495 | 2.40 | -7.426 | -4.654 | | 1 |
| | A99/6496 | 2.30 | -7.539 | -5.494 | | 2 |
| | A99/6497 | 2.20 | -7.006 | -4.560 | | 3 |
| | A99/6498 | 2.10 | -7.558 | -4.843 | 2.10 | 4 |
| | A99/6499 | 2.00 | -8.768 | -5.757 | | 5 |
| | A99/6500 | 1.90 | -8.349 | -5.777 | | 6 |
| | A99/6501 | 1.85 | -8.563 | -5.617 | | 7 |
| | A99/6502 | 1.75 | -7.794 | -5.282 | | 8 |
| | A99/6503 | 1.70 | -7.754 | -5.110 | | 9 |
| | A99/6504 | 1.60 | -7.721 | -4.955 | | 10 |
| | A99/6505 | 1.55 | -8.141 | -5.095 | | 11 |
| | A99/6506 | 1.50 | | | 1.50 | no result |
| | A99/6507 | 1.40 | -7.326 | -5.410 | | 12 |
| | A99/6508 | 1.30 | -7.206 | -5.650 | | 13 |
| | A99/6509 | 1.20 | -7.338 | -6.127 | | 14 |
| | A99/6510 | 1.10 | -7.215 | -6.275 | | 15 |
| | A99/6511 | 1.00 | -7.386 | -6.056 | | 16 |
| | A99/6512 | 0.90 | -7.305 | -5.291 | | 17 |
| | A99/6513 | 0.80 | -8.048 | -5.360 | 0.80 | 18 |
| | A99/6514 | 0.70 | -8.725 | -5.421 | | 19 |
| | A99/6515 | 0.65 | -8.295 | -5.540 | | 20 |
| | A99/6516 | 0.50 | -8.763 | -6.312 | | 21 |
| | A99/6517 | 0.30 | -9.002 | -7.025 | | 22 |
| | A99/6518 | 0.20 | -8.832 | -6.529 | | 23 |
| | A99/6519 | 0.10 | -8.969 | -6.316 | | 24 |
| | | Average | -7.960 | -5.602 | | |
| | | Maximum | -7.006 | -4.560 | | |
| | | Minimum | -9.002 | -7.025 | | |
| | | Median | -7.774 | -5.517 | | |

Table 6.2

with a VG Prism mass spectrometer attached to an on-line VG Isocarb preparation system under standard conditions (Corfield 1995). Daily calibration to the Peedee (PDB) standard, via NBS 19 and Carrera marble, used the Oxford in-house (NOCZ) marble standard. Replication and interlaboratory comparisons are within 0.05‰ for $\delta^{18}O$ and 0.08‰ for $\delta^{13}C$ (Table 6.4).

The growth lines on the shell indicate that specimen 4003.2 was approximately 4 years old when it died. Corbicula rarely live beyond 5 years so this is a young adult. Figure 6.8 shows that the $C^{13}$ and $O^{18}$ isotope curves are frequently in phase with each other (go up and down at approximately the same time), especially in the later years, which is interpreted as a strong temperature signal. The undulating pattern suggests that these curves record a seasonal signal. A shift towards less negative values of $O^{18}$ and $C^{13}$ coincide with growth lines on the shell and is interpreted as a winter growth hiatus (Clark 1974). This particular mollusc also showed a shift to more positive values at the time of its death which was probably in the winter. During the second summer the isotope curves are not in phase, the $O^{18}$ values are more negative whereas the $C^{13}$ values become more positive. The carbon isotope curve may have been affected by changing vegetation or water conditions in this part of the channel during the summer season.

| Species | Sample No. | $\delta^{18}O_{ar}$ max | $\delta^{18}O_{ar}$ min | $\delta^{18}O_{ar}$ av | $\delta^{18}O_{water}$ | Min °C | Max °C | Av °C |
|---|---|---|---|---|---|---|---|---|
| *Corbicula fluminalis* | 4003.2 | -4.560 | -7.025 | -5.602 | -5.3 | 18.3 | 29.9 | 23.2 |
| *Corbicula fluminalis* | 4003.2 | -4.560 | -7.025 | -5.602 | -7.6 | 7.5 | 19.1 | 12.4 |
| *Corbicula fluminalis* | 4003.2 | -4.560 | -7.025 | -5.602 | -8.5 | 3.3 | 14.9 | 8.2 |
| *Corbicula fluminalis* | 4003.2 | -4.560 | -7.025 | -5.602 | -10.2 | -4.7 | 6.9 | 0.2 |
| *Corbicula fluminalis* | 1069.2 | -4.946 | -6.147 | -5.591 | -5.3 | 20.1 | 25.8 | 23.2 |
| *Corbicula fluminalis* | 1069.2 | -4.946 | -6.147 | -5.591 | -7.6 | 9.4 | 15.0 | 12.4 |
| *Corbicula fluminalis* | 1069.2 | -4.946 | -6.147 | -5.591 | -8.5 | 5.1 | 10.8 | 8.2 |
| *Corbicula fluminalis* | 1069.2 | -4.946 | -6.147 | -5.591 | -10.2 | -2.8 | 2.8 | 0.2 |
| *Corbicula fluminalis* | 1069.3 | -4.999 | -7.087 | -6.096 | -5.3 | 20.4 | 30.2 | 25.5 |
| *Corbicula fluminalis* | 1069.3 | -4.999 | -7.087 | -6.096 | -7.6 | 9.6 | 19.4 | 14.7 |
| *Corbicula fluminalis* | 1069.3 | -4.999 | -7.087 | -6.096 | -8.5 | 5.4 | 15.2 | 10.5 |
| *Corbicula fluminalis* | 1069.3 | -4.999 | -7.087 | -6.096 | -10.2 | -2.6 | 7.2 | 2.6 |
| *Corbicula fluminalis* | 1065a | -5.778 | -8.271 | -6.748 | -5.3 | 24.0 | 35.7 | 28.6 |
| *Corbicula fluminalis* | 1065a | -5.778 | -8.271 | -6.748 | -7.6 | 13.3 | 24.9 | 17.8 |
| *Corbicula fluminalis* | 1065a | -5.778 | -8.271 | -6.748 | -8.5 | 9.0 | 20.7 | 13.6 |
| *Corbicula fluminalis* | 1065a | -5.778 | -8.271 | -6.748 | -10.2 | 1.1 | 12.8 | 5.6 |

Minimum temperatures marginal for survival of *Corbicula fluminalis*

Minimum temperature too low for survival or maximum temperature too low for reproduction for *Corbicula fluminalis*

Temperature calculations using the aragonite temperature curve for molluscs (Grossman & Ku, 1986)

$T°C = 21.8 - 4.69 (\delta^{18}O_{aragonite} - \delta^{18}O_{water})$

Note: Water isotopic values are the maximum and minimum values deduced from all the vertebrate teeth studied by A.Jones (2000)

Calculations using the average water value obtained from mammoth and straight-tusked elephant teeth from the Stanton Harcourt Channel

Table 6.3

The same method of sampling and analysis was used for the other *C. fluminalis* from Bed 58. The average isotope $\delta^{18}O$ values for 1069.2 and 1069.3 are close to that of 4003.2 (Table 6.3). Corbicula 4003.2 and 1065a are the same size and there are some similarities in their isotope curves (1065a is not illustrated). Both specimens have a faint growth hiatus between 1.4 and 1.5cm followed by a strong swing to more negative $\delta^{18}O$ and $\delta^{13}C$ values, then a marked growth line at 2.1cm. However, the $\delta^{18}O$ values for 1065a are consistently more negative than those of 4003.2, possibly implying warmer conditions especially in the summer, although the seasonal range of $\delta^{18}O$ values are comparable. It is possible that they are from different layers within Bed 58 as this stratigraphic unit is composite and probably represents several seasons. Alternatively, the shallow pool situation, but close to the active *thalweg*, may have led to varying bottom conditions. There may have been periodic influxes of fresh, oxygenated water containing more $O^{16}$ which may have affected some individuals more than others, depending on their aspect and depth of burial. The maximum, average and minimum values for these two *C. fluminalis* are compared with a modern *Corbicula fluminalis* from the Oyo River, Indonesia (Figure 6.9). 1065a most closely resembles the average and minimum $\delta^{18}O$ values seen for the Indonesian Corbicula. All three Corbicula have a similar seasonal range. The carbon isotope values are much more variable reflecting different water and vegetation conditions. The Indonesian sample has a marked trend to more negative $C^{13}$ values during the monsoon season when there is a large influx of meteoric water, which may have changed the bottom conditions of the river. The $\delta^{18}O$ values seem less affected by the monsoon but $O^{16}$ from the extra rainwater may be compensated for by lower temperatures during this season. The *Corbicula fluminalis* 1065a from Bed 58 is shown (after the removal of remnants of the periostracum) alongside the Indonesian *C. fluminalis* (Figure 6.10). The carbon isotope curves for some molluscs from the Stanton Harcourt Channel also show occasional strong swings to more negative values

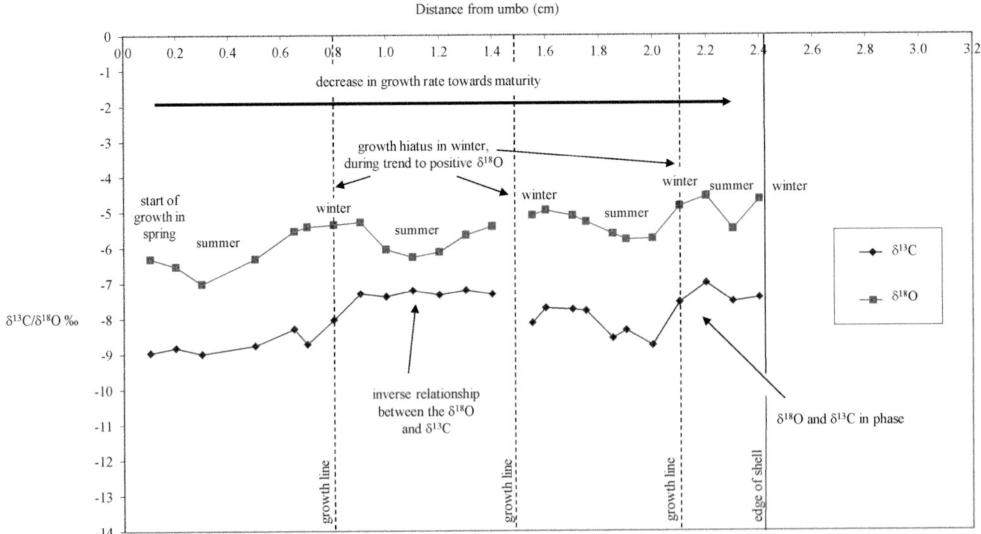


Figure 6.8 Graphs showing variations in the $^{18}$O and the $^{13}$C of shell carbonate of *Corbicula fluminalis* 4003.2 from Bed 58.

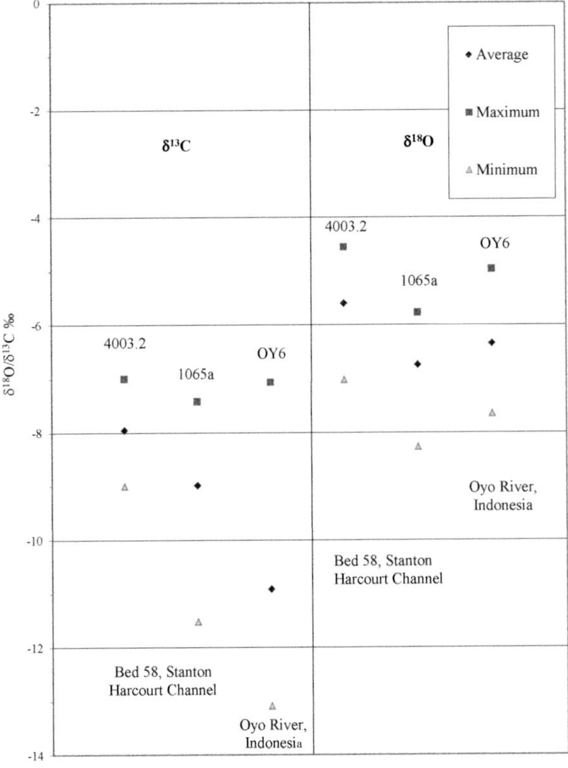


Figure 6.9 Comparison of the Maximum, Average and Minimum values of $\delta^{18}$O and $\delta^{13}$C of shell carbonate of *Corbicula fluminalis* from Bed 58 compared to a modern *Corbicula fluminalis* from Indonesia

Figure 6.10 *Corbicula fluminalis* from the excavation (left) compared to a modern Corbicula from Indonesia

compared to the $O^{18}$ and these are interpreted as seasonal flood events.

Converting the isotope values into actual estimated temperatures is complex because any formulas for the fractionation of shell carbonate rely on an estimate of the water values in which the mollusc lived. Grossman and Ku (1986) studied live molluscs from the continental margins off southern California and Texas, U.S.A. and Mexico. They found a marked temperature dependence of biogenic aragonite and developed an aragonite-water palaeo-temperature formula to apply to molluscs with aragonite shells.

$$T\ (^{\circ}C) = 21.8 - 4.69\ (\delta^{18}O_{aragonite} - \delta^{18}O_{water}$$

The values were recorded as per mille variations from the Chicago Pee Dee Belemnite (PDB) standard. There are more variables in a fresh-water system compared to a marine environment but this formula is still applicable if the mollusc is mainly responding to changes in temperature.

Estimated isotopic values of oxygen in the water from the Stanton Harcourt Channel during MIS 7 were provided by A. M. Jones. As part of a doctoral thesis (Jones 2001) studied the isotopic composition of enamel phosphate and carbonate from the teeth of modern elephants from Amboseli National park, Kenya. She then compared them to mammoth and straight-tusked elephant teeth from the Stanton Harcourt Channel excavation. Amino acid profiles from the Stanton Harcourt teeth are very similar to the modern elephants, indicating well preserved collagen. Bison teeth from the Stanton Harcourt Channel also formed part of her study.

Seasonal fluctuations in $\delta^{18}O_{PO4}$ were observed in the individual teeth records from Stanton Harcourt and the calculated water range was from -5.3 to –10.2‰. The average elephant and mammoth data for ingested water values ranged from -7.6 to -10‰ (Jones *et al.* 2001). Calculated water values were compared to modern precipitation data from the IAEA monitoring station at Wallingford on the River Thames. The Stanton Harcourt teeth fell within the precipitation $\delta^{18}O$ range but towards the negative end of this range. The mammoth teeth tested had slightly more positive values than the straight-tusked elephant. River water values would not be exactly the same as precipitation.

Bison teeth had a $\delta^{18}O_{PO4}$ value of 15.3 %, similar to Scottish cattle today and gave a calculated water value of -8.1 to -8.5‰.

Of the teeth sampled by Jones, the mammoth tooth SH1/353 from Site 1 is in a bed which, stratigraphically, is nearest to the mollusc bed with samples 1065a, 1069.2, 1069.3 and 4003.2 (Bed 58). Although in a bed which was deposited before these molluscs (Bed 48) and about 8m SE, the scenario in the channel setting is similar. The tooth was excavated from the top of a trough of silty gravel with wood and *P. littoralis* just to the north of an Oxford Clay mound believed to be a remnant of the right bank at the channel margin position 3R (Chapter 2 Figure 2.4). It partially rested in a lens of sandy silt of Bed 50, immediately southwest of a mammoth tusk SH1/351. Bed 48 is also one of the layers with *P. littoralis and C. fluminalis* that banks up with *Gryphaea* shells against the SW side of the upside-down skull SH4/124 in Group B (see Chapter 2).

The average $\delta^{18}O_{water}$ calculated from the mammoth tooth SH1/353 was -7.6‰ (Jones *et al.* 2001). The present day $\delta^{18}O_{water}$ of the River Thames averages about -7 ‰.

Using the average water isotope value of -7.6 (estimate from tooth SH1/353) the estimated average temperature for *C. fluminalis* 4003.2 is 12.4 with a temperature range of 7.5 to 19.1 (Table 6.3). These values are comfortably in the temperature range that *C. fluminalis* is known to inhabit. However, if the seasonal fluctuations in the $\delta^{18}O_{water}$ recorded by Jones for all the teeth across the site are taken into account, the temperature range may be much higher (Table 6.3). Estimated summer temperatures for sample 4003.2 are then only just within the tolerance of *C. fluminalis* but winter temperatures are close to the limit for survival. Some mortality in modern Corbicula populations is known to occur when temperatures drop below 4 °C and there is considerable mortality within 1 week at 0°C. Juveniles grow quickly at 25° C (about 1mm a week) but growth stops below 15°C (Meijer and Preece 2000). Reproduction rates also decrease when spring and summer temperatures fall. Specimens 1069.2, 1069.3 and 1065a show a similar picture to 4003.2. All estimated temperatures of these Corbicula fall within the range of the temperatures experienced in Britain at the present day. Water temperatures are likely to be slightly less extreme in the channel compared to those experienced on the land.

*Corbicula fluminalis* is a good indicator of climate because it is an opportunist species. Recently it has been expanding its range in Europe. It is a pest at Hydro Electric Power Station outlets in the River Rhine. It prefers warm water but can tolerate slightly brackish water and most water temperatures above freezing. It has a strong shell to cope with fluctuating energy conditions and can tolerate different types of sediment. *C. fluminalis* is not evenly distributed over the Stanton Harcourt Site. It is present both below and above the relatively undisturbed bone groups A, B and C, D and E in Chapter 2. In later beds across the site, *Potomida littoralis* is present but few Corbicula, which may indicate a less suitable climate at that time, possibly an increase in cold winters. The three *P. littoralis* shells sampled in the isotope study have similar average $\delta^{18}O$ values to all the *C. fluminalis* sampled, as does the tiny *Pisidium moitessarianium* specimen, which was sampled as a whole shell. One specimen of *P. littoralis* gives the most positive $\delta^{18}O$ winter value of all those tested from Stanton Harcourt and is in a similar range to modern Unio sp. from the River Thames near Farmoor.

If climatic changes are rapid, elements of the flora and fauna may not be in total equilibrium with the prevailing climate. All specimens of *C. fluminalis* and *P. littoralis* from the Stanton Harcourt Channel have growth lines which frequently coincide with peaks in the oxygen isotope curves, indicating cool winters. The general decline in the frequency of *Corbicula* and the other warm climate elements higher in the sedimentary sequence in the centre of the excavation area at Dix Pit, is accompanied by a change to moderate or well-sorted gravel, in thicker and more extensive sets of cross-bedding, with subordinate silt lenses. There is also a decrease in the number of derived *Gryphaea* and an increase in the proportion of limestone in the sediment. This is interpreted as a result of deteriorating climatic conditions, with an increase in discharge and/or a decrease in vegetation leading to a subsequent increase in the sedimentation rate. It is likely that the lack of vegetation may have been the most important factor as, without its binding effect, bank instability and sediment mobility would have increased. Occasional thin silt beds with small, articulated, very weathered *Corbicula* are seen within these later gravels. These *Corbicula* may have been reworked or were living at a late stage, when environmental conditions had started to change. One *Corbicula* from this part of the site (6036) has a strong trend to positive $\delta^{13}C$ and $\delta^{18}O$ values near the time of death, probably indicating cold winter temperatures.

In summary Buckingham's study suggests that *C. fluminalis* were not living in their optimum climatic conditions at the time of bone accumulation and were near to their northern limit. Winter growth hiatuses and cold temperatures just before death, experienced by many individuals, imply that winter water temperatures frequently fell below 2 or 3ºC. Estimated water temperatures from the *Corbicula* and *Potomida* profiles are similar or slightly warmer than the present-day Thames, with temperatures mainly between 10ºC and 20ºC but with an annual range of temperature from >25ºC to just below freezing. Seasonal flood events are recorded by changes in sediment type and rapid changes in the carbon isotope record of some molluscs.

## Large vertebrates as climatic indicators

The large vertebrates from the excavations are described in Chapters 4 and 5 and summarised in Table 6.4. The herbivores are steppe mammoth *M. trogontherii*, straight-tusked elephant *P. antiquus*, horse *E. ferus*, red deer *C. elaphus*, and bison *B. priscus*. The latter three are recorded in interglacial assemblages attributed to MIS 9, and bison and red deer (but not horse) in MIS 5e. However, they also occur in assemblages associated with true cold climate obligates such as woolly mammoth *M. primigenius* and woolly rhino *C. antiquitatis*.

The straight-tusked elephant supports all other evidence at Stanton Harcourt for interglacial conditions as it is a consistent faunal element of all post-Anglian interglacials and usually found in association with evidence for temperate forest (Stuart 1982). The steppe mammoth is of particular interest, as described in Chapter 4. Occurring at Eurasian sites of late Early and Middle Pleistocene age, steppe mammoths are generally, but not exclusively, from interglacial context. During this period, they reduced significantly in size (Lister and Scott in press). It is the most commonly occurring vertebrate at Stanton Harcourt and of the 29 British vertebrate assemblages attributed to MIS 7, it is almost invariably the most commonly represented species (Scott in prep.). It might even be considered a marker species for the MIS 7 interglacial (Schreve 1997).

At Stanton Harcourt, some of the vertebrate remains themselves corroborate other data that suggest a shift from fully interglacial to cooler conditions. Jones *et al.* (2001) carried out isotopic analyses of collagen and enamel fractions from mammoth, elephant, horse and bison molars from Stanton Harcourt. The oxygen isotopes were used to calculate palaeotemperatures and the results indicate average annual temperatures to have ranged from similar to the present to cooler by 6 -7 degrees C. Of particular interest here is that the nitrogen isotope data from the collagen fractions of the mammal molars are indicative of water stress within the animal,

| Species | NISP | % |
|---|---|---|
| | | |
| *Canis lupus*, wolf | 1 | 0.09 |
| *Ursus arctos*, bear | 10 | 0.9 |
| *Felis spelaea*, lion | 3 | 0.3 |
| *Palaeoloxodon antiquus*, straight tusked elephant | 57 | 4.9 |
| *Mammuthus trogontherii*, steppe mammoth | 922 | 80 |
| *Equus ferus, horse* | 34 | 2.9 |
| *Cervus elaphus*, red deer | 4 | 0.3 |
| *Bison priscus*, bison | 125 | 10.8 |
| | **1156** | 100 |

Table 6.4 Large vertebrates from Stanton Harcourt: the number of identifiable bones of each species (NISP) and the percentage they represent of all identifiable remains

usually associated with aridity. As cooling and increased aridity are often associated with the onset of glacial periods, the isotopic data may indicate that the final stages of MIS 7 are represented at Stanton Harcourt.

The carnivores (lion, bear and wolf) reveal little about the climate as all occur in British Pleistocene assemblages associated with both cold and temperate climatic indicators.

## The local environment - wood and other vegetation

As described in Chapter 3, well preserved wood was abundant across the site and logs and branches were readily identifiable to species such as oak, willow and dogwood. As the excavation area expanded over the decade, vegetation samples from different parts of the excavation were analysed (see Chapter 3 Table 6).), and Robinson (2012). As with the climatic interpretations based on wood and other vegetation from the site, these analyses reflect differences in the local terrestrial *vs.* aquatic environments. The findings of the various specialists are presented below and illustrated in Figures 6.2, 6.3 and 6.4.

### *Wood*

Gourlay (2000) identified individual pieces of well-preserved wood from Sites 1,4,7,8 and 15. He concluded that most of the trees were mature: oak, hazel, willow, yew and other conifers. The oak ranged from 2 - 120 years. The yew and willow were more than 20 years old. The conifer (probably yew) ranged from 23-50 years. They are mostly the remains of trees after their tops (trunk stems) had been broken off. Some pieces were roots (lack of clear vessel patterns, oval or flattened forms and other features). Others were pieces of wood close to the root/stem collar with characteristic swirling and confusing growth patterns. Some would appear to have been torn off a larger piece such as a main stem or root/collar. This is likely the result of flood water pulling already dead trees out of banks. The wood around the branch/ root/stem junction, being of a tougher structure withstands wear and tear better.

Hather (1993) had bulk samples as well individual pieces of wood in late 1993 and early 1994 (Chapter 3 Table 6). Of the 254 pieces, most were twigs between 3 and 7 years old and a few pieces of mature wood more than 10 years old. As he counted many pieces of wood from bulk samples, it is possible that some species represented mostly by twigs might be disproportionately common. Giving equal weight to fragments whether twigs or mature wood and ignoring the minor taxa (*Betula* and *Juniper*), it would appear that the environment was dominated by *Fagus* beech (47.4%), *Carpinus* hornbeam (31.8%) and *Quercus* oak (24.5%). Consideration of only the twigs works out about the same but percentages based upon the much smaller number of fragments of mature wood reveal *Fagus* beech to be equal to *Quercus* oak (41.6% each) with *Carpinus* hornbeam a less dominant component (16.8%). JH also had mature oak of 80-90 years old (Sample 14001).

The overall picture is of mature deciduous woodland similar to that found in southern Britain or mid France today. As the region is calcareous it is unlikely that it supported dense woodland. Normally, oak/hornbeam and beech are found on acid soils with oak and hornbeam preferring wetter areas and beech drier conditions. Juniper can tolerate a calcareous substrate. The birch might be hairy birch *Betula pubescens* rather than silver birch as the former has a preference for wetter conditions at a riverside association.

### *Pollen, seeds and spores*

Campbell (1993) took bulk samples early in the excavation from Sites 1 and 2. He was specifically looking for pollen at Site 1 (1993) and seeds at Site 2 (1992). This must be borne in mind when making comparisons between the sites based on his data alone.

#### *Site 1 (pollen and spores)*

Pollen was not abundant in this sample, and there were many corroded grains which could not be identified. A bulk sediment sample was processed until 500 grains could be counted. The paucity of pollen in this sample may in part be due to the reworking of the sediment at this locality. The bed which Campbell sampled was early bar top sediment in a meandering river (see Chapter 2). This bed was very thinly laminated and moderately well sorted. The early bar sediments would have been only semi-emergent, with vegetation and wood generally appearing later in the sedimentary pile. A bar is very mobile until it becomes vegetated. Stratigraphically, the sample pre-dates the bones and most of the wood deposits at this locality but there is no doubt that the pollen sample is of MIS7 age as the bed sampled overlies and underlies sediments containing *Corbicula fluminalis.* Wood and a hazelnut were excavated in the gravel bed above the silty sand containing the described pollen assemblage which showed a dominance of tree pollen.

There are two possibilities to explain the predominance of tree pollen. One is that the bar was not vegetated at this time and pollen was blown into the river from trees on the floodplain. The other possibility is that the focus on pollen rather than seeds from this sample may have given undue emphasis to trees because grasses and other plants would have been better represented by seeds.

Trees from this sample are both deciduous (oak, alder and hazel) and coniferous (*Abies* sp.). The dominance of the latter in the arboreal pollen may be attributable to the well-known characteristic of bi-saccate coniferous pollen to be over-represented in water-lain deposits.

Aquatics and *Juncaceae* are poorly represented. One reason for this may be that they produce fragile grains which are often poorly preserved. This may explain their absence from a site where we might expect to find them i.e. a damp riverine environment. However, this area of the site may have been part of the floodplain during the summer months and thus an unsuitable habitat for aquatic species.

*Site 2 (seeds and spores)*

Campbell took 6 bulk sediment samples from this site which was known as 'the wood site' and counted 500 grains from each sample. The majority of identifiable remains were seeds from a silty sand which contained many large fragments of wood. Campbell grouped the seeds and spores into 5 ecological classes: trees and shrubs, herbs of dry land, helophytes (marsh plants), aquatics, and unclassified (no ecological interpretation possible). Taken as a whole, the assemblage would appear to indicate a lightly wooded environment with open areas of both dry and wet grassland typical of riparian vegetation adjacent to a slow-moving river with waterside, emergent and fully aquatic habitats present.

*The trees and shrubs.* Apart from elder, the seeds of trees and shrubs occur in only one of Campbell's samples, despite the fact that the samples contained fragments of wood. In order of abundance are elder, dogwood, blackthorn, alder, pine and birch. Both blackthorn and elder are comparatively intolerant of dense shade and occur infrequently in closed communities. Dogwood is common in chalk scrub vegetation.

*The herbs of dry land group* indicate a moist grassland with such species as buttercup, marsh thistle, autumnal hawkbit and meadow rue. The occurrence of woody nightshade and ragged robin supports the picture of a damp meadow with some woodland. The presence of small scabious, annual knawel and narrow fruited corn salad may indicate areas of dry land. A certain amount of disturbance (possibly trampling by large herbivores) may be indicated by the presence of goosefoot, orache, and knotweed.

*The helophytes* are represented by great sedge and bulrush along with gypsywort which are common along the margins of rivers. The presence of nodding burr marigolds merits the mention that it favours standing water in the winter but not in the growing season.

*The aquatic species* are limited; pondweed dominates the overall assemblage from Site 2 (58%) along with oospores of stonewort which reflects the aquatic nature of the depositional environment. The more extensive aquatic elements characterised by horned pondweed, mare's tail, arrowhead, branched burr reed and water plantain indicates a slow flowing river with a muddy substrate.

### *Seeds and plant fragments*

Robinson (2012) analysed a 500-gram bulk sample from Site 1. This sample was from a *Corbicula* bed and had been sieved through a .5-micron sieve for molluscs. This a larger mesh diameter than is normally used for seed analysis therefore there are likely to have been many more seeds in the sample.

The trees represented are oak, hazel, dogwood, and blackthorn, confirming other wood specialist identifications, with the addition of acer and hawthorn. There is a variety of weeds and aquatic species similar to those found in the sample from Site 2. Robinson also identified a species of bramble and *Filipendula ulmaria*, a herb from Rosacea family

## Insects and the environment

Coope (2000, 2006) provided the following report on the environmental significance of the insect assemblages from the site. See Chapter 3 Table 4.

'The insect species obtained from the organic deposits in filling the Stanton Harcourt channel have two different origins. Firstly, the local species that inhabited the river itself and its immediate surroundings. Secondly, species from more distant habitats that had probably been brought together amongst flood debris.

The insect fauna describes an environment dominated by a rich herbaceous flora growing along the margins of a fairly large, mature river meandering between rapid riffles, shallows and placid pools. In the quieter backwaters oak trunk logs had come to rest submerged in the deeper water. The channel was in places fringed with reed beds where the soil was organic and sulphurous. Between these there were bare mud banks on which there was some growth of algae. On the floodplain there were segments of abandoned channels; pools of standing water in which thrived a variety of pond weeds.

Further afield the landscape was dominated by weedy vegetation growing on both sandy and clayish soils, grading into drier, better drained ground where there were bare patches between the plants. Beside the river there were probably sparse willows and further away stands of oaks. There is evidence of abundant dung in the area from the frequency of *Aphodius carpetanus*, most species of which are obligate dung feeders. Other carnivorous species such as *Oxytelus gibbulus* feed on small insects in dung or decaying vegetable refuse. Further evidence of the presence of large herbivorous mammals in the form of dried out carcases in various stages of decomposition comes from the presence of the beetle *Dermestes murinus* that lives on desiccated carcasses.

There is no significant difference between the faunas from the five samples and the environmental interpretation is therefore based on the fauna as a whole. The beetle species indicate that a wide variety of habitats were available in the immediate neighbourhood and it is likely therefore that much of the fauna was washed into the deposit off the adjacent land surface. A mosaic picture of the local palaeoenvironment can be built up out of the present-day ecological preferences of the various species in these fossil assemblages.

*Aquatic habitats*

Two distinct environments are indicated by the water beetles in these assemblages

a. Running water.

Rapidly flowing water is indicated by the dryopids *Helichus substriatus, Esolus parallelepipedus, Oulimnius* and *Limnius volkmari* which live on micro-organic detritus amongst stones and moss in well oxygenated riffles in rivers and stream. *Platambus maculatus* is an active predator that inhabits submerged vegetation in the shallows of rivers and streams (Balfour-Browne 1950). *Orectochilus villosus* is a whirligig beetle that typically lives in large rivers where it hunts on the water surface for drowning insects. All these species can also be found at the margins of lakes where the well aerated lapping water mimics their more normal habitat. *Macronychus quadrituberculatus* lives in larger rivers in fissures in submerged logs or cracks in stones (Olmi 1976).

The Trichoptera are represented in this assemblage only by larval sclerites of species all of which are aquatic. The majority indicate a large, slowly flowing river. The larvae of *Hydropsyche angustipennis, H.contubernalis* and *Hydropsyche bulbifer* spin regular, cone shaped nets across the current to catch any edible material that drifts by. The latter species lives in small rivers but only in deep places (Wilkinson 1980). The larva of *Cyrnus trimaculatus* is predaceous, building capture nets in lakes and slowly moving rivers (Lemneva 1970). The larvae of *Anabolia nervosa* feed on diatoms and filamentous algae. According to Lepneva (1970, 1971) the larvae need 'slow clear water with a current speed of 0.05 - 0.5 m/sec with summer temperatures of 18°C - 20°C. *Lepidosoma hirtum* is also a species of large rivers. The larvae of *Apatania* are found in cool fast-moving streams and in this respect its habitat resembles those dryopid species mentioned above.

b. Standing or slowly moving water.

In contrast to the indicators of running water, there are a considerable number of beetles that live in standing or slowly moving water. *Agabus nebulosus* is found in ponds without vegetation where the bottom is sand, silt or clay. *Colymbetes fuscus* is also a pond species but found in more vegetated habitats. Most of the species of *Helophorus* are also found in grassy pools that may be very small and shallow. The majority of the hydrophilid species are also found amongst decomposing vegetation in or beside standing water.

There are only a few phytophagous water beetle species in this assemblage. *Haliplus* of the *confinis* group live in clear stationary or slowly moving water where they feed on various characeans (Holmen 1987). *Macroplea appendiculata* spends almost its entire life underwater where it feeds chiefly on *Potamogeton* but also on *Myriophylum* and other pond weeds. *Donacia dentata* feeds on *Sagittaria sagitifolia* and species of *Alisma. Donacia semicuprea* feeds on *Glyceria* [aquatic grass]. The larvae of these species obtain their oxygen by tapping the air channels in the underwater stems. The weevil *Notaris acridulus* feeds on the emergent leaves on *Glyceria.* It is noteworthy that this aquatic sweet-grass is also attractive to large grazing mammals. *Anisosticta novemdecimpunctata* (the water Laybird) feeds principally on Aphids on reedy plants

(Majerus 1994). All these species may all have been living in, or close to, the standing water in abandoned river channels on the floodplain or else in quiet backwaters of the main river.

The juxtaposition of water beetles indicative of both running and still water is not ecologically inconsistent since they may indicate that the river was meandering between a succession of riffles and pools. Because many of these beetles spend almost all their life histories sub aqueously, it is likely that the river at this time continued to flow throughout the year.

*Transitional habitats*

In this category are included those species that inhabit environments which naturally grade from almost fully aquatic to almost dry ground habitats.

Many of the carabid species in this fauna are characteristic of swamps and moist habitats. Most of the data in this section is taken from Lindroth (1992). As far as is possible the species are discussed in the same order as they appear in Table 3.4. *Carabus clathratus* is the most humidity loving of all our species, living beside ponds and lakes where the vegetation is rich and extends into the water. *Notiophilus palustris* and *Trechus secalis* require both humidity and shade which may be provided by deciduous trees or by tall vegetation. *Dyschirius globosus* is very eurytopic but is often found in sparse woodlands, swamps, meadows and peat bogs (though rarely with *Sphagnum*). *Elaphrus riparius* is found on the shores of still or slowly flowing water where the vegetation is poor or even completely absent. *Bembidion properans* is likewise a species of open sun-exposed habitats often in the vicinity of water, usually where the soil is clayish. *Bembidion varium*, *Bembidion articulatum* and *Bembidion octomaculatum* live in moist habitats on soft clayish soils by pools that are exposed to the sun and which dry up in the summer. Similarly, *Bembidion articulatum, Bembidion gilvipes, Bembidion clarki* and *Bembidion doris* are found in damp habitats often in sparse woodland where they live beside ponds. *Patrobus atrorufus* occurs on clayish soils that are mixed with humus and moderately to very moist and where there is some shade from rich vegetation. *Pterostichus strenuus* is also found in damp deciduous woodland where it is found amongst leaf litter and moss. *Oodes helopiodes* is partly subaquatic, living amongst reedy vegetation where the water is frequently dirty and smells of hydrogen sulphide $H_2S$ and where there are bare patches between the plants. *Polistichus connexus* is found on damp river banks where it seems to prefer clayish soils.

Most of the staphylinid species in this fauna are predators on small arthropods and worms in accumulations of decomposing vegetation. *Corylophus cassidoides* is also a predator in rotting detritus of *Phragmites*, *Carex* and grasses.

Both the adult beetles and the larvae of *Heterocerus* live in shallow burrows which they excavate in damp soil where they probably feed on organic particles in the soil (Clarke 1973). *Limnichus pygmaeus* feeds on algae, its larvae borrowing in the damp mud.

Many of the phytophagous species in this fauna indicate the presence of swampy or damp habitats. Reedy vegetation is indicated by *Donacia bicolor* which feeds on *Sparganium ramosum* and *Carex* and by *Plateumaris affinis* which feeds on *Carex*. The weevil *Notaris scirpi* feeds on the leaves of *Scirpus*, and *Carex* and less often of *Typha*. *Limnobaris pilistriata* lives on various *Juncaceae* and Cyperaceae, its larvae feeding on the roots, particularly of *Scirpus lacustris* (Koch

1992). *Hydrothassa marginella* lives on various species of *Ranunculus* and on *Caltha palustris.* The large weevil *Liparus germanus* lives in damp habitats usually on *Petasites* and less often on *Heracleum* and *Angelica.* Both the larvae and adults of the byrrhid species *Porcinolus murinus* and *Byrrhus* feed exclusively on mosses.

*Drier ground habitats*

There are several species amongst the Carabidae that indicate open environments where there is only a moderate vegetation cover and where the soil is rich in humus; conditions similar to those of cultivated areas at the present day. Such species include *Nebria brevicolli, Bembidion quadrimaculatum, Bembidion obtusum, Pterostichus melanarius, Calathus fuscipes, Calathus melanocephalus, Amara sp.* and *Syntomus truncatellus.*

The larvae of the click-beetles *Agiotes, Adelocera murins* and *Athous haemorrhoidalis* are the familiar 'wireworms' feeding underground on roots often causing considerable damage. The larvae of *Dacillus cervinus* live in meadow-like places on roots, but the adult beetle feeds on a variety of flowers notably of Umbelliferae. *Opatrum sabulosum* also feeds on grass roots on dry, sandy waste ground. *Rhyssemus germanus* is found at the foot of plants in dry sandy places, typically beside rivers (Jessop 1986). *Heptaulacus villosus* is a subterranean species living amongst humus on dry sandy or chalky soils.

The chrysomelid and curculionid (weevil) species are all phytophagous [insects that feed on plants]. Species in this assemblage indicate a diverse open ground flora. Thus *Phyllotreta atra* feeds on the leaves of Cruciferae. Most species of *Chaetocnema* live in the stems and roots of grasses. The numerous species of *Apion* feed on a wide range of herbaceous plants. *Otiorhynchus ligneus, Trachyphloeus, Barynotus obscurus* and *Alophus triguttatus* are polyphagous on a wide variety of herbs, their larvae living underground on roots. *Cleonus piger* lives on various thistles particularly on *Cirsium* and *Carduus. Sitona* species feed on various Papilionaceae and have larvae that live underground on the roots of their host plants. *Stomodes gyrosicollis* feeds on the leaves of *Trifolium arvense* and *Medicago lupulina* (Hoffman1950). The larvae of *Bruchus rufimanus* develop inside the seeds of similar plants. *Orthochaetes setiger* inhabits dry sun-exposed sandy banks where it usually feeds on various Compositae but also on other weedy plants. Its larvae mine the leaves of its hosts.

Several carabid species such as *Elaphrus riparius, Poecilus sp.* and *Agonum sexpunctatum* are pronounced heliophiles, requiring bare soils between the plants that are exposed to the sun.

*Evidence from the insects for the presence of trees*

Several species indicate the local presence of deciduous trees. The weevil *Rhynchaenus quercus* feeds exclusively on *Quercus*, the larvae mining the leaves. The scolytid *Xyleborus dryographus* drills deep galleries into oak wood. *Rhynchaenus rufimanus* attacks *Populus* and *Salix.* The anobiid species *Dorcatoma chrysomelina* lives in fungi such as *Polyporus* on dead or dying trees. Finally, as mentioned above, *Macronuchus quadrituberculatus* is found in the fissures of submerged logs.

*Specialised habitats.*

The relatively large number of specimens of *Aphodius* indicate the abundant presence of dung deposited on land. Species of the *Oxytelus gibbulus* group are predators and are today found in dung (Hammond et al. 1979). The histerid species prey upon dipterous larvae (maggots) often in decomposing carcases. *Dermestes murinus* lives in dried out carcases. The silphid species *Phosphuga atrata* is a specialist predator on snails'.

## Molluscs and the environment

Keen (1992) and Gleed-Owen (1998) reported that the mollusc species (Chapter 3 Table 3) indicate a fairly typical river assemblage of temperate character. The water was evidently well oxygenated and there would have been a variety of channel and associated flood-plain habitats in the vicinity, with both flowing water and the occasional flood-plain pool.

The main elements are indicative of moving water (*A. fluviatilis V. piscinalis, C. fluminalis, P. amnicum, E. henslowana, and O. moitessierianum*). Weed and mud-dwelling species (*A. crista, A.lacustris, L. peregra*), abundant in backwaters of the Thames today, are less well represented, so it is probable that this stretch of the river was not very weedy or stagnant. The larger mollusc *P. littoralis* prefers moderately large rivers with depths of about two metres.

The number of taxa and species of land snail show that land surfaces were in the close vicinity. The land fauna is mostly of grassland or marsh species (*P. muscorum, Vallonia* sp.) but shade-demanding elements (*E. montana, A. nitidula. N. hammonis. C. bidentata*) are also present. The woodland obligate gastropod *Spermodea lamellata* lives in deep leaf under trees whose dead leaves are durable and slow to break down, such as beech and oak and indicates the presence of an open but mature woodland. *Valvata piscinalis* has an operculm and can tolerate some drying out such as on bar surfaces.

Overall, the evidence from the molluscs supports the other environmental analyses in presenting a picture of a moderate river flowing through a predominantly grassy landscape with stands of woodland close by.

## Vertebrates and the environment

### *Species representation*

The fish species found in the excavated material (Table 6.5) are typical of clean reaches of the Thames today (B. Irving 1995). Other small vertebrates were insufficient to use as environmental indicators.

Of the large vertebrates in Table 6.4, steppe mammoth *M. trogontherii*, bison *B. priscus* and horse *E. ferus* are predominantly grazers and their presence suggests extensive grassland. The preferred habitat of the lion *P. spelaea* is also open terrain. However, straight-tusked (forest) elephant *P. antiquus*, red deer *C. elaphus*, brown bear *U. arctos* and wolf *C. lupus* are all indicators of the proximity of woodland.

| | |
|---|---|
| *Gasterosteus aculeatus* | three-spined stickleback |
| *Esox lucius* | pike |
| *Perca fluviatilis* | perch |
| *Anguilla anguilla* | eel |
| *Abramis brama* | bream |
| *Gobio gobio* | gudgeon |
| *Leuciscus leuciscus* | dace |
| *Squalius cephalus* | chub |
| *Rutilus rutilus* | roach |
| *Barbatula barbatula* | stone loach |
| *Salmo salar* | salmon |

Table 6.5 Icthyofaunal species identified in the Stanton Harcourt channel (B. Irving pers. comm.)

Overall, the picture is of a predominantly grassland environment with some deciduous woodland nearby. Two isotopic analyses of large vertebrate dentition served to corroborate the faunal and other biological data.

### *Isotopic analyses of mammalian teeth*

Jones *et al.* (2001) carried out isotopic analyses of collagen and enamel fractions from mammoth, elephant, horse and bison molars from Stanton Harcourt. While the oxygen isotopes were used to calculate palaeotemperatures, the carbon isotopic data from both collagen and enamel fractions indicate a 100% C3 diet (both grazing and browsing) which is analogous to the vegetation evidence for a mixed grassland and deciduous woodland habitat.

Lee-Thorp (2020) also carried out isotopic analyses of tooth enamel from the site. This Chapter concludes with her report as follows:

*Context and aims*

The primary purpose for analysing Stanton Harcourt tooth specimens for carbon and oxygen isotopes was to establish at least two in-house isotopic standards, for application in all our mass spectrometry runs of fossil enamels. For mass spectrometry, an important part of the quality control process involves the regular assessment against standards with a material of similar matrix (i.e. fossilised enamel in this case), produces enough clean enamel material for use as a standard, and preferably shows distinct values for $\delta^{13}C$ and $\delta^{18}O$. The fossils from Stanton Harcourt and Sutton Courtenay were likely to satisfy at least two of these criteria, in that there were numerous megafaunal teeth, and they had been established to be either from MIS7 (Interglacial) or MIS3 (Glacial). The second aim was to establish their isotopic composition and compare against the values obtained by Jones *et al.* (2001).

*Materials and Methods*

Three teeth were provided, consisting of (1) the lower right molar of *Palaeoloxodon antiquus* (straight-tusked elephant), from MIS7 Interglacial channel deposits (specimen number SH4/45); (2) Upper right molar of a *Mammuthus trogontherii* (steppe mammoth), from

Interglacial MIS7 deposits (specimen number SH6/138); and (3) a part molar of a Mammuthus primigenius (woolly mammoth) from MIS3 deposits (specimen number SC 707). The first two are from Stanton Harcourt, and the third from Sutton Courtney.

Although rather little isotopic work has been published on British mega-fauna from these periods, one can predict from first principles that the flora eaten by these herbivorous animals consisted entirely of plants following the C3 pathway because of the northerly latitude. We understand well the range of values from such C3 environments; and tooth collagen and tooth enamel $\delta^{13}C$ should reflect this composition, as shown for dentine collagen and enamel bioapatite of 7 Stanton Harcourt specimens by Jones et al. (2001), that indicate very standard C3-feeder values. There are isotopic differences amongst C3 plant taxa growing in different habitats, most particularly where plants grow in dense forest stands and giving low $\delta^{13}C$ due to the canopy-effect, which are passed onto the tissues of animals (e.g. van der Merwe and Medina 1991). No canopy effects were visible in the Jones (op. cit.) dataset, however. For $\delta^{18}O$, values reflect a mix between drinking water values (dependent largely on patterns of precipitation) and oxygen from plant water and carbohydrates. Our first expectation was that values should vary between the Interglacial and the Glacial due to precipitation under cooler conditions (Dansgaard 1964), although the magnitude of any shift is hard to estimate. Values for bioapatite CO3 shown in Jones (op. cit.) suggested $\delta^{18}O$ values are very similar to those found today. Whatever the case, these $\delta^{13}C$ and $\delta^{18}O$ values should be isotopically distinct from other fossils available to us from African sites and provide a much lower endpoint for 2- or 3- point reference standards regressions.

On examination of the three specimens, we found that only the first two were suitable for our purposes, because their enamel plates could be removed and crushed. The shape and state of fossilization of the 3rd specimen, unfortunately precluded this.

Thus, for specimens (1) and (2) (above) the enamel plates were carefully and manually removed from the dentine and cementum matrix, cleaned of all adhering non-enamel material, and reduced to small fragments. The fragments from each sample were then ground in a Spex-Mill under liquid nitrogen temperatures in aliquots, and then re-united and mixed. Each powdered and thoroughly mixed sample was divided again into suitable sized aliquots for chemical pre-treatment to remove possible contaminants. We followed our normal laboratory process, based on Sponheimer (1999), up-scaled to account for the higher mass of powder. This uses a sodium hypochlorite solution (1.75%) as a precaution against organic remnants (although unlikely in enamel of this age) for 30 minutes, followed by centrifugation and rinsing in de-ionised water. Weak 0.1M acetic acid is used to remove any possible extraneous carbonates and diagenetic (altered) enamel, which tends to be more soluble. Following thorough rinsing again, powders were dried in a freeze-drier, and again each sample was reunited and thoroughly mixed.

We measured small aliquots (1.5mg) of the two samples SH4/45 and SH6/138 by mass spectrometry multiple times, in two laboratories, in order to establish (i) the values, and (ii) their reproducibility. One set was measured at the University of Bradford's Stable Light Isotope Facility by Andrew Gledhill, and another set subsequently in our own Laboratory at Oxford.

*Results and Discussion*

One set of data is shown in Table 6.6 and serves to illustrate the results. $\delta^{13}C$ is expressed versus the international PDB standard, and $\delta^{18}O$ versus the international SMOW standard, according to convention (although the latter can also be expressed versus PDB).

Based on the means and standard deviations, the two individuals show very similar isotopic signatures. The pure C3 signature shown by both individuals is to be expected for the reasons given above – Stanton Harcourt was then, as now, a cool, temperate environment with only C3 plants. The value for SH4/45 is consistent with its collagen $\delta^{13}C$ value of -20.8‰ shown in Jones et al. (2001). However their enamel carbonate $\delta^{13}C$ value differs - at -11.3‰ with only a standard analytical error provided compared to the value shown here. The difference is likely due to two factors. One is that they used only a Hydrogen peroxide pre-treatment that has been shown to be completely ineffective in removing organics (Snoeck and Pellegrini 2015), and no acid pre-treatment. Secondly, the measurements were carried out on an instrument accustomed to measurement of relatively enriched marine carbonate samples, not bioapatites. They do not report whether appropriate (negative value) standards were used, nor the values, so one has to assume that the standards used were not matched in matrix or in isotopic composition. This is a problem where the instrumental response slope is not quite 1 since the values of the unknown are very far from the standard values. We believe that our result of -12.01 +/-0.04‰, now confirmed on two instruments, is correct.

The $\delta^{18}O$ values, which reflect ultimately the isotopic composition of precipitation and hence environmental drinking and plant waters, are similar to what we would expect to see today and indeed the value for SH4/45 is close to the value obtained by Jones *et al.* (2001). A small survey of horse values gave similar values also gave very similar values for the south of England (Bell *et al.* 2006). Values for the two individuals do differ slightly – by 0.88‰ for $\delta^{13}C$ and 0.7‰ for $\delta^{18}O$. Based on just one individual of each species, we cannot know the range of variability for each taxon in this environment. However, putting aside inter-individual variation within a taxon for the moment, these small differences *may* reflect subtle distinctions in vegetation consumed and drinking behaviour by these two taxa, and/or their digestive physiology. We would require a more significant sized sample to ascertain whether this is the case.

The value of both specimens as reference standards is demonstrated by their reproducibility for which the standard deviations serve as a good indicator. Analytical errors of 0.04 and 0.06‰ for $\delta^{13}C$ and 0.10 and 0.11‰ for $\delta^{18}O$ are very precise. We continue to use the *Mammuthus* tooth as one of our essential standards for enamel bioapatite work, to provide a good C3-feeder reference endpoint.

| | $\delta^{13}C_{VPDB}$ ‰ | $\delta^{18}O_{VSMOW}$ ‰ |
|---|---|---|
| SH4/45 | -12.01 | 24.55 |
| SH4/45 | -12.06 | 24.52 |
| SH4/45 | -11.98 | 24.53 |
| SH4/45 | -11.98 | 24.34 |
| **mean** | **-12.01** | **24.49** |
| **std dev** | **0.04** | **0.10** |
| | | |
| SH6/138 | -12.70 | 25.35 |
| SH6/138 | -12.77 | 25.22 |
| SH6/138 | -12.84 | 25.08 |
| SH6/138 | -12.76 | 25.15 |
| **mean** | **-12.77** | **25.20** |
| **std dev** | **0.06** | **0.11** |

Table 6.6 Results of isotopic analyses of tooth enamel of straight-tusked elephant (SH4-45) and steppe mammoth (SH6-138) from Stanton Harcourt (Lee-Thorp 2020). See text for discussion.

# Chapter 7

# The Artefacts

A total of 36 artefacts came from the Channel: eleven flint handaxes (not all complete), four pointed quartzite handaxes, eleven flint flakes, four flint cores, five pieces of flint debitage and one quartzite flake (Table 7.1 and Figure 7.14).

In the preliminary report on the excavation (Buckingham *et al.* 1996), Derek Roe described the nine artefacts recovered by that date. He also made notes on some of the later material. In this Chapter, Terry Hardaker reviews and expands upon Roe's initial report and his notes to describe the total assemblage and discusses the Channel artefacts in the light of other lithics from surrounding gravel pits. In the concluding section of the Chapter, Nick Ashton considers the Stanton Harcourt assemblage in the context of other MIS 7 assemblages in Britain.

## Descriptions of the artefacts

It should be noted that some of the artefacts were numbered as they were described rather than with their find numbers in the 1996 *Journal of Quaternary Science* report. The correct numbers are allocated here and the *JQS* numbers follow in brackets, where relevant.

### *Artefact A1* (numbered 1 in *JQS*)

This is a small pointed flint handaxe with dark yellow patina (76 x 46 x 25 mm). It is crudely made with irregular surfaces and poor-quality edge working, and is moderately rolled on both faces with edge damage all round. A more recent break reveals that the natural colour of the flint is yellow not black: this is characteristic of the local gravel flint, which is often hard to work, and is often found as relatively small nodules. This would explain the small size and crude morphology of this piece.

### *Artefact A2* (numbered 2 in *JQS*)

This is a cordiform flint handaxe with grey-yellow patination (88 x 60 x 31mm). A modern break shows a deep patination (3mm) with grey interior punctuated by lighter grey inclusions. The artefact retains a substantial amount of cortex around the butt which is unworked. The maker has had trouble reducing one of the edges to an acute angle resulting in numerous hinge fractures which, in the end, failed to make headway. This edge would have been almost useless as a cutting or slicing tool; the other edge was finished satisfactorily. The artefact is quite heavily abraded including smoothing of the arêtes, suggestive of long-term exposure or, more likely, time spent in a fluvial environment. The small size of the flint nodule and interior colour of the flint suggests a local source for this artefact.

### *Artefact A3* (numbered 7 in *JQS*)

A medium sized broad oblique flint flake (52 x 81 x 21mm) in deep yellow patinated flint. It is sharp but with many minute edge spalls characteristic of fluvial abrasion rather than retouch. The natural break of this flake makes a regular gentle arc which could have served as a cutting

| No | Description | Material |
|---|---|---|
| A1 | Handaxe | Flint |
| A2 | Handaxe | Flint |
| A3 | Flake | Flint |
| A4 | Large handaxe | Flint |
| A5 | Flake | Flint |
| A6 | Small Middle Palaeolithic core | Flint |
| A7 | Small ovate | Flint |
| A8 | Core? | Flint |
| A9* | Handaxe | Quartzite |
| A10 | Broken handaxe | Flint |
| A11 | Retouched flake fragment | Flint |
| A12 | Core | Flint |
| A13 | Handaxe tip | Flint |
| A14 | Denticulate flake | Flint |
| A15 | Small flake | Flint |
| A16 | Broken handaxe tip | Flint |
| A17 | Debitage | Flint |
| A18 | Flake | Flint |
| A19 | Handaxe on a flake, possibly Middle Palaeolithic | Flint |
| A20 | Handaxe | Flint |
| A21 | Debitage | Flint |
| A22 | Small handaxe | Flint |
| A23 | Crude handaxe | Flint |
| A24* | Handaxe | Quartzite |
| A25 | Handaxe thinning flake | Flint |
| A26 | Large flake | Flint |
| A27 | Handaxe | Quartzite |
| A28 | Flake | Flint |
| A29 | Broken flake tip | Quartzite |
| A30 | Handaxe | Quartzite |
| A31 | Flake with facetted platform | Flint |
| A32 | Debitage | Flint |
| A33 | Flake | Flint |
| A34 | Debitage | Flint |
| A35 | Debitage | Flint |
| A36 | (deleted) | - |
| A37 | Core | Flint |
| A38 | Handaxe | Quartzite |

*Artefacts A9 and A24, although both quartzite handaxes, are omitted from the discussion as they were surface finds from the vicinity prior to the excavation. A36 has been deleted as it was misinterpreted as an artefact.

Table 7.1 Artefacts from the Stanton Harcourt Channel.

tool or scraper without further retouch. The origin of the flint is uncertain but its intense yellow colour suggests a Chiltern source.

***Artefact A4*** (numbered 3 in *JQS*)

A large complete pointed flint handaxe (177 x 85 x 45mm), symmetrical with straight edges and thin flaking suggesting soft hammer finish. It has been worked more crudely round the butt. There is a large cortical méplat towards the butt (a méplat is a flat unworked area possibly retained to make it easier to hold in the hand). The yellow patination is identical to the Gravelly Guy handaxes which it resembles in size, workmanship and quality. The edges are almost entirely damaged (both naturally and with some deliberate retouch) so little of the original knapped edge remains. While the original surfaces of the handaxe show typical glossy patina characteristic of the Gravelly Guy series, many of the edge spalls lack this gloss, suggesting subsequent edge damage eating into a thick patination that formed after the handaxe was made.

On the faces of the handaxe, the arêtes between the negative scars are quite sharp, suggesting that the artefact was not present in a streambed for long, if at all, and was not long on the land surface before becoming buried. On the surface of the artefact there are slight indications that the unpatinated interior is of black flint. There are no modern breaks to verify this. An artefact of this size in flint is most likely to have been sourced from Chiltern flint some 20km distant, as were many of the Gravelly Guy handaxes. This is the finest of the Stanton Harcourt handaxes.

Roe's description was: 'Well made from a substantial nodule of chalk flint, to judge from the nature and distribution of the surviving cortex patches. The whole of the implement is patinated and stained yellowish brown; although there are some ancient damage scars, there are no recent ones to reveal the internal colour of the flint though a hint of underlying dark grey or black can be observed in one or two places where the patina is thinnest. This is a large pyriform handaxe, with a linguate tip that has suffered a certain amount of ancient damage (revealed as scars with much lighter patination) and may formerly have been more pointed than now appears. The butt is fully worked, but a large patch of cortex affects one edge just above it. Much battering at this point suggests determined but unsuccessful attempts on the knapper's part to detach flakes that would have made the implement flatter, perhaps with a view to getting rid of the cortex subsequently. The opposite edge has only very minor patches of remaining cortex. The flaking is of high quality on both faces, and some of the scars certainly look like soft-hammer work. The condition is not quite fresh: the ridges have been slightly dulled, and the delicate edges have suffered a little battering in places, but the implement is certainly less worn than artefacts 1 and 2.'

***Artefact A5*** (numbered 8 in *JQS*)

An irregular yellowish patinated flint flake (124 x 63 x 34mm) with a prominent bulb and platform. This is good quality 'Chiltern' flint removed from a large nodule perhaps in the course of making a handaxe. The arêtes are not abraded but the edges are. However there appears to be retouch along one edge (11 tangent removals all from the same side: see Figure 7.1). This is different from the erratic edge damage seen on the rest of the artefact. The

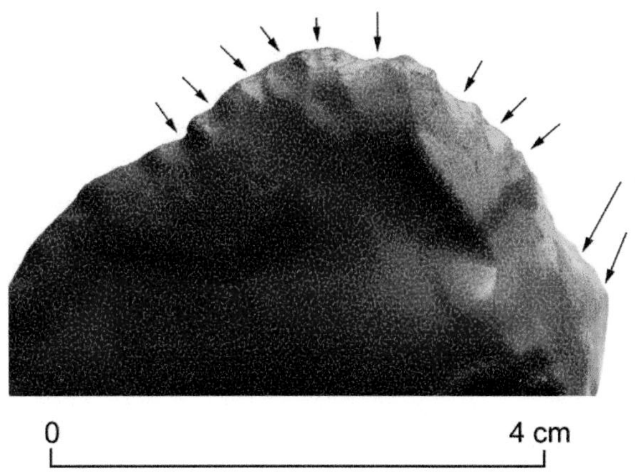


Figure 7.1 Artefact A5 detail showing retouched edge

artefact may thus be described as 'Possibly a nose scraper'. Such forms are rare in the Lower Palaeolithic, although recognition is difficult. The abrasion of the edges while the arêtes are left fresh simply indicates that acute angles on flint are much more vulnerable to damage that obtuse ones.

***Artefact A6*** (numbered 6 in *JQS*)

This is a cortical flake fragment converted to a core (67 x 41 x 19mm). The platform and bulb are missing. The flint is cream and patinated with traces of grey or black interior. Roe commented that it began as a cortical flake, but was then adapted by simple preparatory flaking to become a small core, evidently specifically designed to allow the extraction of a single, thin, elongated flake from what had been the original flake's bulbar surface. The preparation of the core included the creation of what might pass for a facetted striking platform, from which the intended flake was subsequently struck. In its final form, the piece is accordingly classifiable as a small struck Levalloisian core of a simple kind. The flake it yielded would have carried few primary dorsal scars and, at its distal end, was diverted into a somewhat irregular shape by the presence of a small hinge-fracture feature on the core. The whole operation might be seen as a neat piece of opportunism by a knapper who was familiar with the Levallois technique.

The raw material is also of some interest, as it comprises banded flint which is probably of Chiltern origin. This core is one of the freshest in the collection. The ridges are virtually unworn, and a number of small sharp projections survive on the edges, including that of the prepared striking platform, which would quickly have disappeared if the artefact had been deposited in a gravel bed under high-energy conditions.

***Artefact A7*** (numbered 4 in *JQS*)

This is a small bifacially worked ovate tool (48 x 39 x 19mm) is of yellowish patinated flint retaining 70% cortex on one side. It appears to be a deliberately shaped tool rather than a core. It has edge damage but the arêtes are sharp. A small tool suggests that flint was a precious commodity and every little piece was used.

***Artefact A8*** (numbered 9 in *JQS*)

This is a large nodule (core?) in greyish flint (105 x 83 x 54mm). Two long removals and one broader removal on the dorsal side look humanly struck, but there are no platforms or bulb scars, and the surfaces of these features are not smooth (as is common with a human-struck removal) but rough. There are other small chips from the edges. It compares with Artefact 23, a handaxe with similar rough surfaces on the removal scars. Possibly this raw material was particularly intractable and attempts to work it resulted in failure and it was abandoned. This is unlikely to be Chiltern flint and was probably sourced from local river gravels.

***Artefact A10***

This is probably a crude handaxe with the tip broken off (112 x 80 x 34mm). It is of yellow patinated flint with substantial cortex (75% on ventral side as white areas, 60% on the dorsal side as white or yellow areas). This artefact has been frost-cracked. Further cracks are evident on the cortex of both sides. The original flint nodule was quite flat and the maker failed to thin the middle part to obtain the 'weighted butt' normally needed for a balanced handaxe. The workmanship may have been hampered by the possibility that the frost cracks were antecedent to the knapping thus rendering it delicate, and the handaxe may have broken in manufacture. It is slightly rolled on the edges and the surfaces. The original blank may have been about 140mm long and thus probably not of local gravel origin but brought from the Chilterns. The broken surface is whitish but this may be a patina developed since the break; no indication of the interior colour of the flint is apparent.

***Artefact A11***

A small unpatinated flake fragment in brown-grey flint with soft grey mud adhering (57 x 36 x 14mm). There is extensive edge damage but it appears mostly random and the arêtes are sharp. However, a long line of 18 small spalls along one edge is curious; they are mostly from the same side (Figure 7.2). The platform and bulb of the flake are missing but the edge break here is facetted suggesting that the artefact may have been deliberately broken.

***Artefact A12***

A triangular flat core in glossy greenish flint (73 x70 x 20 mm). It is unpatinated and rolled. The flat shape is unusual and there are about 15 removals on one side, while the opposite side is mainly natural.

***Artefact A13***

This flint handaxe tip (63 x 63 m) with yellow patination is the same as those from nearby Dix Pit and Gravelly Guy. The edges are very battered; none of the original edges are present. The edge-battered surfaces are a lighter colour than the rest of the artefact and the same colour as the broken section across the artefact. Down one of the edges, the battering is all from the same side suggesting that this was a broken tip that was subsequently resharpened to use as a scraper. The surfaces still have small traces of the mud and sand matrix in which it was found.

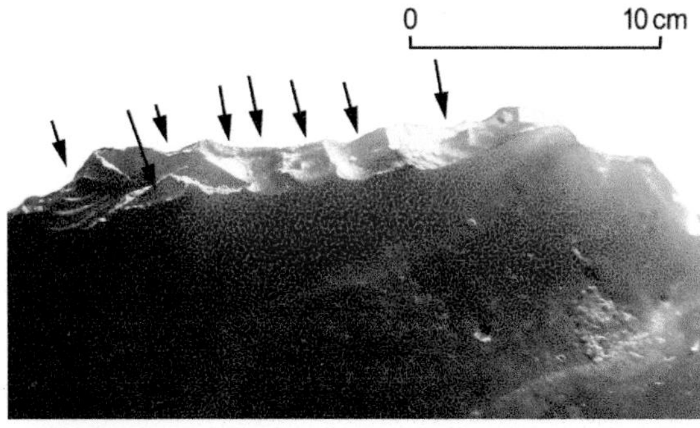


Figure 7.2 Artefact A11: part of the row of small spalls on one edge

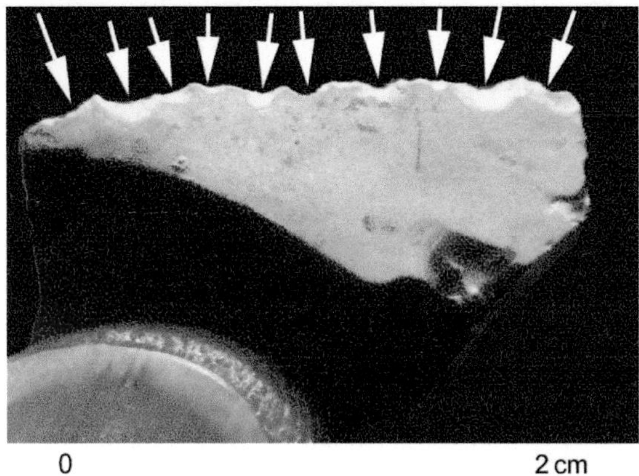


Figure 7.3 Enlargement of Artefact A14 showing ten spalls all from the same side along one edge

The flaking on what remains of the main artefact is thin suggesting soft hammer work. This artefact seems to have had two lives.

***Artefact A14***

A small yellowish patinated flake (15 x 22 x 5mm) with one denticulate edge (ten spalls from one side and three from the other). Making a denticulate from such a small flake is not characteristic of the Lower Palaeolithic, so the denticulation is probably natural (Figure 7.3).

***Artefact A15***

This is an extremely small yellowish patinated flint flake (16 x13 x 5mm) with a small area of cortex.

***Artefact A16***

A broken (ovate?) handaxe tip (62 x 43 x 20mm) of glossy, multi-coloured flint. This artefact shows fine soft hammer work at the tip. Its edges have been chipped rather than worn; the unchipped edges are sharp.

### *Artefact A17*

This small piece of debitage (35 x 23 x 10mm) is of unpatinated black flint. It is quite rolled with edge damage.

### *Artefact A18*

This flake of yellow stained flint (48 x 38 x 13mm) has a row of small retouch scars along the long edge, some with whiter patination which, if intentional, would suggest that the flake was adapted to be used as a tool. It is slightly rolled.

### *Artefact A19*

A glossy unpatinated cordiform flint handaxe made on a flake; grey with yellow inclusions (82 x 66 x 20mm). The flake presented steep edges towards the point which the maker has struggled to reduce, producing numerous small step fractures but never penetrating the unwanted mass at the tip. However, with a smooth ventral surface and all removals from the dorsal side, it was possible to make a serviceable tool with straight edges. A more ambitious approach would have been to try thinning the tip with removals from the dorsal side, but with the risk of reducing the handaxe size and losing its functionality.

The tool superficially resembles a bout coupé handaxe. Roe suggested it might be classed as a Middle Palaeolithic discoidal core, but the very deliberate working to form a point despite challengingly steep angles suggests that the maker was more intent on a handaxe form.

It is difficult to surmise where this flint came from. If it is river gravel flint, it is exceptionally fine and lacking in frost damage. More likely it is imported from the Chilterns.

### *Artefact A20*

This is a flat-butted pointed flint handaxe (102 x 66 x 32mm). The butt is entirely cortical with no attempt at working. Despite its inelegantly shaped butt, this handaxe is a masterpiece with long thin removals penetrating to the centre of the artefact on both sides, a well-thinned point, and straight edges. As is common in this assemblage, the edges also show many small step fractures presumably made while refining the tool, along with yet smaller spalls which may be use wear or subsequent fluvial battering. The flint has deep yellow glossy patination comparable to the larger Chiltern handaxes from Gravelly Guy (also *cf.* Artefact 4, which could have been made by the same hand).

### *Artefact A21*

This is a yellow-stained piece of debitage (42 x 38 x 17mm) with cortex remaining on the ventral side. It is sharp but there is some edge damage. There are discoidal removals on the dorsal side but this item is too small to be regarded as a discoidal core.

*Artefact A22*

A small flint handaxe very similar to Artefact A1 (68 x 51 x 36mm). Roe describes it as 'tending towards cordiform, made from a pebble of cherty flint surely not of Chiltern origin. The shape is determined by the pebble. A fair amount of cortex has been left: at least some more could have been removed but only at the cost of reducing what was always going to be a small tool. The maker has been able to contrive a robust if somewhat blunt point, a chopper-like butt and irregular but usable edges. This is an enterprising use of a locally acquired rock'. It has muddy yellow patination. The removal scars have unusually rough surfaces similar to Artefact 8 (which is stained grey), suggesting that they are both of the same type of low-grade flint and difficult to work. The surfaces are slightly rolled, and the edges damaged all round.

*Artefact A23*

This very crude ovate handaxe is made of the same kind of flint as Artefact 8 (118 x 62 x43mm). One edge was completely beyond reduction and remains as a 'third side' up to 40mm across, tapering towards the tip. The butt has a very large concave piece missing, perhaps convenient as a mêplat. This is a rolled item with wear on the edges as well as on the faces. The surfaces have iron-stained concretions. The material may be locally sourced.

*Artefact A25*

A flint handaxe thinning flake, ivory/yellow stained (36 x 30 x 5mm). It is slightly rolled and soft hammer work is implied.

*Artefact A26*

This is a large crude pointed flint flake, part cortical (121 x 69 x 40mm). It has an ivory patina and is probably of Chiltern origin. The large bulb suggests hard hammer work. The ventral side consists solely of the removal surface, with no further working or retouch. The dorsal side shows only two main removals but remnants of hinge fractures near the dorsal ridge tell of previous removals when the artefact was much larger. There have also been three rather feeble attempts to thin the large mass of flint in the middle of the artefact on the other side of the dorsal ridge. This artefact is interesting because it is a large primary flake taken off a very large nodule. Practicalities would suggest that early humans, rather than carrying heavy stones of excess weight, would have made their handaxes at the raw material source. As to its purpose, the long line of retouch along one edge, entirely from one side along some 85 mm, surely suggests its use as a scraper or knife. Both surfaces and edges are sharp.

*Artefact A27*

A quartzite handaxe (117 x 91 x 40mm; Figure 7.4). It is a particularly interesting piece because it was made on the most intractable raw material one can imagine! In knapping terms, quartzitic sandstones come in a range of qualities from the impossible to the superb, the latter almost matching flint. There is evidence that Lower and Middle Palaeolithic humans were capable of discriminating quartzite raw material qualities by testing cobbles before knapping (Hardaker and MacRae 2000: 57). In this case however the raw material was so unsuitable that

Figure 7.4 Quartzite handaxe Artefact A27

traces of the flaking scars are virtually all subsumed into the uneven surfaces produced by the removals. The only way one can tell it is a handaxe is from the unmistakable profile and plan shapes, the cortical butt (paradoxically the only smooth surface on the tool) and the beautifully straight cutting edges that the maker has managed to achieve on both edges.

This is a masterpiece of execution. It demonstrates that amongst the population there were occasionally people who developed outstanding knapping skills. It also suggests that there was indeed an inadequate supply of good flint at least for some of the time during the human occupation of the area in MIS 7. There was also a shortage of good quartzite!

***Artefact A28***

This is a thin pointed flint flake in yellow/grey material (43 x 29 x 8mm). It is possibly a handaxe trimming flake but the edges are damaged all round.

***Artefact A29***

A broken tip of a large quartzite flake (86 x 60 x 25mm). This is the most rolled artefact in the collection on edges and surfaces; it has clearly seen time in a fluvial context. There are only three removals on the dorsal side which also has a slither of the original cobble, and the opposite side seems to be 'natural' rather than flaked. It is difficult to surmise its purpose which was possibly to get long blades off a long flake, or to use as a backed knife.

***Artefact A30***

This is a quartzite handaxe on a cobble (98 x 56 x 41mm). Minimal working has achieved a useful straight edge on one side and a tapered point but the cortex still extends across much of the second edge, which was impossible to penetrate. The surfaces and edges are moderately rolled. Roe was non-committal on this piece, saying it was 'perhaps best classed as a side chopper, but the presence of a point at each end gives it many of the features of a thick handaxe'.

### *Artefact A31*

This very fresh flint flake, with faceted platform, is broken at the distal end (25 x 23 x 5mm). The edges almost free of damage. This, and its facetted platform might suggest a Middle Palaeolithic date although it is certainly from a secure context within the excavation.

### *Artefact A32*

This very small piece of debitage is of yellow stained flint with some cortex (28 x 22 x 11mm).

### *Artefact A33*

A small pointed flint flake with deep yellow staining (36 x 23 x 10mm).

### *Artefact A34*

A very small piece of flint debitage, yellow stained (28 x 20 x 7mm). The platform and bulb are preserved but it is somewhat rolled and with some edge damage. None of the flaking scars shows point of origin, thus the possibility remains of it being natural.

### *Artefact A35*

This is a black, unpatinated piece of flint debitage with several natural (thermal) surfaces (28 x 28 x 8mm).

### *Artefact A37*

A dark yellow flint core - the only true core in the whole collection (76 x 52 x 38mm). It is a bifacial pyramid core with characteristics not unlike an unstruck Levallois core, but the working on one side has not progressed sufficiently to verify this - there are only two removals. There is also a large area of cortex on the opposite side, so the pyramid may just be fortuitous although, by MIS 7, Levallois technology is present in Britain. The artefact is well-rolled on both surfaces and edges. The question arises: what was the artefact for if not to make a Levallois flake? It must be greatly reduced from its original size and ended up with an edge that could be used for processing animal hides.

### *Artefact A38*

This quartzite handaxe (100 x 56 x 47mm) is made on a small cobble with 30% cortex still remaining. It is quite rolled and has been flaked to provide a single good straight cutting edge with good thinning at the tip but less attention to the butt. The maker has achieved a useful tool with minimal flaking effort in material that is extremely hard to work. It is typical of the quartzite items in the wider area, where practical considerations seem to have taken precedence over the quest for symmetry.

## Artefacts from the wider context near Stanton Harcourt

The Channel artefacts need to be viewed in the context of the additional finds, collected mainly by R.R. MacRae from the 1980s at pits adjacent to the Channel excavation. Most came from Gravelly Guy and Dix Pit (also known as Smiths Pit) which are close to the Stanton Harcourt Channel (Figure 7.5). These non-Channel artefacts were analysed by Woo Lee (2001) but have been reviewed by the present author. A summary is shown in Table 7.2.

The relative proportions of different typologies in the Channel and the adjacent pits are generally in line with each other except that there were relatively more flakes discovered in the Channel. That is to be expected since the Channel was excavated meticulously while the other pits were not excavated at all. The artefacts listed in Table 2 resulted from scouring reject heaps and the floor of the quarries after gravel extraction.

### *Selected items from Dix and Gravelly Guy pits*

The numbering on these artefacts is from several different hands at different times. Probably they were left unnumbered originally but MacRae added numbers on some in pencil. Subsequently coloured labels were added, and two sets of coloured labels also appear on some.

1. *Ficron - Gravelly Guy.* Figure 7.6.

269x119x49mm. Made from a flat flint nodule as seen from the cortex remaining on both sides at the butt, and traces of cortex further up the artefact. White patina shows small areas of later edge damage. Long thin removals suggest soft hammer work. Slightly rolled. Good symmetry with fairly straight cutting edges. The butt is unworked and attempts at thinning on one side of the butt have resulted in multiple step fractures. This indicates that the maker was trying to produce a bisymmetrical balance but gave up. One edge closer to the point could have done with further thinning but at the risk of breaking off the tip. The maker was fortunate a large flat natural nodule and the exceptional length of this artefact is surely due to the excellent suitability of the raw material. This artefact is the clearest evidence that raw material was sourced from the base of chalk downs and not from an Upper Thames fluvial context.

2. *Large ovate flint handaxe (Also labelled 22 and SH 1) Dix Pit SU41205.* Figure 7.7.

191x112x53mm. Finely thinned tip with soft hammer flakes extending to apex. Irregular butt partially worked. Thick ivory coloured patination with yellow stains. Slightly rolled. Extremely straight cutting edges but with small edge spalls (edge damage).

3. *Cordiform flint handaxe No 22 - Smiths Pit.* Figure 7.8.

109x79x36mm. Well finished with soft hammer work, deep yellow colour with thick patination, straight cutting edges. Quite sharp, some small edge damage.

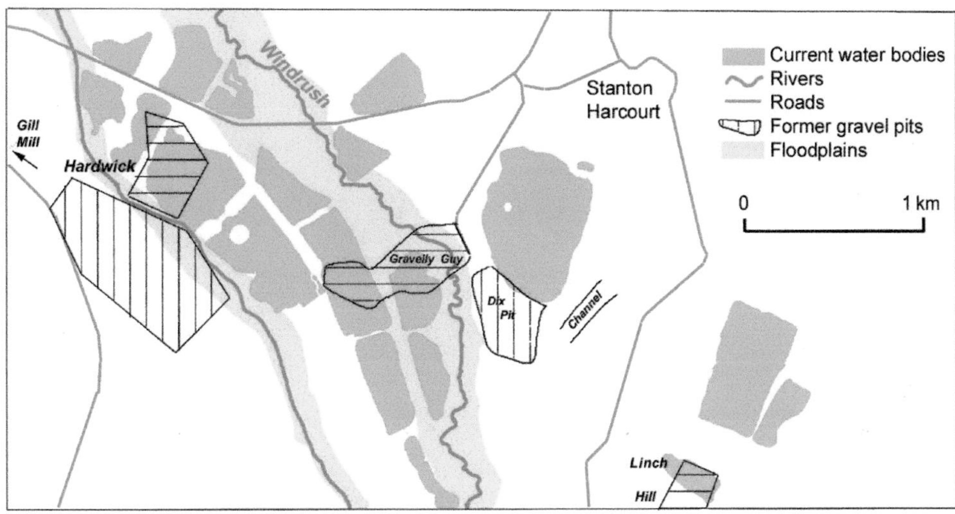


Figure 7.5 Map of Stanton Harcourt with artefact zones as mapped by MacRae in 1990 (pers. comm.). 'Dix Pit' was also known as Smiths Pit and was later extended into the 'Channel' area. Hardwick is probably not lithologically parallel to the Channel and any finds from there are omitted.

Figure 7.6 Gravelly Guy Ficron

Figure 7.7 Dix Pit handaxe 2 - finely worked tip

Figure 7.8 Smiths Pit Cordiform Flint handaxe 22

Figure 7.9 Smiths Pit Flint handaxe 7

Figure 7.10 Smiths Pit flint handaxe 11

Figure 7.11 Stanton Harcourt quartzite handaxe unnumbered

4. *Pointed flint handaxe No 7 (Also labelled 9 and 34) - Smiths Pit.* Figure 7.9.

139x77x38mm. Tip missing. Thick, ivory patination. Straight-edged with méplat on side of butt.

5. *Ovate flint handaxe (Pencil number 11) - Smiths Pit.* Figure 7.10.

152x81x45mm. Invasive soft hammer flaking, thick ivory patination, méplat or break on side of butt. Slightly less successful edge working but carefully worked thin tip.

6. *Thick cordiform quartzite handaxe (Unnumbered) labelled Stanton Harcourt, SU412056, dated 1984.* Figure 7.11.

10x6x43mm. There is an error with the grid reference, which should be SP412056, and is thus located in what is now the industrial estate about 1.2km north of the Channel. The handaxe is very rolled on both faces but the edges are only moderately rolled and with no edge damage. Flaking was extremely skilled including the butt area producing sharp straight cutting edges with a slight twist (not a 'twisted ovate'). Only a small area of cortex remains on one face.

| Material | Artefact type | Number found in pits adjacent to Channel | Number found in Channel excavation |
|---|---|---|---|
| Flint | Pointed handaxe | 36 | 3 |
| Flint | Ovate handaxe | 10 | 5 |
| Flint | Cordiform handaxe | 1 | 3 |
| Flint | Cleaver | 2 | 0 |
| Flint | Ficron | 4 | 0 |
| Flint | Handaxe fragment | 6 | 0 |
| Flint | Chopper | 2 | 0 |
| Flint | Core/core tool | 10 | 4 |
| Flint | Flake/flake tool | 16 | 12 |
| Flint | Debitage | 10 | 4 |
| Quartzite | Pointed handaxe | 11 | 3 |
| Quartzite | Cordiform handaxe | 0 | 1 |
| Quartzite | Cleaver | 2 | 0 |
| Quartzite | Ovate handaxe | 1 | 0 |
| Quartzite | Chopper | 4 | 0 |
| Quartzite | Chopper-core | 3 | 0 |
| Quartzite | Core | 1 | 0 |
| Quartzite | Debitage | 7 | 0 |
| Quartzite | Flake/flake tool | 3 | 1 |
| Quartzite | Retouched flake | 1 | 0 |
| Chert? | Pointed handaxe | 1 | 0 |
| TOTAL | | 131 | 36 |

Table 7.2 Artefacts from adjacent pits compared with those from the Stanton Harcourt Channel

7. *Quartzite cleaver from 'Stanton Harcourt' (Numbered 66 in red).* Figure 7.12

If correctly identified this is the first quartzite cleaver from the upper Thames area (It does seem to have all the relevant characteristics: chisel tip, tranchet removal at tip, and heavier butt than tip, but clearly this was a tough task to achieve in quartzite.

### *Discussion*

Useful information can be gained from a comparison of the Channel artefacts and those in the vicinity of the site. This includes raw material sources, stylistic groupings, influence of raw material on artefact typology and form, relative proportions of different typologies, and the possible effect of all these on hominid lifestyles.

From studies of wide-area surface scatters of Palaeolithic artefacts in the UK (Hardaker 2017), Africa (Hardaker 2011, 2019; Sampson 2006) and India (e.g. Paddayya 2001) it is clear that Lower Palaeolithic communities did not live solely in isolated groups alongside rivers, but tended to form lattices of occupation across the wider terrain where the different communities were able to contact one another over quite large distances. Hence the find of a cluster of artefacts in the Stanton Harcourt Channel would mean that other artefact scatters could be expected

Figure 7.12 Quartzite cleaver from Stanton Harcourt (pit not recorded)

Figure 7.13 (a) Handaxe of local flint with modern break showing very thin white cortex under the brown stained surface. (b) Remote flint with break showing very thick white cortex below the surface. (c) Local flint handaxe showing typical brown colour and rough cortical surfaces

in the vicinity, and so it has proved. It may reasonably be hypothesised that these wider finds belong to the same general community as those in the Channel. Their lithological context, within a thin bed of coarse gravels and clays overlain by MIS 6 gravels, corresponds generally with the lithologies of the Channel zone. The only caveat is that very few of the artefacts from the surrounding pits were recovered *in situ*, although many had clays adhering which betrayed their context. Most came from the reject heaps that are part of the gravel quarrying process. However, MacRae observed that these were invariably found at the base of the pit and were subsequently buried by the sand and gravels of the Summertown-Radley Member of the Upper Thames Formation (MIS 6). At the time of their collection, MacRae attributed them to the early part of this cold stage but, as Scott and Buckingham (2001) point out, none was recovered from the overlying MIS 6 'cold' gravel and there is a family likeness between the excavated Channel and surrounding artefacts in terms of typology and patination which further augments the case that they all belong to the same period. Furthermore, from these areas MacRae also collected teeth of mammoth *Mammuthus trogontherii* and elephant *Palaeoloxodon antiquus* similar to those from the Channel (Scott and Buckingham *op. cit.*), thus confirming the likelihood of a temporal link. If this premise is correct, the assemblages from adjacent pits greatly enhance what can be read into the Channel archaeology.

The question of the origin of the raw material for the flint assemblages can now be better understood. Linch Hill, Gravelly Guy and Dix Pit (also known as Smiths Pit) yielded predominantly flint material, but Hardwick and Gill Mill west of the Channel are almost entirely quartzite. These latter two quarries also yielded remains of woolly mammoth *Mammuthus primigenius* and reindeer *Rangifer tarandus* (K. Scott *pers. comm.*). These factors combine to indicate that they are not of the same age as the Channel deposits, hence their omission from the present discussion.

The closest sources of flint to Stanton Harcourt are the White Horse Hills, over 20km to the south, or the Wallingford Fan gravels, 13 km to the east (MacRae 1988b). Stray flint nodules

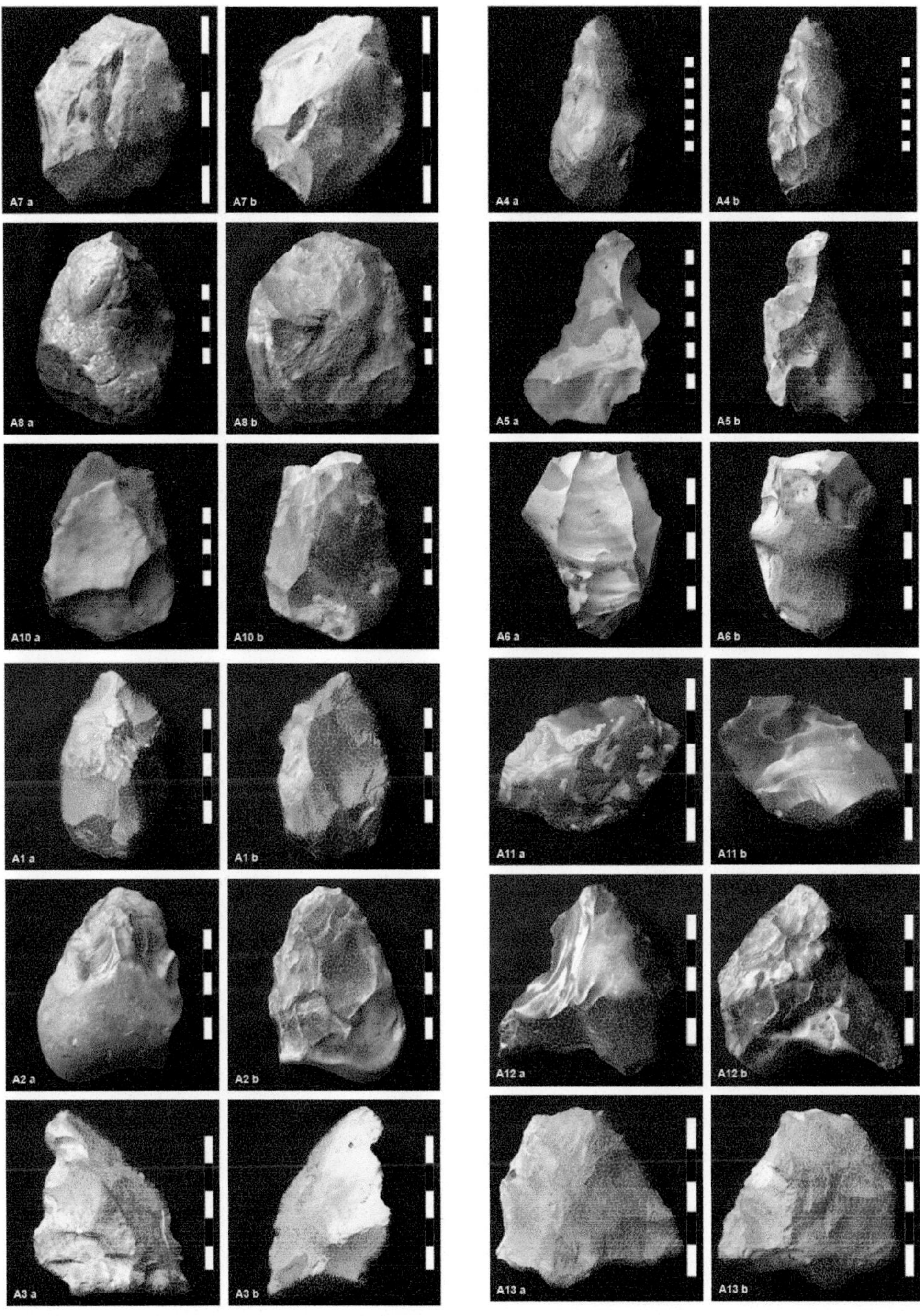


Figure 7.14 The complete assemblage of artefacts from Stanton Harcourt

A17 a
A17 b
A14 a
A14 b
A18 a
A18 b
A15 a
A15 b
A19 a
A19 b
A16 a
A16 b
A23 a
A23 b
A20 a
A20 b
A25 a
A25 b
A21 a
A21 b
A26 a
A26 b
A22 a
A22 b

Figure 7.14 Continued, The complete assemblage of artefacts from Stanton Harcourt

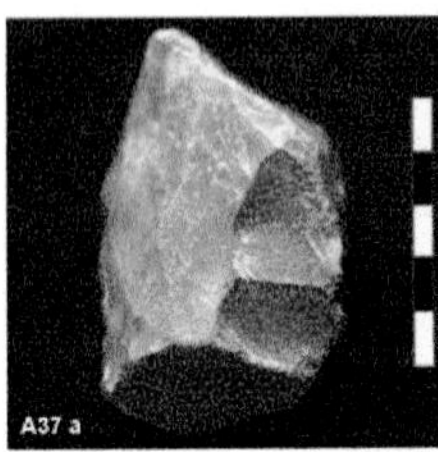


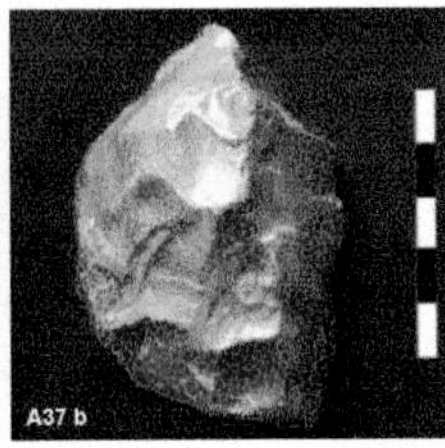


occur in some of the Quaternary gravels in the vicinity but these were seldom large enough to fashion handaxes and moreover, if they had previously been frost-cracked by exposure to intense cold, they may have readily fractured in knapping (MacRae *op. cit.*). However, the handaxes recovered are often sturdy and large, notably the remarkable ficron from Gravelly Guy, 269mm in length, which was the third largest handaxe in Britain when found in 1986 (MacRae 1986). Such examples have clearly come from the chalk bedrock and thus indicate the distances travelled by MIS 7 humans to obtain good quality flint. MacRae discussed the possible methods of transport (MacRae 1988b), suggesting that at Stanton Harcourt, because of the distance involved, finished handaxes rather than nodules or roughouts might have been transported from the raw material source. How many such items could have been transported in one journey? And how did they carry them? It seemed likely to MacRae that some kind of carrier was employed: a bag made of skins or a belt of fibres. Speculation, of course, but it appears unlikely that they were carried one by one in the hand. Such efforts to get flint handaxes on to home territory tells of the importance attached to efficient cutting tools. The only alternative would have been the local quartzite and flint from the rivers or from the Northern Drift surface deposits; these were occasionally used, but only in the ratio of 1:4.

Distinguishing remote from local flint is based on colour, surface morphology and, where it can be seen, the thickness of the cortical layer. The latter is evident when there are recent breaks in the flint (Figure 7.13). The local flint, sometimes called 'Drift' flint, is found in gravels throughout the Upper Thames. Its origin is uncertain, possibly from the north, and it is frequently seen as small nodules in the fluvial gravels, often frost-cracked indicating it has been through a very cold climatic phase. Seldom could a tool be made larger than about 120mm in length. It would have been fairly easy to work but liable to fracture unpredictably. Quartzite, on the other hand, was available from fluvial deposits in the Stanton Harcourt area because of the tongue of the Northern Drift (the river deposit of the pre-Anglian proto-Thames) that brought water-worn cobbles south from the Triassic Bunter beds of the Midlands. It would have been common but very hard to work.

Faced with these raw material options, early humans evidently considered it worth the effort to travel long distances, possibly following animal tracks as pathways, to obtain better cutting tools. Their knowledge of topography and geology would have extended far beyond what the eye could see. That in turn points to the high priority they might have attached to their stone tools, and possibly also the importance of large animal carcasses in their diet. They could probably have made out with a local diet of small animals and plant food and saved the expeditionary risks. They chose not to. This is the picture these stone tools paint of the level of sophistication and common-sense solutions of British hominids during MIS 7.

Some of the flake/core artefacts in flint from the Channel show signs that they are not waste material but seem to have had small retouch to the edges to render them steeper and thus capable of use as scrapers or more robust cutting tools (see Figures 7.1 and 7.2). Fewer of the flakes from the wider area were identified as retouched: Woo Lee actually identified 19 items from this group as miscellaneous 'tools' that had retouch applied, but current inspection suggests that it is hard to distinguish natural damage from human retouch. As flint was a scarce commodity at Stanton Harcourt, it is likely that humans would have found a use for all shapes and sizes. The natural mechanism by which small flakes can become so damaged along their edges probably results from tumbling in fast-moving water, possibly intensified by seasonal chill. It is an open question whether the gradient of the Channel was sufficient to create the energy needed to produce these damaged edges, although in the *thalweg* (the fastest moving water in a channel) bed load material of sand, gravel and cobbles could have abraded the artefacts. Those artefacts located at the interface of the channel sediments and the overlying MIS 6 sediment could have suffered some frost damage or the edge damage could have been caused as a result of a consistent natural force working on any exposed flint, such as when MIS 6 ice extended across the Midlands. Ice wedge casts at the excavation, post-dating the Channel sediments, indicate periglacial conditions in the Stanton Harcourt area during MIS 6

Conventional terminology (such as side choppers, backed blades or knives) was used by Roe to describe some of the items in the Stanton Harcourt gravels. The present author wonders whether such precise interpretations are stretching the terminology beyond what is justified. Whether we call it a side chopper or a crude handaxe (e.g. Artefact A30) is less important than getting into the mind of the maker. The humans who made the Stanton Harcourt tools were clearly used to working in a range of raw materials from good chalk-based flint through poor local flint to quartzite. They did have certain templates embedded in their thinking, one of which was the handaxe shape, and they only fell short of the ideal when the raw material prevented it. Thus Artefact A30 is, in the present author's opinion, a handaxe because it has a thinned pointed tip, an attempt to work both sides to a straight edge, and a butt heavier than the tip. A 'side chopper' would not share these characteristics. There are plenty of 'choppers' amongst the Cassington/Hardwick/Gill Mill collections that one could compare to test this idea.

There is no clear evidence of Prepared Core Technology (PCT) in this assemblage. Yet the best of the handaxes from the adjacent pits are slim, well-made artefacts that could have been struck from flakes – a hallmark of late Acheulian/early Middle Stone Age workmanship, but the definitive tell-tale signs are missing (see Figure 7.7). At the same time, many are definitely not struck on flakes because of the disposition of cortex remaining on the artefact. However, in the wider context of the finds from adjacent pits, we certainly have flint cleavers and ficrons, both indicative of the Later Acheulian in Britain on the cusp of Mode 3 technology (White *et al.* 2011, 2018). There is also what appears to be a quartzite cleaver (Figure 7.12). Apart from Artefact A6 (a flake converted into a core), the only Mode 3 contender from the Channel is Artefact A37, purportedly a Levallois core, but to my thinking a fortuitous shape which is actually just a core. No other prepared core examples have come from the gravel pits nearby, except a single doubtful flake from Cassington. Given the cumulative evidence from all these finds, it seems that PCT is absent from the Channel area, being confined to a smallish cluster in the Abingdon region (Hardaker 2006). This situation adds another piece to the jigsaw of typological change that archaeologists are currently building for the British marine isotope stages.

The more restricted area of the Channel yielded only one fine handaxe (Artefact A4) to match the best from the adjacent pits. This shares a common feature of most of the adjacent handaxe repertoire in elegant soft-hammer working especially towards the tip. It seems there was a deliberate policy to save the best workmanship for the tip (see for example Figure 7.7). White and Foulds (2018), in their discussion on handaxe symmetry, have suggested that Acheulian workmen took pleasure in shaping their handaxes and the best of the present assemblage would certainly chime with this comment.

This preference has not transferred to the quartzite handaxes although a few are carefully worked with long thin removals. It is suggested that this is simply because of the difficulty of introducing elegance into the knapping procedure as applied to quartzite material. The rather greater rolling seen on the quartzite items is puzzling and just might be a hint that some of them belong to a much earlier period.

The flint handaxes from the surrounding pits vary in quality; not all are finely worked. Whether this represents two separate populations working at different times, or just variability in human skills amongst the same population, it is impossible to be sure. But on the premise that early humans varied in their abilities just as all mammals do, it seems the latter is more likely.

Whereas some of the Channel artefacts are undoubtedly from a fluvial context, others were originally lying on open ground and were subsequently buried by fluvial gravels. Their generally sharp condition (apart from some of the quartzites) suggest they have not travelled far and that they represent roughly the places where early humans worked, lived and dropped their tools. As would be expected, this extends beyond the immediate environs of the river as the people ranged widely in their home environment.

### The Stanton Harcourt artefacts and other British assemblages

A postscript is added to this Chapter by Nick Ashton (pers. comm.). He notes that the assemblages from Stanton Harcourt and surrounding pits are unusual in the context of other MIS 7 assemblages in Britain. The absence of clear Levallois technology stands in contrast to the MIS 8 to 7 assemblages in the Lower and Middle Thames valley and East Anglia (Scott 2010). At Ebbsfleet, Crayford and Creffield Road, Levallois technology dominates the assemblages with little evidence of handaxe manufacture, and likewise in Suffolk at sites such as Stoke Tunnel and Brundon. There are also several Levallois assemblages in the Solent, such as Warsash, which probably date from MIS 8 to 7 (Davis *et al.* 2016). It has been suggested that there is a regional pattern in Britain with Levallois being mainly limited to the south and east of Britain, but with handaxe manufacture surviving in the west at sites such as Harnham in Wiltshire in MIS 8, or Pontnewydd in north Wales in MIS 7 (Aldhouse-Green *et al.* 2012; Bates *et al.* 2014; Ashton *et al.* 2018). Stanton Harcourt adds another important site to the pattern that seems to be emerging of different populations and cultural traditions in the south and east, compared to the west of Britain from MIS 8 to MIS 7.

# Chapter 8

# Neanderthals in the Thames Valley

Although some 160,000 artefacts are attributed to the Middle and Upper Pleistocene of Britain, the skeletal remains of their makers are extremely rare. The earliest hominin remains for this period date to c.500,000 years ago and come from Boxgrove, West Sussex. Here a partial shin bone and two teeth are ascribed to *Homo heidelbergensis* (Stringer *et al.* 1998; Roberts and Parfit 1999). Dated to c.400,000 years ago is the partial skull of a young female Neanderthal *Homo neanderthalensis* from Swanscombe, Kent. No further hominin remains are known in Britain until MIS 7, represented by the teeth and partial jaws of several individual Neanderthals excavated in Pontnewydd Cave, Wales (Compton and Stringer 2012).

Whereas the Continental archaeological evidence shows a more or less continuous Neanderthal presence in the Middle and Upper Pleistocene, various factors combined to ensure that the Neanderthal occupation of Britain was likely to have been limited and sporadic. During cold intervals such as MIS 8 (and especially MIS 6), when sea level was at its lowest and Britain was part of the Continental mainland, western and northern regions of Europe were probably inhospitable environments for people for long periods. As the climate ameliorated at the MIS 8/7 transition, a range of animal species of temperate adaptation evidently moved westwards into Britain. Vertebrate assemblages considered to date to the earlier part of MIS 7 indicate a temperate environment with abundant vegetation that supported a range of large herbivores and attendant predators. Species with a preference for open grassland predominate, but a significant woodland component is also evident (Schreve 1997, 2001). Ashton and Scott (2016) assign 4 lithic assemblages (dominated by Levallois technology) to this late MIS 8/early MIS 7 period (Table 8.1).

As sea levels rose during MIS 7e, access to Britain would have become increasingly impossible. As reviewed by Pettitt and White (2012), opinion varies regarding the extent to which Britain remained an island for the rest of MIS 7. Sea-level fell during the two cool intervals (MIS 7b and 7d) of the interglacial but it seems that only during MIS 7d did sea-levels become sufficiently depressed for Britain to have become a peninsula and thus accessible again from the Continental mainland. Although the range of species that entered Britain at this time is not significantly different from those earlier in MIS 7, those of woodland preference are less well represented than in MIS 7e (Schreve 1997, 2001). Schreve ascribed the majority of MIS 7 vertebrate faunal assemblages to the latter part of the interglacial. Since her analysis, further assemblages that can be attributed to the period MIS 7d-7a have become available for study (Scott in prep.). However, of 29 fossil vertebrate assemblages, only 10 have artefacts, the majority of which were collected rather than excavated (Table 8.1). While this might indicate that human presence in Britain was extremely rare in MIS 7, it might also reflect the nature of the context in which the vertebrate remains and artefacts have been found. As Pettitt and White (*op. cit.*) point out, they have been recovered principally from fluvial gravels because these have been the focus of gravel extraction. Early people, although periodically visiting river margins, may have been conducting their activities more extensively on higher ground, the evidence for which is simply not being found.

| SITE | NUMBER of ARTEFACTS | EXCAVATED or COLLECTED | MIS |
|---|---|---|---|
| LATTON | 8 | excavated | Late MIS 7 |
| STANTON HARCOURT | 36 | excavated | Late MIS 7 |
| STOKE TUNNEL, MAIDENHALL | 20 | excavated | MIS 7 |
| BRUNDON | 34 | collected | MIS 7 |
| CRAYFORD | 265 | collected | MIS 7 |
| PONTNEWYDD CAVE | 1037 | excavated | MIS 7 |
| WEST THURROCK | 229 | old excavation | MIS 8/early MIS 7 |
| EBBSFLEET | 579 | collected | MIS 8/early MIS 7 |
| CREFFIELD RD, ACTON | 343 | collected | MIS 8/early MIS 7 |
| YIEWSLEY, WEST DRAYTON, HILLINGDON | 290 | collected | MIS 8/early MIS 7 |

Table 8.1 British sites of MIS 7 age with artefacts. With the exception of Pontnewydd, Wales, all sites are from within or on the surface of fluvial sediments laid down by the Thames or its tributaries. The proposed date for the assemblages is mainly dependent on the interpretation of terrace stratigraphy. The data for Stanton Harcourt and Latton are from Scott and Buckingham (2001) and for all other sites from Ashton & Scott (2016).

Figure 8.1 Flint handaxe Artefact A4 from Stanton Harcourt compared with one of the remarkable bifaces made of elephant bone from the Middle Pleistocene site of Castel di Guido (Boschian and Saccà 2015)

Although the British evidence is scarce, the artefacts from Stanton Harcourt show the undoubted presence of Neanderthals in the Upper Thames Valley in the latter part of MIS 7. The richer Continental records, apart from showing that the Neanderthals were intelligent, innovative tool makers, also enable us to know what they looked like (Humphrey and Stringer 2018). Adults ranged from about 1.50 - 1.75m tall and weighed about 64-82kg. They had strong, muscular bodies, and wide hips and shoulders. Their skulls were long and low (compared to the more globular skull of modern humans) with a characteristically prominent brow ridge above their eyes. Their faces were also distinctive. The central part of the face protruded forward and was dominated by a very big, wide nose. The chin of a Neanderthal was far less prominent than that of a modern human. Their front teeth were large, and scratch-marks show they were regularly used like a third hand when preparing food and other materials.

One can only speculate on the day-to-day existence of these Neanderthals in the Upper Thames Valley but the environmental data from Stanton Harcourt describe a temperate climate, rich

Figure 8.2 Top: Mammoth skull SH4-124.
Below: Elephant skull, log and associated artefacts at the Middle Pleistocene Acheulian site of Gesher Benot Ya'aqov, Israel (Goren-Inbar *et al.* 1994)

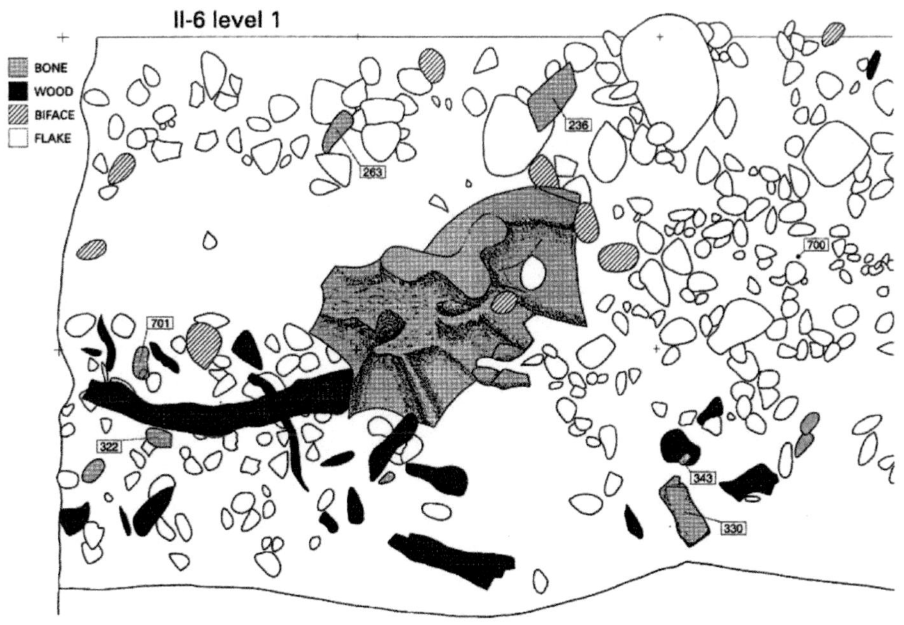

resources of food (both animal and plant), and the ready availability of raw materials such as stone and wood. As regards the availability of food, the site yields plenty of evidence. Whether hunted or scavenged (or both) the meat of bison, horse, red deer, mammoth and elephant was potentially available, perhaps seasonally. Apart from the large herbivores, there was a variety of other food available. If the Neanderthals were able to catch or trap fish, there were numerous species in the river: bream, stickleback, pike, chub, eel, gudgeon, dace, cyprinid, perch, dace, roach, stone loach, and salmon (Chapter 5). Freshwater molluscs were abundant and larger species such as *Potomida littoralis* (about the size of a black mussel) may have been a source of protein. Seeds and nuts (including acorns and hazelnuts) could have been collected seasonally and many of the plants listed in Table 6.1 have edible leaves, roots and bulbs.

There are no certain cut-marks on the Stanton Harcourt bones that might identify butchery. This could be explored further but it is possible that the necessity to fibreglass or plaster larger bones will have obscured such damage. It is also possible that with the mass of flesh available on dead animals such as mammoth or bison, the bone surfaces were seldom reached in the cutting process. Modern observations also indicate that cut-marks on elephant bones occur infrequently in cases where human activity is ascertained, mostly because of the thickness of the periosteum, which protects the underlying bone (Crader 1983, Haynes 1991). As Zutovski and Barkai (2016) describe, the archaeological record reveals that Lower Palaeolithic Acheulian early humans exploited elephants by hunting and/or by collecting carcasses, and butchered and processed elephants apparently for meat and fat consumption, and possibly also to extract bone marrow. This pattern of behaviour and adaptation was evidently practiced over three continents of the Old World for hundreds of thousands of years. At many sites where elephant bones occur with artefacts, there is evidence that the artefacts were manufactured and resharpened at the site of the elephant carcass. Interestingly, in some instances where suitable raw material (flint and other hard stone) was perhaps unavailable, elephant bone was used to make artefacts, including handaxes (Boschian and Saccà 2015; Zutovski and Barkai *op. cit.*). One might imagine that any piece of bone with a sharp edge would have been sufficient to use as a cutting tool, but to have replicated an Acheulian stone handaxe in bone is remarkable (Figure 8.1).

No such bone artefacts occurred at Stanton Harcourt but of possible archaeological interest is the mammoth skull SH4-124 described in Chapter 2. This skull was frontal side down on an oak log. The surface condition of the occipital area, the premaxillae and the zygomatic bones (tusk sockets and cheek bones) is good, but the top of the skull (frontal and parietal regions) had been severely damaged in antiquity. Goren-Inbar *et al.* (1994) describe excavations at Gesher Benot Ya'aqov, a Middle Pleistocene Acheulian site in Israel. Here the skull of a straight-tusked elephant lies on an oak log. The underside of the skull faces upwards and although the surface condition of the skull is generally good, the parietal and occipital regions have been smashed. The authors suggest that the log might have been used to turn the skull over to reveal the occipital region allowing the most direct access to the brain. While the role of the log in the possible butchery of skull SH4-124 can only be surmised, the similarities between it and the Israeli elephant skull are striking (Figure 8.2).

It would not be unusual for a mammoth skull to be smashed to access the brain. Various authors cite ethnographic literature describing particular preference on the part of humans for certain body parts of their prey including the brain. The brain of an elephant is protected

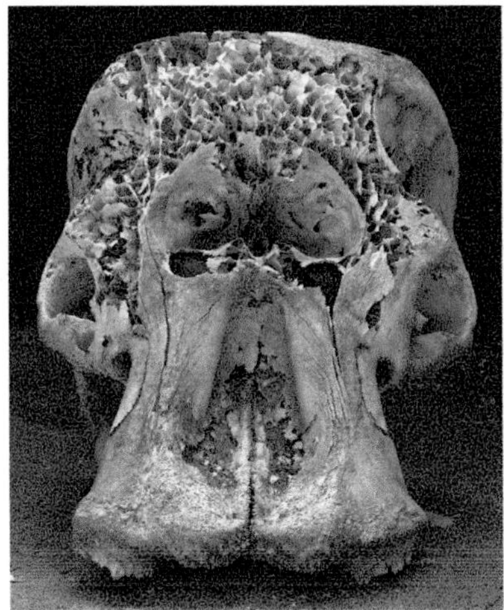

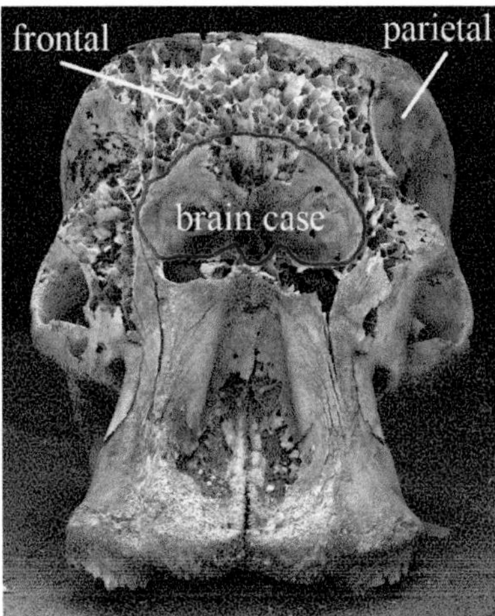


Figure 8.3 Damaged skull of African elephant revealing the honeycomb-like tissue that comprises the crown of the skull (frontal and parietal regions). As the braincase is behind the nasal cavity, the entire frontal and parietal region would have to be smashed to access the brain. Photograph by kind permission of Erika Gouws Publishing Services & Photography

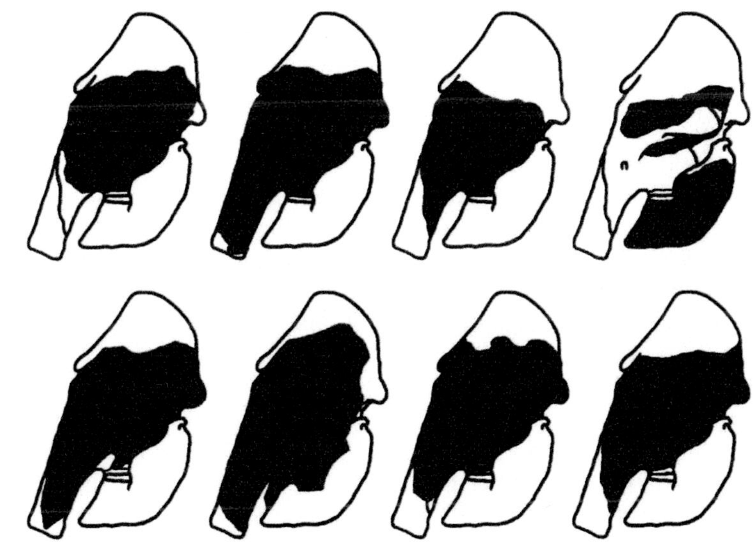

Figure 8.4 Surviving parts (shaded) of mammoth skulls from La Cotte de St. Brelade, Jersey, Channel Islands (modified from Scott 1989)

by a mass of honey-comb air cells (Figure 8.3) so extensive damage is necessarily done to the skull to access the brain. Agam and Barkai (2016) describe 13 sites with elephant or mammoth skulls in an archaeological context and suggest that the available data clearly indicate that Palaeolithic hominins exploited elephant meat and fat routinely and abundantly. The smashing of skulls to access the brain is frequently mentioned in the literature. At La Cotte de St Brelade, for example, excavations revealed late Middle Palaeolithic artefacts and animal remains including cranial material representing at least 11 mammoths (Scott 1986, 1989). The skulls were generally complete but, invariably, the frontal and parietal regions were broken revealing the braincase (Figure 8.4).

Ethnographic accounts of the exploitation of an elephant head also describe a preference for the trunk and tongue, and the abundant meat and fat in the hollow above the eye. An interesting 19th C account, for example, of an African tribe butchering an elephant: '*The skull itself is broken up for the sake of the oily fat which fills the honeycomb-like cells which intervene between the plates of the skull*' (Wood 1868: 138 as referenced by Agam and Barkai *op. cit.*). Other ethnographic accounts mention the use of brain tissue in the tanning of hides (Richter and Dettloff 2002). This activity usually takes place near a water source as water is required in the tanning process. Stone tools have frequently been used to scrape off the flesh, muscle and hair to expose the dermis (skin layer) before soaking the hides for several days. North American Indians combine deer or bison brains with water which makes the hides much softer.

The climatic and sedimentary evidence from the excavation at Stanton Harcourt indicates that there were distinct annual seasons with average winter and summer temperatures probably very similar to those in southern Britain today (Chapter 6). This raises the question of whether the Neanderthals were likely to have been summer visitors or were able to remain in the area on a year-round basis. There are no caves in the area that might have provided shelter but the abundant evidence for mature woodland in the vicinity of the site would have facilitated the erection of canopy-type structures over which to hang skins to provide shelter. The hides of any of the large mammals in the region could have been used for such shelters. All the mammals recorded at the site could have provided skins for clothing, in particular, the fur of the brown bear. On the basis of much older evidence at Boxgrove, for example (Roberts and Parfit 1999), hominins were already adept at processing animal hides, possibly for clothing and shelter. Cut-marks on bear bones are reported from Neanderthal-occupied caves in southern Europe (Romandini *et al.* 2018) and, although the evidence is scarce in Britain, cut-marks made with flint tools on the articulated metatarsals of a brown bear from Grays Thurrock suggest butchery and skinning (Schreve and Currant 2003). While the brown bear would have been a formidable target for a Neanderthal, it is known that even in historical times, the indigenous Japanese Ainu hunted bears for meat and fur armed only with spears (Isabella 2017). There is every likelihood that MIS 7 Neanderthals were doing the same.

There is little in the Pleistocene record to indicate the frequency or duration of hominin visits to Britain but at Stanton Harcourt the number of different layers in which the stone tools were excavated certainly demonstrates Neanderthal presence in the valley on more than one occasion. As detailed in Chapter 2, the deposits in which the bones and artefacts occurred is not a vertical sequence but rather a gradual lateral infilling as the river migrated sideways across the Oxford Clay surface. More than 100 very thin beds were recorded in the southern part of the excavation area even though the full depth of the deposit is only about 1 metre.

Warm water molluscs throughout the sequence at this locality indicate that all the bone beds and at least 10 of the excavated artefacts accumulated here during MIS7. The majority of the artefacts are in layers which also have bones.

The sedimentary evidence in this part of the excavation site suggests that a point bar with an undulating surface built near the left bank while the Oxford Clay was being undercut at the right bank. As a bar grows the potential for environmental material to collect increases when it becomes semi-emergent and vegetation starts to colonize the surface. The earliest signs of human presence and the first bones appear at approximately the same time in the sedimentary sequence as grassy vegetation, when the bar had grown enough for it to be an extension of the floodplain at times of low water. This bar would have provided easy access for people and animals to reach the water, especially in the summer months. The clay banks would not have been high (up to 1m) on the right bank, providing no obstacle to human activity. The water would also have been deeper nearer the right bank and the mature trees here would have been useful in many ways.

A migrating channel also provides an environment for the quick burial of material that has collected on the bar surface or near the right bank without long distance transport of those items. A few unrolled artefacts on the bar surface, articulated bones and possible in-situ mammoth skulls in different layers suggest repeated visits by people to this site. Ready-made stone tools were brought from further afield but a small number of excavated flakes suggest that they were occasionally re-sharpened here. Riffle sections of the river, such as at Site 7, where there was a shallower section near the boulder-like 'turtle' stones, may also have provided a local crossing point for animals and people. Artefacts found at other sites in the region, such as 'Gravelly Guy' (see Chapter 7) also indicate people in the wider landscape in the Thames Valley during MIS7.

As described in Chapter 6, later sediments in the sequence across the site indicate less optimal climatic conditions than in the main body of the deposits. There is less organic material in these better sorted sands and gravels and although the mollusc *Potomida littoralis* is still present, the more temperature sensitive *Corbicula fluminalis* becomes scarce. As the climate became increasingly cool, humans possibly moved south and thus it is perhaps not surprising that no artefacts were recovered from these later beds. Finally, the MIS 7 channel sediments were buried by limestone dominated braided river deposits during the periglacial conditions of MIS 6.

Despite the renewed connection between Britain and the Continental mainland that occurred when sea-levels fell with the onset of MIS 6, the extremely cold conditions appear to have discouraged people from remaining in or visiting Britain. To our knowledge, no artefacts can certainly be attributed to MIS 6 sediments in this area. Even when warm conditions returned in MIS 5e, the associated rapid rise in sea-level seems to have precluded access to Britain once again. Some species evidently crossed before the channel flooded; species such as red deer and forest elephant from the interglacial are indicative of warm climate and extensive forestation. Whether or not some people also made the crossing during MIS 6 or MIS 5 is unknown, but no human remains or artefacts have been found from Britain for the period of 180,000 to 60,000 BP.

# References

Aldhouse-Green, S. H. R., Peterson, R. and E.A. Walker 2012. *Neanderthals in Wales. Pontnewydd and the Elwy Valley Caves.* Oxford, Oxbow Books.

Agam, A. and R. Barkai 2016. Not the brain alone: The nutritional potential of elephant heads in Paleolithic sites. *Quaternary International* 406 Part B: 218-226.

Agenbroad, L.D. 2004. Hot Springs, South Dakota: entrapment and taphonomy of Columbian mammoth, in D.L. Agenbroad and J.I. Mead (eds) *The Hot Springs Mammoth Site: a Decade of Field and Laboratory Research in Paleontology, Geology and Paleoecology*: 113-127. Hot Springs, SD: Mammoth Site Inc.

Agenbroad, L.G. and L. Nelson 2002. *Mammoths: Ice Age Giants.* Mineapolis: Lerner Publications Company.

Aldhouse-Green, S. H. R., Peterson, R. and E.A. Walker 2012. *Neanderthals in Wales. Pontnewydd and the Elwy Valley Caves.* Oxford, Oxbow Books.

Allman, M and D.F. Lawrence 1972. Staining Techniques, in *Geological Laboratory Techniques*: 95-112. London, Blandford Press.

Álvarez-Lao, D.J. and M. Méndez 2011. Ontogenetic changes and sexual dimorphism in the mandible of adult woolly mammoths (*Mammuthus primigenius*). *Géobios* 44(4): 335-343.

Anderson, R and B. Rowson 2020. Annotated list of the non-marine Mollusca of Britain and Ireland. Viewed 2 February 2021, https://www.conchsoc.org/sites/default/files/admin/British revised NMList_2020.pdf.

Ashton, N. and B. Scott 2016. The British Middle Palaeolithic. *Quaternary International* 411: 62-76

Ashton, N. M., Harris, C. R. E. and S.G. Lewis, S. G. 2018. Frontiers and routeways from Europe: the Early Middle Palaeolithic of Britain. *Journal of Quaternary Science* 33(2): 194–211.

Averianov, A.O. 1996. Sexual dimorphism in the mammoth skull, teeth and long bones, in Shoshani, J. and P. Tassy (eds) *The Proboscidea. Evolution and Palaeoecology of Elephants and Their Relatives*: 260–267. New York: Oxford University Press.

Azzaroli, A., de Giuli, C., Ficcarelli, G. and D. Torre 1988. Late Pliocene to early mid-Pleistocene mammals in Eurasia: faunal succession and dispersal events. *Palaeography, Palaeoclimatology, Palaeoecology* 66: 77-100.

Barnett, R., Mendoza, M.L.Z., Soares, A.E.R., Ho, S.Y.W., Zazula, G., Yamaguchi, N., Shapiro, B., Kirillova, I.V., Larson, G. and M.T.P. Gilbert 2016. Mitogenomics of the Extinct Cave Lion, Panthera spelaea (Goldfuss, 1810). Resolve its Position within the Panthera Cats. *Open Quaternary* 2: 4.

Barron, E. and D. Pollard 2002. High resolution climate simulations of oxygen isotope stage 3 in Europe. *Quaternary Research* 58: 296-309.

Bates, M. R., Wenban-Smith, F. F., Bello, S. M., Bridgland, D. R., Buck, L. T., Colllins, M. C., Keen, D. H., Leary, J., Parfitt, S. A., Penkman, K. E. H., Rhodes, E., Ryssaert, C. and J. E. Whittaker 2014. Late Middle Pleistocene human occupation and palaeoenvironmental reconstruction at Harnham, Salisbury (UK). *Quaternary Science Reviews* 101: 159–76.

Bennett, D. and R. S. Hoffmann 1999. *Equus caballus.* Mammalian Species. *American Society of Mammologists* 628:1-14.

Bere. R.M. 1966. *The African Elephant.* London: Littlehampton Book Services Ltd.

Bojarska, K. and N. Selva 2012. Spatial patterns in brown bear *Ursus arctos* diet: the role of geographical and environmental factors. *Mammal Review* 42 (2): 120-143.

Bond, G., Kromer, B., Beer, J., Muscheler, R., Evans, M.N., Showers, W., Hoffmann, S., Bond, R.L., Hajdas, I and G. Bonari 2001. Persistent solar influence on north Atlantic climate during the Holocene. *Science* 294 (5549): 2130-2135.

Boschian, B. and D. Saccà 2015. In the elephant, everything is good: Carcass use and re-use at Castel di Guido (Italy). *Quaternary International* 361: 288-296.

Bowen, D.Q., Hughes, S., Sykes, G.A. and G.H. Miller 1989. Land-sea correlations in the Pleistocene based on isoleucine epimerization in non-marine molluscs. *Nature* 340: 49–51.

Bowen, D.Q. (ed.) 1999. *A Revised Correlation of Quaternary Deposits in the British Isles.* Geological Society, London, Special Reports, 23.

Breda, M., Collinge, S.E., Parfitt, S.A. and Lister A.M. 2010. Metric analysis of ungulate mammals in the early Middle Pleistocene of Britain, in relation to taxonomy and biostratigraphy. *Quaternary International* 228: 136-156.

Bridgland, D.R. 1994. *Quaternary of the Thames.* Joint Nature Conservation Committee, Geological Conservation Review, Series 7. London: Chapman and Hall.

Bridgland, D. R. 2014. Lower Thames terrace stratigraphy: latest views, in D R. Bridgland, P. Allen and T.S. White (eds) *The Quaternary of the Lower Thames & Eastern Essex*: 1-10. Cambridge: Field Guide, Quaternary Research Association.

Bridgland, D.R., D.H. Keen and D. Maddy 1989. The Avon Terraces: Cropthorne, Ailstone and Eckington, in D.A. Keen (ed.) *West Midlands*: 51-67. London: Field Guide, Quaternary Research Association.

Bridgland, D.R., Allen, P., Austin, L., Irving, B., Parfitt, S., Preece, R.C. and R.M. Tipping 1995. Purfleet interglacial deposits: Bluelands and Greenlands Quarries (TQ 569586), in D.R. Bridgland, P. Allen and B.A. Haggert (eds) *The Quaternary of the Lower Reaches of the Thames*: 167-184. Durham: Field Guide, Quaternary Research Association.

Bridgland, D.R., Maddy, D. and M. Bates 2004. River terrace sequences: templates for Quaternary geochronology and marine-terrestrial correlation. *Journal of Quaternary Science* 9: 203-218.

Bridgland, D. R., Howard, A. J., White, M. J., White, T. S. and R. Westaway 2015. New insight into the Quaternary evolution of the River Trent, UK. *Proceedings of the Geologists' Association* 126: 466-479.

Briggs, D.J., Coope, G.R. and D.D. Gilbertson 1985. *The Chronology and Environmental Framework of Early Man in the Upper Thames Basin: A New Model.* BAR British Series 137.

Buckingham, C.M. 2004. The Application of Sedimentological and Isotopic Studies to the Environmental Interpretation of the Stanton Harcourt Channel Deposit. Unpublished PHD Dissertation, Oxford Brookes University, Oxford.

Buckingham, C.M. 2007. The context of mammoth bones from the Middle Pleistocene site of Stanton Harcourt, Oxfordshire, England. Quaternary International 169-170: 137-148.

Buckingham, C.M., Roe, D.A. and K. Scott 1996. A preliminary report on the Stanton Harcourt Channel Deposits (Oxfordshire, England): geological context, vertebrate remains and palaeolithic artefacts. *Journal of Quaternary Science* 11: 397-415.

Burger, J., Rosendahl, W., Loreille, O., Hemmer, H., Eriksson, T., Gotherstr, A., Hiller, J., Collins, M.J., Wess, T. and K.W. Alta 2004. Molecular phylogeny of the extinct cave lion *Panthera leo spelaea. Molecular Phylogenetics and Evolution* 30: 841–849.

Carr, N.J. 1950. Elephants in the eastern province. *Northern Rhodesia Journal* 2: 25-28.

Candy, I. and D. Schreve 2007. Land-sea correlation of Middle Pleistocene temperate sub-stages using high-precision uranium-series dating of tufa deposits from southern England. *Quaternary Science Reviews* 26:1223–1235.

Carne, P. 2000. *Deer of Britain and Ireland: their origins and distribution.* Shrewsberry: Swan Hill Press.

Cartledge, K.M. 1985. Non-marine Mollusca, in Briggs, D.J., Coope, G.R. and D.D. Gilbertson (eds) *The Chronology and Environmental Framework of Early Man in the Upper Thames Basin: A New Model*: 118-131. British Archaeological Reports British Series 137, Oxford.

Chrószcz, A., Janeczek, M., Pasicka, E. and J. Klećkowska-Nawrot 2011. Height at the withers estimation in the horses based on the internal dimension of cranial cavity. *Folia Morphologica* 73 (2): 143–148.

Clark, G.R. 1974. Growth lines in invertebrate skeletons. *Annual Review of Earth and Planetary Sciences* 2: 77-99.

Coe, M. 1978. The decomposition of elephant carcasses in the Tsavo (East) National Park, Kenya. *Journal of Arid Environments* 1: 71-86.

Compton, T. and C.B. Stringer 2012. The human remains, in S.H.R. Aldhouse-Green, R. Peterson and E.A. Walker (eds). *Neanderthals in Wales. Pontnewydd and the Elwy Valley Caves*: 118-230. Oxford: Oxbow Books.

Coneybeare, A. and Haynes, G. 1984. Observations on Elephant Mortality and Bones in Water Holes. *Quaternary Research* 22: 189-200.

Coope, G. R. 1986. Coleoptera analysis, in B. E. Berglund (ed.) *Handbook of Holocene Palaeoecology and Palaeohydrology*: 703-713. Chichester: J. Wiley and Sons.

Coope, G.R. 2001. Biostratigraphical distinction of interglacial coleopteran assemblages from southern Britain attributed to oxygen isotope stages 5e and 7. *Quaternary Science Reviews* 20: 1717-1722.

Coope, G.R. 2006. Age and environment of the (MIS 7) channel at Stanton Harcourt, Oxfordshire, interpreted from insect analysis. Unpublished report.

Coope, G. R., Shotton, F. W. and I. Strachan 1961. A Late Pleistocene fauna and flora from Upton Warren, Worcestershire. *Philosophical Transactions of the Royal Society of London* B244: 379-421.

Corfield, R.M. 1995. An introduction to the techniques, limitations and landmarks of carbonate oxygen isotope palaeothermometry, in D.W.J. Bosence and P.A. Allison,P.A. (eds) *Marine palaeoenvironmental analysis from fossils.* Geological Society, London, Special Publication 83: 27-42.

Corfield, T.F. 1973. Elephant mortality in Tsavo National Park, Kenya. *African Journal of Ecology* 11 (3-4): 339–368.

Cox, B.M., Hudson, J.D. and D.M. Martill 1992. Lithostratigraphic nomenclature of the Oxford Clay (Jurassic). *Proceedings of the Geologists' Association* 103: 343-345.

Crader, D.C. 1983. Recent single-carcass bone scatters and the problem of 'butchery' sites in the archaeological record, in J. Clutton Brock and C. Grigson (eds) *Animals and Archaeology: Hunters and their Prey* :107-141. BAR International Series 163 Volume 1.

Currant, A.P. 1986. Man and the Quaternary Interglacial Faunas of Britain, in S.N. Collcutt (ed.) *The Palaeolithic of Britain and its Nearest Neighbours: Recent Trends*: 50-52. Department of Archaeology and Prehistory, Sheffield University: J.R. Collis Publications.

Currant, A.P. 1989. The Quaternary origins of the modern British mammal fauna. *Biological Journal of the Linnean Society* 38: 22-30.

Davies, P. 2002. The straight-tusked elephant *(Palaeoloxodon antiquus)* in Pleistocene Europe. D. Phil. Dissertation. University of London. ProQuest U642959.

Davis, R. J., Hatch, M., Ashton, N. M., Hosfield, R. T., and S.G. Lewis. 2016. The Palaeolithic record of Warsash, Hampshire, UK: implications for late Lower and early Middle Palaeolithic occupation history of Southern Britain. *Proceedings of the Geologists' Association* 127: 558-574.

Davison, J., Ho, S.Y.W., Bray, S.C., Korsten, M., Tammeleht, E., Hindrikson, M., Østbye, K., Østbye, E., Lauritzen, S-K., Austin, J., Cooper, A. and U. Saarma 2011. Late-Quaternary biogeographic scenarios for the brown bear (*Ursus arctos*), a wild mammal model species. *Quaternary Science Reviews* 30 (3-4): 418-430.

De Rouffignac, C., Bowen, D.Q., Coope, G.R., Keen, D.H., Lister, A.M., Maddy, D., Robinson, J.E., Sykes, G.A. and M.J.C. Walker 1995. Late Middle Pleistocene interglacial deposits at Upper Strensham, Worcestershire, England. *Journal of Quaternary Science* 19: 15-31.

Dickinson, M. R. 2018. Enamel Amino Acid Racemisation Dating and its Application to Building Proboscidean Geochronologies. Unpublished PhD Dissertation, University of York.

Dickinson, M. R., Lister, A. M. and K.E.H. Penkman 2019. A new method for enamel amino acid racemization dating: A closed system approach. *Quaternary Geochronology* 50: 29-46.

Diedrich, C.G. 2012. Late Pleistocene steppe lion *Panthera leo spelaea* (Goldfuss 1810) skeleton remains of the Upper Rhine Valley (SW Germany) and contributions to their sexual dimorphism, taphonomy and habitus. *Historical Biology* 24(1):1-28.

Dixon, E.J. 1999. *Bones, Boats and Bison. Archeology and the First Colonization of Western North America.* University of New Mexico Press.

Douglas-Hamilton, I. and Douglas-Hamilton, O. 1975. *Among the Elephants.* New York: Viking.

Eltringham, S.K. 1991. The Illustrated Encyclopaedia of Elephants. London: Salamander Books.

Eltringham, S.K. *Elephants.* Dorset, Blanford Press.

Flower, L.O.H. 2014. Canid evolution and palaeoecology in the Pleistocene of western Europe, with particular reference to the wolf *Canis lupus* L. 1758. Unpublished D. Phil. Dissertation. Royal Holloway, University of London.

Flower, L.O.H. 2016. New body mass estimates of British Pleistocene wolves: Palaeoenvironmental implications and competitive interactions. *Quaternary Science Reviews* 149: 230-247.

Gee, H.E. 1993. The distinction between postcranial bones of *Bos primigenius* Bojanus, 1827 and *Bison priscus* Bojanus, 1827 from the British Pleistocene and the taxonomic status of Bos and Bison. *Journal of Quaternary Science* 3-4: 175-192.

Gibbard, P.L. 1995. The formation of the Strait of Dover, in R.C. Preece (ed.) *Island Britain: A Quaternary Perspective.* Vol. 96. Geological Society Special Publication: 15–26.

Gleed-Owen, C.P. 1998. Quaternary herpetofaunas of the British Isles: Taxonomic descriptions, palaeoenvironmental reconstructions and biostratigraphic implications. Unpublished PhD Dissertation. Coventry University.

Godwin, H. 1975. *The History of the British Flora: a factual basis for phytogeography.* Cambridge: Cambridge University Press.

Goren-Inbar, N, Lister, A., Werker, E. and M. Chech 1994. A butchered elephant skull and associated artifacts from the Acheulian site of Gesher Benot Ya'aqov, Israel.
*Paléorient* 20: 99-112.

Goren-Inbar, N., Werker, E. and C.S. Feibel 2002. *The Acheulean Site of Gesher Benot Ya'aqov, Israel: the Wood Assemblage.* Oxford: Oxbow Books.

Green, C.P., Coope, G.R., Jones, R.L., Keen, D.H., Bowen, D.Q., Currant, A.P., Holyoak, D.T., Ivanovich, M., Robinson, J.E., Rogerson, R.J. and R.C. Young 1996. Pleistocene deposits at Stoke Goldington in the valley of the Great Ouse, UK. *Journal of Quaternary Science* 11: 59-87.

Grossman, E.L. and T. Ku 1986. Oxygen and carbon isotope fractionation in biogenic aragonite: temperature effects. *Chemical Geology (Isotope Geoscience Section)* 59: 59-74.

Guthrie, R.D. 1984. Mosaics, allelochemics, and nutrients: An ecological theory of Late Pleistocene megafaunal extinctions, in P.S. Martin and R.G. Klein (eds) *Quaternary Extinctions: A prehistoric* revolution: 259-298. University of Arizona Press.

Hanks, J. 1979. *The Struggle for Survival: the Elephant Problem*. New York: Mayflower Books.

Hardaker, T. 2001. New Lower Palaeolithic Finds from the Upper Thames, in S. Milliken and J. Cook (eds) *A Very Remote Period Indeed: Papers on the Palaeolithic Presented to Derek Roe*: 180-198. Oxford, Oxbow Books.

Hardaker, T. 2006. *The Solent Thames Research Framework: The Lower and Middle Palaeolithic of Oxfordshire.* English Heritage.

Hardaker, T. 2017. Lower Palaeolithic artefacts in surface contexts in Britain with special reference to the Northern Drift (Oxfordshire) and Warren Hill (Suffolk). *Proceedings of the Geologists' Association* 128 (3): 303-325.

Hardaker, T. 2020. A geological explanation for occupation patterns of ESA and early MSA humans in Southwestern Namibia? An interdisciplinary study. *Proceedings of the Geologists' Association* 131(1):8-18.

Hardaker, T. and R.J. MacRae 2000. A lost river and some Palaeolithic surprises. *Lithics* 21: 52-59.

Haynes, G. 1987. Proboscidean Die-offs and Die-outs: Age Profiles in Fossil Collections. *Journal of Archaeological Science* 14: 659-668.

Haynes, G. 1988. Longitudinal Studies of African Elephant Death and Bone Deposits. *Journal of Archaeological Science* 15: 131-157.

Haynes, G. 1991. *Mammoths, Mastodonts, and Elephants: Biology, Behavior, and the Fossil Record.* Cambridge: Cambridge University Press.

Haynes, G. 2006. Mammoth landscapes: good country for hunter-gatherers. *Quaternary International* 142–143: 20–29.

Haynes, G. 2017. Finding meaning in mammoth age profiles. *Quaternary International* 443 (A): 65-78.

Haynes, G., Klimowicz, J. and P. Wojtal 2018. A comparative study of woolly mammoths from the Gravettian site Kraków Spadzista (Poland), based on estimated shoulder heights, demography, and life conditions. *Quaternary Research* 90 Special Issue 3: 1–20.

Hemmer, H. 2011. The story of the cave lion - *Panthera leo spelaea* (Goldfuss, 1810) - A review. *Quaternaire*, Supplement Issue 4: 201-208.

Hinton, M.A.C. 1909-10. British Fossil Voles & Lemmings. *Proceedings of the Geological Association* XXI: 491.

Hollingworth, N.T.J. and P.B. Wignall 1992. The Callovian-Oxfordian boundary in Oxfordshire and Wiltshire, based on two new temporary sections. Proceedings of the Geologists' Association 103 (1): 15-20.

Humphrey, L. and C. Stringer 2018. *Our Human Story*. London: Natural History Museum.

Irving, B. (1995) Unpublished identifications of fish remains from Stanton Harcourt.

Isabella, J. 2017. From Prejudice to Pride: How Japan's Bear-Worshipping Indigenous Group Fought Its Way to Cultural Relevance. Viewed 5th September 2019, www.hakaimagazine.com, Smithsonian.

Jachmann. H. 1988. Estimating age in African elephants: a revision of Laws' molar evaluation technique. *African Journal of Ecology* 26 (1): 51–56.

Jones, A.M. 2001. Environmental signals in proboscidean molars: understanding the isotopic variations in enamel and collagen. Unpublished D.Phil. Dissertation, University of Oxford.

Jones, A.M., Koch, P.L. and E.D. Young 1998. In-stu UV laser ablation and irm-GCMS measurements of sub annual oxygen isotope variations in proboscidean enamel phosphate-prospects for seasonal palaeoclimatological and paleoecological studies. *Geological Society of America Abstracts Program.*

Jones, A., O'Connell,T., Young, E., Scott, K., Buckingham, C., Iacumin, P. and M. Brasier 2001. Biochemical data from well preserved 200ka collagen and skeletal remains. *Earth and Planetary Science Letters* 5976: 1-7.

Keen, D.H. 1990. Significance of the record provided by Pleistocene fluvial deposits and their included molluscan faunas for palaeoenvironmental reconstruction and stratigraphy: cases from the English Midlands. *Palaeogeography, Palaeoclimatology, Palaeoecology* 80: 25-34. Amsterdam, Elsevier Science Publishers, B.V.

Keen, D.H. (1992) Unpublished report on the molluscs from Stanton Harcourt

Keen, D.H. 1995. Raised beaches and sea-levels in the English Channel in the Middle and Late Pleistocene: problems of interpretation and implications for the isolation of the British Isles, in R.C. Preece (ed.) *Island Britain: a Quaternary Perspective*: 63-74. London: Geological Society Special Publications 96.

Keen, D.H. 2001. Towards a late Middle Pleistocene non-marine molluscan biostratigraphy for the British Isles. *Quaternary Science Reviews* 29: 1657-1665.

Kennedy, W.J., Taylor, J.D. and A. Hall 1969. Environmental and biological controls on bivalve shell mineralogy *Biological Reviews* : 44(4):499-530

Klein, R.G. and K. Cruz-Uribe 1984. The Analysis of Animal Bones from Archeological Sites. University of Chicago Press.

Krasińska M. and Z.A. Krasiński 2002. Body mass and measurements of the European bison during postnatal development. *Acta Theriologica* 47: 85–106.

Kurtén. B. 1968. Pleistocene Mammals of Europe. London: Weidenfeld and Nicholson.

Larramendi, A. 2014. Shoulder Height, Body Mass, and Shape of Proboscideans. *Acta Palaeontologica Polonica* 61(3): 537-574.

Laws, R.M. 1966. Age criteria for the African elephant Loxodonta a. africana. *African Journal of Ecology* 4 (1): 1-37.

Lee, P.C., Sayialel, S., Lindsay, W.K. and C.J. Moss 2012. African elephant age determination from teeth: validation from known individuals. *African Journal of Ecology* 50: 9–20.

Lewis, S.G., Maddy, D., Buckingham, C. Coope, G.R., Field, M.H., Keen, D.H., Pike, A.W.G., Roe, D.A. and K. Scott 2006. Pleistocene fluvial sediments, palaeontology and archaeology of the Upper River Thames at Latton, Wiltshire, England. *Journal of Quaternary Science* 21(2): 181-205.

Lindeque, M. 1989. Population dynamics of elephants in Etosha National Park, S.W.A./Namibia. Unpublished PhD dissertation, University of Stellenbosch.

Lindeque, M. 1991. Age structure of the Elephant population in the Etosha National Park, Namibia. *Madoqua* 18 (1): 27-32.

Lipecki, G. and P. Wojtal 1996. Mammoth population from Cracow Spadzista Street (B) site. *Acta zoologica Cracoviensia* 39 (1): 289-292.

Lister, A. 1990. Taxonomy and biostratigraphy of Middle Pleistocene deer remains from Arago (Pyrénées-Orientales, France). *Quaternaire*: 225-230.

Lister A.M. 1996. Sexual dimorphism in the mammoth pelvis: an aid to gender determination, in J Shoshani and P. Tassy (eds) *The Proboscidea. Evolution and Palaeoecology of Elephants and their Relatives*: 254-259. Oxford, Oxford University Press.

Lister, A.M. 1999. Epiphyseal fusion and postcranial age determination in the woolly mammoth *Mammuthus primigenius,* in G. Haynes, J Klimowicz and J.W.F. Reumer (eds) *Mammoths and the Mammoth Fauna: Studies of an Extinct Ecosystem.* DEINSEA 6: 79-88.

Lister, A.M. 2009. Late-glacial mammoth skeletons (*Mammuthus primigenius*) from Condover (Shropshire, UK): anatomy, pathology, taphonomy and chronological significance. *Geological Journal* 44 (4): 447-479.

Lister, A. M 2014. *Mammoths: Ice Age Giants.* London: Natural History Museum.

Lister, A.M. 2017. On the type material and evolution of North American mammoths. *Quaternary International* 43: 14-31

Lister A.M. and L.D. Agenbroad 1994. Gender determination of the Hot Springs mammoths, in L.D. Agenbroad and J.I. Mead (eds) *The Hot Springs Mammoth Site: a Decade of Field and Laboratory Research in Paleontology, Geology and Paleoecology,*). Mammoth Site: Hot Springs, South Dakota: 208 - 214.

Lister, A.M. and A.V. Sher 2001. The origin and evolution of the woolly Mammoth. *Science* 294: 1094–1097.

Lister A. and P. G. Bahn 2007. *Mammoths: Giants of the Ice Age.* London: Marshall Editions.

Lister, A.M. and A.J. Stuart (eds) 2010. The West Runton Freshwater Bed and the West Runton Mammoth. *Quaternary International* 228 (1-1).

Lister A.M. and A.V. Sher 2015. Evolution and dispersal of mammoths across the Northern Hemisphere. *Science* 350 (6262): 805 - 809.

Lister, A.M. and K. Scott *in press* Mammoth evolution in the Late Middle Pleistocene of Europe.

Lucht, W.H. 1987. *Die Käfer Mitteleuropas: Katalog.* Krefeld: Goecke and Evers.

Ludwig, A., Pruvost, M., Reissmann, M., Benecke, N., Brockmann, G.A., Castaños, P., Cieslak, M., Lippold, S.,Llorente, L.,Malaspinas, A-S.,Slatkin, M. and M. Hofreiter 2009. Coat color variation at the beginning of horse domestication. *Science* 324:485.

Luedtke, B.E. 1992. *An Archaeologist's Guide to Chert and Flint.* Archaeological Research Tools 7: 107-109. Los Angeles: University of California, Institute of Archaeology.

MacRae, R.J., 1986. The great handaxe stakes. *Lithics* 8, 15-17.

MacRae, R. J. 1987. Palaeolithic artefacts from Stanton Harcourt. *Oxoniensia* 52: 179-181.

MacRae, R. J. 1988a. The Palaeolithic of the Upper Thames and its quartzite implements, in R.J. MacRae and N. Moloney (eds*) Non-flint stone tools and the Palaeolithic occupation of Britain*: 123-54. BAR British Series189.

MacRae, R.J. 1988b. Belt, Shoulder-bag or Basket? An Enquiry into Handaxe Transport and Flint Sources. *Lithics* 9: 2-8.

Maddy, D. 1997. Uplift-driven valley incision and river terrace formation in southern England. *Journal of Quaternary Science* 12 (6): 539-545.

Meijer, T. and R.C. Preece 2000. A review of the occurrence of *Corbicula* in the Pleistocene of North-west Europe. *Geologie en Mijnbouw/Netherlands Journal of Geosciences* 79: 241– 255.

Marciszak, A., Schouwenburg, C. and R. Darga 2014. Decreasing size process in the cave (Pleistocene) lion *Panthera spelaea* (Goldfuss, 1810) evolution – a review. *Quaternary International* 339–340: 245-257.

Marsolier-Kergoat, M-C, Palacio, P., Berthonaud,V., Maksud, F., Stafford, T., Bégouën, R. and J-M. Elalouf 2015. Hunting the Extinct Steppe Bison (*Bison priscus*): Mitochondrial Genome in the Trois-Frères Paleolithic Painted Cave. PLoS One Jun 17, 10(6): 1-16.

Mecozzi, B and Lucenti, S.B. 2018. The Late Pleistocene Canis lupus (Canidae, Mammalia) from Avetrana (Apulia, Italy): reappraisal and new insights on the European glacial wolves. *Italian Journal of Geosciences* 137 (1): 138–150.

Meijer, T. and R.C. Preece 2000. A review of the occurrence of *Corbicula* in the Pleistocene of North-West Europe. *Geologie en Mijnbouw/Netherlands Journal of Geosciences* 79 (2/3): 241-255.

Mitchell, G.F., Penny L.F., Shotton, F.W. and R.G. West 1973. A correlation of Quaternary deposits in the British Isles. *Geological Society of London*, Special Report 4: 1-99.

Mol, D., de Vos, J. and J. van Der Plicht 2007. The presence and extinction of Elephas antiquus Falconer and Cautley, 1847, in Europe. *Quaternary International* 169:149-153.

Murton, J.B., Baker, A., Bowen, D.Q., Caseldine, C.J., Coope, G.R., Currant, A.P., Evans, J.G., Field, M.H., Green, C.P., Hatton, J., Ito, M., Jones, R.L., Keen, D.H., Kerney, M.P., McEwan, R., McGregor, D.F.M., Parish, D., Robinson, J.E., Schreve, D.C. and P.L. Smart 2001. A Late Middle Pleistocene temperate–periglacial-temperate sequence (Oxygen Isotope Stages 7–5e) near Marsworth, Buckinghamshire, UK. *Quaternary Science Reviews* 20: 1787–1825.

Murton, J.B., Bowen, D.Q., Candy, I., Catt, J. A., Currant, A., Evans, J.G., Frogley, M., Green, C.P., Keen, D.H., Kerney, M.P., Parish, D., Penkman, K., Schreve, D.C., Taylor, S., Toms, P.S., Worsley, P. and L. York 2015. Middle and Late Pleistocene environmental history of the Marsworth area, south-central England. *Proceedings of the Geologists' Association* 126 (1): 8-49.

Oakley. D. O. S., Kaufamn. D. S., Gardner, T. W., Fisher, D. M. and R.A. VanderLeest 2017. Quaternary marine terrace chronology, North Canterbury, New Zealand, using amino acid racemization and infrared-stimulate luminescence. *Quaternary Research* 87: 151-167.

Olsen, S. J. 1960. Postcranial skeletal characters of *Bison* and *Bos*, *Papers of the Peabody Museum of Archaeology and Ethnology* 35(4) Harvard University.

Pacher, M. and A.J. Stuart 2009. Extinction chronology and paleoecology of the cave bear *Ursus spelaeus. Boreas* 38 (2): 189-206.

Paddayya, K. 2001. The Acheulian culture project of the Hunsgi and Baichbal Valleys, Peninsular India, in L. Barham and K. Robson-Brown (eds) *Human Roots: Africa and Asia in the Middle Pleistocene*: 235-258. Western Academic and Specialist Press Ltd, UK.

Palombo, M. R. 2001. Endemic Elephants of Mediterranean Islands: Knowledge, Problems and Perspectives, in G. Cavarretta, P.Giola, M.Mussi and M.R.Palombo (eds) *La Terra degli Elefanti*: 486-491. Rome: Consiglio Nazionale della Ricerche.

Palombo, M.R., Filippi, M.L., Iacumin, P., Longinelli, A., Barbieri, M., and A. Maras 2005.
Coupling tooth microwear and stable isotope analyses for palaeodiet reconstruction:
the case study of Late Middle Pleistocene *Elephas* (*Palaeoloxodon*) *antiquus* teeth from Central Italy (Rome area). *Quaternary International* 126-128: 153-170.

Pappa, S. 2014. Preliminary morphometrical analysis of Pleistocene bear specimens from caves sites in south-west England. *Quaternary Newsletter* 134: 40-43.

Parfitt, S.A., Barendregt, R.W., Breda, M., Candy, I., Collins, M.J., Coope, G.R., Durbridge, P., Field, M.H., Lee, J.R., Lister, A.M., Mutch, R., Penkman, K.E.H., Preece, R.C., Rose, J., Stringer, C.B., Symmons, R., Whittaker, J.E., Wymer, J.J. and A.J. Stuart 2005. The earliest record of human activity in northern Europe. *Nature* 438, 108-1012.

Penkman, K. E. H., Preece, R. C., Bridgland, D. R., Keen, D. H., Meijer, T., Parfitt, S. A., White, T. S. and M.J. Collins 2011. A chronological framework for the British Quaternary based on Bithynia opercula. *Nature* 476 (7361): 446-449.

Penkman, K. E. H., Preece, R. C., Bridgland, D. R., Keen, D. H., Meijer, T., Parfitt, S. A., White, T. S. and M.J. Collins 2013. An aminostratigraphy for the British Quaternary based on Bithynia opercula. *Quaternary Science Reviews* 61, 111-134.

Perry, J.S. 1954. Some observations on growth and tusk weight in male and female African elephants. *Proceedings of the Zoological Society of London* 124 (1): 97-104.

Pettitt, P.B. and M.J. White 2012. *The British Palaeolithic: Human Societies at the Edge of the Pleistocene World*. London: Routledge.

Pike, A.W.D., Hedges, R.E.M. and P. van Calsteren 2002. U-series dating of bone using the diffusion-absorption method. *Geochimica et Cosmochimica Acta* 660: 4273-4286.

Rees-Jones, J. 1995. Optically Stimulated Luminescence (OSL) dating report on sediment from Stanton Harcourt, Oxfordshire. Unpublished report.

Preece, R.C. 1995. *Island Britain: a Quaternary Perspective.* Geological Society, London.

Pruvosta, M., Bellone, R., Benecke, N., Sandoval-Castellanos, E., Cieslak, M., Kuznetsova, T., Morales-Muñiz, A., O'Connor, T., Reissmann, M., Hofreiter, M., and A. Ludwig 2014. Genotypes of predomestic horses match phenotypes painted in Paleolithic works of cave art. *Proceedings of the National Academy of Sciences* 108 (46): 18626-18630.

Rees-Jones, J. 1995. Optical dating of selected British archaeological sediments. Unpublished D.Phil Dissertation, University of Oxford.

Richter, M. and D. Dettloff 2002. Experiments in brain tanning with a comparative analysis of stone and bone tools. *Journal of Undergraduate Research* 5: 301-318.

Roberts, M.B. and S.A. Parfitt 1999. *Boxgrove. A Middle Pleistocene hominid site at Eartham Quarry, Boxgrove, West Sussex*. London: English Heritage Archaeological Report 17.

Robinson, M. 2012. Unpublished identifications of seeds from Stanton Harcourt.

Romandini, M., Terlato, G., Nannini, N., Tagliacozzo, A., Benazzi, S. and M. Peresani 2018. Bears and humans, a Neanderthal tale. Reconstructing uncommon behaviors from zooarchaeological evidence in southern Europe. *Journal of Archaeological Science 90*: 71-91.

Roth, V.L. 2001. Ecology and evolution of dwarfing in insular elephants, in P. G. Cavarretta, M. Mussi and M. R. Palombo (eds) *The World of Elephants: proceedings of the 1st International* Congress: 507–9. Roma: Consiglio Nazionale delle Ricerche.

Roth, V.L. 1984. How elephants grow: Heterochrony and calibration of developmental stages in some living and fossil species. *Journal of Vertebrate Paleontology* 4 (1): 126-145.

Sandford, K.S. 1925. The Fossil Elephants of the Upper Thames Basin. *Quarterly Journal of the Geological Society* 81: 62-86.

Sala, B. 1986. *Bison schoetensacki* Freud from Isernia la Pineta (early Mid-Pleistocene - Italy) and revision of the European species of bison. *Palaeontographia* Italica 74: 113-170.

Sampson, G. 2006. Acheulian quarries at hornfels outcrops in the Upper Karoo region of South Africa, in N. Goren-Inbar and G. Sharon (eds) *Axe Age: Acheulian toolmaking from quarry to discard*: 75-107. London, Equinox.

Schreve, D.C. 1997. Mammalian biostratigraphy of the later Middle Pleistocene in Britain. Unpublished Ph.D. Dissertation, University of London.

Schreve, D.C. 2001a. Differentiation of the British late Middle Pleistocene interglacials: the evidence from the mammalian biostratigraphy. *Quaternary Science Reviews* 20: 1693-1705.

Schreve, D. C. 2001b. Mammalian evidence from Middle Pleistocene fluvial sequences for complex environmental change at the oxygen isotope substage level. *Quaternary International* 79: 65-74.

Schreve, D.C. and A.P. Currant 2003. The Pleistocene history of the Brown Bear (*Ursus arctos* L.) in the western Palaearctic: a review, in B. Kryštufek, B. Flajšman and H.I. Griffiths (eds) *Living with Bears. A Large European Carnivore in a Shrinking World*. Ljubljana, Ekološki forum LDS, pp 27-39. ISBN: 9789619052242.

Scott, B. 2010. *Becoming Neanderthal: The Earlier British Middle Palaeolithic*. Oxford: Oxbow Books.

Scott, K. 1980. Two hunting episodes of Middle Palaeolithic age at La Cotte de Saint-Brelade, Jersey (Channel Islands). *World Archaeology* 12 (2): 137-152. London: Routledge and Kegan Paul.

Scott, K 1986a. The Large Mammal Fauna, in P. Callow and J. Cornford (eds) *La Cotte de St. Brelade* 1961-1978: Excavations by C.B.M.McBurney: 109-138. Norwich: Geo Books.

Scott, K. 1986b. The Bone Assemblages of Layers 3 and 6, in P. Callow and J. Cornford (eds) *La Cotte de St. Brelade* 1961-1978: Excavations by C.B.M.McBurney: 159-184. Norwich: Geo Books.

Scott, K. 2001. Late Middle Pleistocene Mammoths and Elephants of the Thames Valley, Oxfordshire, in G. Cavarretta, P.Giola, M.Mussi and M.R.Palombo (eds) *La Terra degli Elefanti*: 247-254. Rome: Consiglio Nazionale della Ricerche.

Scott, K. 2007. The ecology of the late middle Pleistocene mammoths in Britain. *Quaternary International* 169-170: 125-136.

Scott, K. in press La Cotte de St Brelade: in pursuit of the mammoths. In *Peopling La Manche* (C. Gamble ed.) Special Volume of the Prehistoric Society.

Scott, K. in prep. Large vertebrates in Britain during MIS 7: the fossil evidence reviewed

Scott, K. and C.M. Buckingham 1997. Quaternary Fluvial Deposits and Palaeontology at Stanton Harcourt Oxfordshire, in S.G. Lewis and D. Maddy (eds) *The Quaternary of the South Midlands and the Welsh Marches*: 115-126. London: Field Guide, Quaternary Research Association.

Scott, K. and C.M. Buckingham 2001a. Preliminary report on the excavation of Late Middle Pleistocene deposits at Latton, near Cirencester, Gloucestershire. *Quaternary Newsletter* 94: 24-29.

Scott, K. and C.M. Buckingham 2001b. A River Runs Through It: A Decade of Research at Stanton Harcourt, in S. Milliken and J. Cook (eds) *A Very Remote Period Indeed*: 206-213. Oxford, Oxbow Books.

Scott, K. and A.M. Lister in prep. Size reduction in *Mammuthus trogontherii* in the late Middle Pleistocene; the evidence from Britain.

Sher, A.V. 1997. An Early Quaternary bison population from Untermassfeld: *Bison menneri* sp. nov. In R.-D. Kahlke (ed.) *Das Pleistozän von Untermassfeld bei Meiningen (Thüringen) Teil 1*: 101-180. Monographien des Römisch-Germanischen Zentralmuseums Mainz vol. 40.

Shpansky, A.V., Svyatko, S.V., Reimer, P.J. and S. V. Titov 2016. Records of *Bison priscus* Bojanus (Artiodactyla, Bovidae) skeletons in Western Siberia. *Russian Journal of Theriology* 15(2): 100–120.

Stansfield, F.J. 2015. A Novel Objective Method of Estimating the Age of Mandibles from African Elephants (*Loxodonta africana Africana*). Viewed 15 October 2019, https://doi.org/10.1371/journal.pone.0124980

Stringer, C.B., E. Trinkaus and M.B. Roberts 1998. The Middle Pleistocene human tibia from Boxgrove. *Journal of Human Evolution* 34 (5): 509–547.

Stout, D., Apel, J., Commander, J. and M. Roberts 2014. Late Acheulean technology and cognition at Boxgrove, UK. *Journal of Archaeological Science* 41: 576-590.

Stuart, A.J. 1982. *Pleistocene Vertebrates in the British Isles.* London: Longman.

Stuart, A.J. 1991. Mammalian extinctions in the late Pleistocene of northern Eurasia and North America. *Biological Science Reviews* 66: 453-562.

Stuart, A.J. 2005. The extinction of woolly mammoth (*Mammuthus primigenius*) and straight tusked elephant (*Palaeoloxodon antiquus*) in Europe. *Quaternary International* 126–128: 171–177.

Stuart, A.J. and A.M Lister 2001. The mammalian faunas of Pakefield/Kessingalnd and Corton, Suffolk; evidence for a new temperate episode in the British early Middle Pleistocene. *Quaternary Science Reviews* 20: 1677-1692.

Stuart A.J. and A.M. Lister 2011. Extinction chronology of the cave lion *Panthera spelaea* *Quaternary Science Reviews* 30: 2329-2340.

Sutcliffe, A.J. 1975. A hazard of interpreting glacial-interglacial sequences. *Quaternary Newsletter* 17: I-4.

Sutcliffe, A.J. 1976. The British glacial-interglacial sequence. Reply to Dr R. West. *Quaternary Newsletter* 18: 1-7.

Sutcliffe, A.J. 1995. Insularity of the British Isles 250 000-30 000 years ago: the mammalian, including human, evidence, in R.C. Preece (ed.) *Island Britain: a Quaternary perspective*: 127-140. Geological Society Special Publication 96.

Tikhonov, A., Agenbroad, L. and Vartanyan, S. 2003. Comparative analysis of the mammoth populations on Wrangel Island and the Channel Islands, in J.W.F. Reumer, J. de Vos, and D. Mol (eds) *Advances in Mammoth Research*. Proceedings of the Second International Mammoth Conference Rotterdam, 16–20 May 1999. *Deinsea* 9: 415–420.

Sukumar, R. 2003. *The Living Elephants. Evolutionary Ecology, Behavior, and Conservation*. Oxford University Press.

Van Asperen, E.N. 2011. Distinguishing between the late Middle Pleistocene interglacials of the British Isles: a multivariate approach to horse biostratigraphy. *Quaternary International* 231: 110-115.

Turner, A., 1984. Dental sex dimorphism in European lions (*Panthera leo* L.) of the Upper Pleistocene: palaeoecological and palaeoethological implications. *Annales Zoologici Fennici* 21 (1): 1-8.

Turner, A and Antón, M. 1997. *The Big Cats and their fossil relatives*. New York: Columbia University Press.

West, R.G. 1963. Problems of the British Quaternary *Proceedings of the Geological Association* 74:147-186.

West, R.G. 1968. *Pleistocene Geology and Biology*. London: Longmans.

West, R.G. 1981. Palaeobotany and Pleistocene stratigraphy in Britain. *New Phytologist* 87: 127-137.

West, R.G. 1984. Interglacial, interstadial and oxygen isotope stages. *Dissertationes Botanicae* 72: 345-357.

White, M., Ashton, N., and R. Scott 2011. The emergence, diversity and significance of Mode 3 (prepared core) technologies, in N. Ashton, S. Lewis and C. Stringer (eds) *The Ancient Human Occupation of Britain*: 53-65. London: Elsevier Science.

White, M. and F. Foulds 2018. Symmetry is its own reward: on the character and significance of Acheulean handaxe symmetry in the Middle Pleistocene. *Antiquity* 92 (362): 304-319.

White, M., Bridgland, D.R., Schreve, D.C., White, T. S, and K.E.H. Penkman 2018. Well-dated fluvial sequences as templates for patterns of handaxe distribution: understanding the record of Acheulian activity in the Thames and its correlatives. *Quaternary International* 480: 118-131.

White, O.A. and C.G. Diedrich 2012. Taphonomy story of a modern African elephant *Loxodonta africana* carcass on a lakeshore in Zambia (Africa). *Quaternary International* 276: 287-296.

Williamson B.R 1975. Seasonal distribution of elephants in Wankie National Park. *Arnoldia (Rhodesia)* 7 (11): 1-16.

Winograd, I.J., Coplen, T.B., Landwehr, J.M., Riggs, A.C., Ludwig, K.R., Szabo, B.J., Kolesar, P.T. and K.M. Revesz 1992. Continuous 500,000-year climate record from vein calcite in Devils' Hole, Nevada. *Science* 258: 255-260.

Wood, A.A. 2004. An Investigation into the Differentiation of *Bos primigenius* and *Bison priscus* with a Description of the Post-cranial Aveley Aurochs and an Analysis of Body Mass Estimates and Palaeoecology. Unpublished MSc Dissertation. University of London.

Woo Lee, H. 2001. *A Study of lower Palaeolithic stone artefacts from selected sites in the Upper and Middle Thames Valley, with particular reference to the MacRae collection.* BAR British Series 319.

Zhou, L.P., McDermott, F., Rhodes, E.J., Marseglia, E.A. and P.A. Mellars 1997. ESR and Mass-spectrometric uranium-series dating studies of a mammoth tooth from Stanton Harcourt, Oxfordshire, England. *Quaternary Science Reviews* (*Quaternary Geochronology*) 16: 445-454.

Zutovski, K and R. Barkai 2016. The use of elephant bones for making Acheulian handaxes: A fresh look at old bones. *Quaternary International* 406: 227-238.